Springer Climate

Springer Climate is an interdisciplinary book series dedicated to climate research. This includes climatology, climate change impacts, climate change management, climate change policy, regional climate studies, climate monitoring and modeling, palaeoclimatology etc. The series publishes high quality research for scientists, researchers, students and policy makers. An author/editor questionnaire, instructions for authors and a book proposal form can be obtained from the Publishing Editor.
Now indexed in Scopus® !

Elena G. Popkova
Editor

Smart Green Innovations in Industry 4.0

New Opportunities for Climate Change Risk Management in the "Decade of Action"

Springer

Editor
Elena G. Popkova
RUDN University
Moscow, Russia

ISSN 2352-0698 ISSN 2352-0701 (electronic)
Springer Climate
ISBN 978-3-031-45829-3 ISBN 978-3-031-45830-9 (eBook)
https://doi.org/10.1007/978-3-031-45830-9

This Springer imprint is published by the registered company Springer Nature Switzerland AG
The registered company address is: Gewerbestrasse 11, 6330 Cham, Switzerland

Climate Action Through Smart Green Innovation in Industry 4.0 (Introduction)

Climate change determines the landscape of society and the economy, which, in turn, affects the quality of life. Climate change risks are global and high. Nevertheless, an unexpected management solution for them is found in the Fourth Industrial Revolution. No previous industrial revolution has shown as much attention to environmental protection as the Fourth Industrial Revolution. This advantage has been achieved through the unity of the global community in implementing the UN initiative in the Decade of Action.

Thanks to this, there was a synthesis of high-tech and environmentally friendly innovations emerging and being put into practice in Industry 4.0. It is scientifically correct to call them smart green innovations. Their special feature is a climate perspective on Industry 4.0 innovation. Innovative technologies 4.0 are examined for environmental safety before they are approved and put into practice, which is why environmental quality certification of high-tech products has become very popular. Additionally, there is specific climate-oriented R&D to develop technology 4.0 specifically to combat climate change.

The described new vision of quality of life, with its inseparable connection to technology 4.0, can be clearly seen by superimposing the climate landscape of Industry 4.0 on Maslow's pyramid of needs. In particular, organic needs of the first level include the universal availability of quality food, which is strongly influenced by climate risks. The low-tech approach to farming provides only limited opportunities to manage these risks. The high-tech approach reduces these risks to a minimum or even completely overcomes them with agriculture 4.0.

Security needs on the second level include the prevention of environmental disasters, which have become more frequent due to climate change. Climate monitoring with machine vision and regenerative natural resource management with the support of AI can predict and reverse climate change. Needs for belonging on the third level in the information society and capitalism 4.0 are increasingly being met through belonging to the green digital workforce. As a continuation of this, the need for respect at the fourth level can be met through the career building of a green digital workforce.

The cognitive needs of the fifth level often manifest themselves in practice in the desire of members of climate-responsible communities to learn more about climate issues and solutions, which is achieved through electronic communication, distance learning, and smart assistants. The aesthetic needs at the sixth level in the Decade of Action are embodied in the Sustainable Development Goals (SDGs) and are manifested in the harmony of society, economy, and nature. These needs are met through the beauty and health of the environment and the restoration of climatic order. Finally, the need for self-actualization at the seventh (highest) level is met through personal growth in the search for one's place in a climate-responsible society 4.0.

In practice, the integration of technology 4.0 into the climate economy gets four manifestations. The first manifestation is entrepreneurial. Its point is that entrepreneurship in digital economy markets exhibits climate responsibility. Influenced by the Fourth Industrial Revolution, climate responsibility 4.0 has emerged, with entrepreneurs adapting the breakthrough technologies available to them to implement their climate change initiatives.

This manifestation encompasses many progressive entrepreneurial practices, including green blockchain for climate optimization of investment flows, uncrewed electric vehicles, and green robots for multifunctional climate applications: from climate change monitoring and energy conservation to automated waste sorting and disposal and recycling.

The second manifestation is technological. It is related to combating climate change by relying on cutting-edge technology and Industry 4.0 green innovations. In this case, technology 4.0 is considered without explicit reference to the subjects of its implementation and with a strict focus on the fight against climate change. Responsible production and consumption in the information society, as well as the development of sustainable yet smart territories in the digital economy, significantly contribute to combating climate change.

The value of smart green innovation in Industry 4.0 is that it combines previously separate practices for digital modernization and economic sustainability. This transforms the essence of the innovation economy, which acquires a new – climate dimension and limitation. This makes it possible to streamline the development of the innovation economy and direct it toward improving the quality of life. Technology 4.0 and innovation cease to be an end in itself and become a tool for implementing the climate agenda in the Decade of Action.

The third manifestation is regulatory. It includes government regulation and corporate management of decarbonization and sustainable economic development based on ESG principles. This manifestation contributes to the revision of the energy economy, which is acquiring climate reference points. High-tech energy (EnergyTech) and smart grids (Smart Grid) significantly contribute to their achievement.

The ESG principles are inspired by the spirit of the Decade of Action. They emerged in the synthesis of society's progressiveness, soft international law, and the UN's global initiatives that unite humanity. Regulation in support of decarbonization means moving toward clean energy and renewable natural resources to preserve the legacy for future generations. This is facilitated by the massive support and adherence to ESG principles in government, society, and business.

The fourth manifestation is social. Sociocultural support for combating climate change is very important. Environmental thinking and climate-responsible lifestyles are embedded in tradition and fashion – together, they shape the outlines of today's progressive and sustainable communities represented by climate change activists.

Green employment is a contemporary reality and the norm, which is encouraged by the state, society, and business. Green personnel is developed by educational institutions at all levels, carefully distributed in the labor market, and nurtured through corporate environmental responsibility at HRM. In an Industry 4.0 environment, a green and digital workforce is often inseparable.

The problem is that the described manifestations are treated separately in the available publications, which is a shortcoming of the existing literature. The lack of a systemic understanding of Climate Economy 4.0, and its prospects is a gap in the literature. This hinders its development and prevents the potential of climate economy 4.0, which makes it necessary to fill the gap identified. This motivated the creation of this book, which demonstrates how business, technology, government, and society are united in fighting climate change, thus achieving harmony with nature.

This systemic vision is useful for each actor in the climate economy 4.0 to understand its role in its development, as well as for forecasting and foresight purposes. This book aims to identify new opportunities to manage the risks of climate change in the Decade of Action with smart green innovations in Industry 4.0. The book is original in that it offers a promising solution to the challenges related to the development of Industry 4.0 and combating climate change in the Decade of Action through the creation and implementation of smart green innovation 4.0.

The book reveals the practical experience of the development of climate economy 4.0 in the countries of Central Asia and Eastern Europe, with particular attention to the experience of the Eurasian Economic Union (EAEU). The book also covers the experience of Zambia. For this reason, the book considers the peculiarities of developed and developing countries. The book is organized as follows: The first part of the book examines climate responsibility in managing entrepreneurship in digital economy markets. The second part focuses on combating climate change by relying on cutting-edge technology and Industry 4.0 green innovations, such as AI, IoT, digital environmental platforms, big data, agrotech, and e-commerce. The third part explores the experience and prospects of decarbonization and sustainable economic development based on ESG principles. The fourth part, which closes the book, explores sociocultural support for climate change and green employment. Particular attention is paid to social management issues in the innovation economy, overcoming cultural differences, and ensuring cultural inclusiveness.

The book is intended for scientists studying climate change and climate action. The book provides them with theoretical insights and methodological recommendations for managing the risks of climate change in the Decade of Action through Industry 4.0 smart green innovations. The book is also of interest to representatives of climate-responsible businesses, climate activists, and state and supranational regulators of the climate economy. In this book, they will find an overview of international experiences and promising applied solutions for managing the risks of climate change in the Decade of Action through Industry 4.0 smart green innovations.

RUDN University
Moscow, Russia

Elena G. Popkova

Contents

Part I
Climate Responsibility in Managing Entrepreneurship in Digital Economy Markets

Competitiveness of Climate-Responsible Entrepreneurship of Central Asia and Eastern Europe in the World Markets of the Digital Economy

Elena G. Popkova

1 Introduction

The development of climate-responsible entrepreneurship in the markets of the digital economy defines the important aspects of world development. It combines economic processes that are to increase the level of competitiveness, innovative mechanisms that are supposed to look for new methods of solving entrepreneurial problems and the tools for fighting climate change and achieving balance within all the above processes. Climate change is usually assessed in the context of CO_2 emissions and search for the ways to reduce them [7] against the background of the necessity to increase economic potential. Innovative achievements are studied from the position of digital technologies' role in the processes of adaptation to climate change [8] and stimulation of low-carbon innovations [4]. Given the achievements in the development of digital technologies, more progressive countries obtain an advantage over other countries, which allows them to achieve a higher level of competitiveness.

In modern conditions, the indicators of economic effectiveness and competitiveness remain the actual indicators of the success of entrepreneurs and countries. However, the growth of the importance of the Sustainable Development Goals and the extension of their influence on the strategic planning systems outlines the topical directions for scientific studies, the essence of which lies in the necessity to reach competitiveness in the combination of climate-responsible and cost-effective activities. Here digital markets are individual platforms with specific features, which allow assessing the role of digital technologies in the achievement of economic and environmental goals. At that, environmental factors are often considered as opposed to economic factors. The presence of a certain conflict between the desire for maximisation of profit and compliance with the adopted norms of responsible behaviour

E. G. Popkova (✉)
RUDN University, Moscow, Russia

E. G. Popkova (ed.), *Smart Green Innovations in Industry 4.0*, Springer Climate,
https://doi.org/10.1007/978-3-031-45830-9_1

often determines the direction of scientific search, which would allow identifying the balance between the desire to ensure high quality of the environment, which requires unproductive costs, and receipt of good economic results [5].

In this context, countries of Eastern Europe and Central Asia are particularly interesting, since they went through a transformation of the economic system from planning to the market economy, which allowed them to form systems of effective functioning of entrepreneurship, achieve competitiveness in certain segments of the world market, develop digital technologies and implement the mechanisms of fighting climate change. Thus, the topical scientific task regarding this group of countries is determining the level of influence of the results of economic progress on the processes of sustainable development, which are considered from the position of success in the world markets of the digital economy.

2 Materials and Methods

The methodology of this research is based on the matrix method, which involves the graphic representation of the studied countries on the criteria of the level of the digital economy development and effectiveness of the policy of fight against climate change. The information base of the analysis includes the data of the US AID, in particular, the Environmental Performance Index, developed by CIESIN and Yale University and the ICT Adoption index, calculated by the World Economic Forum as a component of the Global Competitiveness Report. The Environmental Performance Index characterises the level of assuaging the consequences of climate change and combines nine indicators, which characterise the rates of CO_2 emissions growth, forecasted level of CO_2 emissions, growth in methane emissions, etc. The ICT Adoption index characterises the level of development of the infrastructure of the digital economy – dissemination of mobile Internet, broadband Internet and the number of Internet users.

The research objects are countries of Central Asia and Eastern Europe. The similarity between these regions is the historical period when the countries of the sample were in the sphere of the influence of the USSR and the planned economy. To identify countries by the regional groups, the generally acknowledged regional division of countries of the world was used [10]. According to it, the sample contains 15 countries: 5 countries of Central Asia and 10 countries of Eastern Europe. Some data on these countries are lacking: there are no data on the digital economy for Uzbekistan, Turkmenistan and Belarus. Thus, a sample of 12 countries was studied.

The issue of competitiveness of climate-responsible entrepreneurship in countries of Eastern Europe and Central Asia in the markets of the digital economy was previously considered by scholars within studies of problems, issues and processes of interconnection between the level of economic development of countries and its effect on the environment in the context of Environmental Kuznets Curve [1, 5]; influence of government policy of nature protection activities regulation on the state of development of innovations and competitiveness of countries [2, 4]; specifics of sustainable development in the digital economy [6]; influence of innovations on climate change [8, 12] and on economic development and competitiveness [9]; study of the parameters of economic development from the position of the Green Competitiveness concept [3]; analysis of climate change and reaction to them in countries of Eastern Europe and Central Asia [7].

The purpose of this work is to discover the level of competitiveness of climate-responsible entrepreneurship in the markets of the digital economy in countries of Central Asia and Eastern Europe; for this, the ratio between the level of development of the digital economy and the policy of fighting climate changes in the designated countries is assessed.

3 Results

From the position of economics, competitiveness is an important indicator pointing to the successfulness of functioning of economic subjects compared to others. Most often, competitiveness is assessed through the conquered share of the market, correspondence to certain parameters that characterise demand, etc. From the position of climate responsibility of entrepreneurship in digital markets, an important indicator that characterises competitiveness is the achievement of a high level of compliance with the criteria of sustainable and digital development, which is accompanied by support for a certain balance between these indicators.

With this in mind, an important task of this research is coordinating the assessment of climate-responsible activities and the level of the digital economy development. The matrix method with the graphic representation of results is used for this. According to this method, on one axis of the system of coordinates the values of the Environmental Performance Index are shown (y); on another – the ICT Adoption index (x). The values of countries of the sample are shown at the crossing of the indices' values, which allows stating the domination of activities aimed at the development of the digital economy, fight against climate change or achievement of certain balance by these indicators. Values of the indices are shown in Fig. 1.

Figure 1 shows the position of countries of the sample that belong to the Eastern European region (♦), and Central Asian region. To improve the analytical value of the results, additional lines are used – they allow for a better description of the level of climate-responsible entrepreneurship's competitiveness in the markets of the digital economy.

Given the size of the scale of the assessment of the index, the median of all possible values for each index is 50 points. According to this, the scale was divided into four squares by a value of 50 points. As a result, in the lower left corner, one country (Tajikistan) is located – it is peculiar for a low level of development of ICT and environmental effectiveness. In the upper left corner, there are countries that are peculiar for a high level of ICT development, combined with a low level of environmental effectiveness. Most of the countries and the global average value are in this

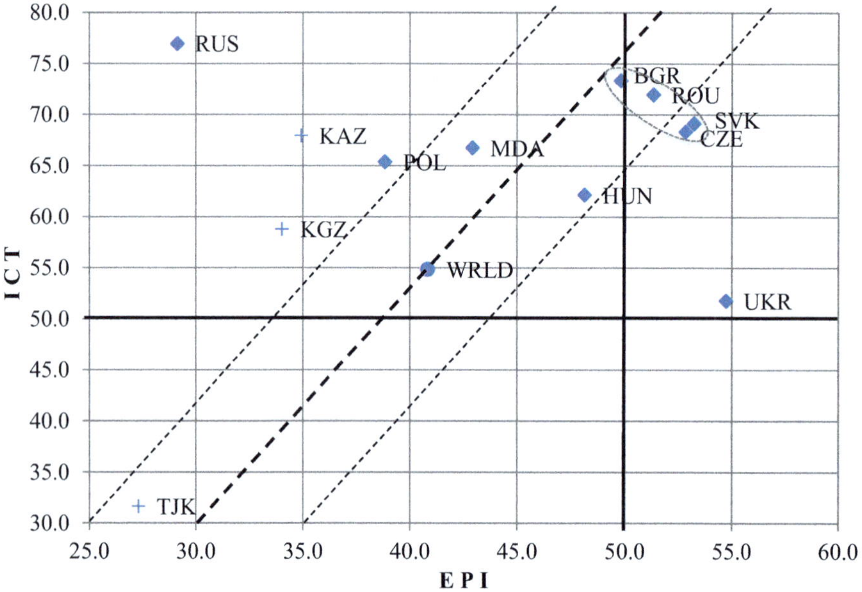

Fig. 1 Matrix of the balance of the development of climate-responsible entrepreneurship and the digital economy in countries of the sample. (*Source*: Created by the authors using [11])

square. In the upper right corner, countries with high values of ICT and environmental effectiveness are located.

The results demonstrate that almost all countries of the sample have a rather high (above the average value – 50 points) level of ICT development. However, only four countries of the sample have also high values of the Environmental Performance Index. That is, in most cases, the development of ICT takes place without consideration of climate-responsible entrepreneurship. Countries of the sample that demonstrate a high level of environmental effectiveness and development of ICT are Romania, Slovakia, the Czech Republic and Ukraine. Still, while the first three countries also have a high level of ICT development (67–73 points), Ukraine is close to the level of 50 points. For all Central Asian countries, the value of the Environmental Performance Index is below 50 points. Almost all Eastern European countries have higher values of the Environmental Performance Index compared to Central Asian countries.

Another tool of analysis in Fig. 1 is diagonal dotted lines, the central of which goes from the start of the system of coordinates through the global average value, and side lines characterise the range of deviation of the given values. This line is connected to the global average level of indicators. It points to the domination of the priority of development of the digital economy or compliance with the principles of climate-responsible activity. Within the outlined range there are five countries, which are peculiar for a certain balance between the priorities of development: Tajikistan, Moldova, Hungary, Bulgaria and Romania. Slovakia, the Czech Republic and Ukraine are characterised by the domination of the priority of climate-responsible entrepreneurship, while Russia, Kazakhstan, Kyrgyzstan and Poland's priority is the development of the digital economy against the background of a low level of climate-responsible entrepreneurship.

The highest level of competitiveness of climate-responsible entrepreneurship in the world markets of the digital economy is peculiar to four countries of Eastern Europe: Bulgaria, Romania, Slovakia and the Czech Republic. These countries demonstrate the highest level of climate effectiveness, combined with a high level of digital economy development. These countries form a certain cluster, the parameters of which could be considered guidelines for other countries of the sample. The competitiveness of climate-responsible entrepreneurship in countries of Central Asia is much lower. A rather significant level of development of the digital economy, peculiar to Kazakhstan, is accompanied by a low level of environmental effectiveness; Kyrgyzstan's position is slightly worse, while the competitiveness of climate-responsible entrepreneurship in Tajikistan in the digital economy is the lowest within the sample.

4 Discussion

The problems considered in this research form an important scientific direction, which covers theoretical and practical aspects of the development of economic systems from the position of their competitiveness by the criteria of climate responsibility in the markets of the digital economy. Therefore, perspectives of the research have a large potential, aimed at improvement of the methodology of assessing the competitiveness of climate-responsible entrepreneurship, determination of the influence of environmental or digital criteria on economic indicators and searching for the methods of achievement of the optimal result from a win-win position.

The research directions could be implemented within several concepts, including the hypothesis of the environmental Kuznets curve [1], which studies how economic growth and access to digital technologies allow reducing the burden on the environment; green competitiveness [3], which defines the key criteria of success of entrepreneurial structures from the position of a well-balanced combination of the environmental and economic interests in the markets of the digital economy; the concept of sustainable development [6], which is based on the Sustainable Development Goals and envisages the search for factors and conditions which would allow for the achievement of the most optimal result of the combination of the economic and digital potential in favour of the interests of fighting climate change.

Additional directions for the research that can substantially expand the results obtained are the search for the possibility to use digital technologies to solve the problems of climate change and the formation of a system of management that can stimulate entrepreneurial structures in the markets of the digital economy for climate-responsible activity on the principles of mutual profit from the position of economic effectiveness and the good for the environment.

5 Conclusions

Achievement of a high level of climate-responsible entrepreneurship's competitiveness in the markets of the digital economy in countries of Central Asia and Eastern Europe involves a balanced combination of both directions of economic development, which is characterised by indices of ICT development and environmental effectiveness. The use of the matrix method to assess the indicators allowed determining the level of climate-responsible entrepreneurship's competitiveness for the group of considered countries and the level of domination of the digital or climate priority in this aspect, as well as grouping them by the analysed indicators. In this context, it was determined, that countries of Eastern Europe achieved a mostly better level of competitiveness of climate-responsible entrepreneurship in markets of the digital economy. Except for Russia, Poland and Ukraine, the position of Eastern European countries in the matrix demonstrates a well-balanced level of environmental effectiveness and development of ICT. Countries of Central Asia demonstrate the priority of the development of the digital economy over the criterion of environmental effectiveness or are characterised by low values of both parameters, which does not allow identifying their position as competitive from the position of climate responsibility in digital markets.

References

1. Al Khars MA, Alwahaishi S, Fallatah MR, Kayal A (2022) A literature review of the Environmental Kuznets Curve in GCC for 2010-2020. Environ Sustainability Indic 14. https://doi.org/10.1016/j.indic.2022.100181
2. Ambec S, Cohen MA, Elgie S, Lanoie P (2010) The porter hypothesis at 20: can environmental regulation enhance innovation and competitiveness? TSE Working Paper Series. Research Group: Environmental Economics and Natural Resources, pp 10–215. http://publications.ut-capitole.fr/3556/1/10-215.pdf. Accessed 18 Jan 2023
3. Chygryn O, Lyulyov O, Pimonenko T, Myronenko N (2021) Key indicators of green competitiveness: the EU and Ukraine's performance. E3S Web Conf 307:03003. https://doi.org/10.1051/e3sconf/202130703003
4. Dechezleprêtre A, Martin R, Bassi S (2019) Chapter 10: Climate change policy, innovation and growth. In: Handbook on green growth. Edward Elgar Publishing, Cheltenham, UK. https://www.elgaronline.com/display/edcoll/9781788110679/9781788110679.00018.xml. Accessed 18 Jan 2023
5. Egorova S, Kistaeva N, Kulachinskaya A, Nikolaenko A, Zueva S (2021) Development of methods for assessing the impact of environmental regulation on competitiveness. Int J Technol 12(7):1349–1358
6. Karpunina EK, Lapushinskaya GK, Arutyunova AE, Lupacheva SV, Dubovitski AA (2020) Dialectics of sustainable development of digital economy ecosystem. In: Popkova E, Sergi B (eds) Scientific and technical revolution: yesterday, today and tomorrow. ISC 2019, Lecture notes in networks and systems. Springer, Cham, p 129. https://doi.org/10.1007/978-3-030-47945-9_54

7. Li R, Jiang H, Sotnyk I, Kubatko O, Almashaqbeh YAI (2020) The CO2 emissions drivers of post-communist economies in Eastern Europe and Central Asia. Atmos 11(9):1019. https://doi.org/10.3390/atmos11091019. Accessed 18 Jan 2023
8. Matos S, Viardot E, Sovacool BK, Geels FW, Xiong Y (2022) Innovation and climate change: a review and introduction to the special issue. Technovation 117:102612. ISSN 0166-4972. https://doi.org/10.1016/j.technovation.2022.102612
9. Miethlich B, Belotserkovich D, Abasova S, Zatsarinnaya E, Veselitsky O (2020) Digital economy and its influence on competitiveness of countries and regions. ESPACIOS 41(12):20. http://ww.revistaespacios.com/a20v41n12/20411220.html. Accessed 18 Jan 2023
10. UN (1999) Standard country or area codes for statistical use, vol 49. United Nations, Statistical Office, New York. https://unstats.un.org/unsd/methodology/m49/. Accessed 18 Jan 2023
11. USAID (2022) Country dashboard. U.S. Agency for International Development. https://idea.usaid.gov/cd. Accessed 18 Jan 2023
12. Yusuf MF, Ashari H, Razalli MR (2018) Environmental technological innovation and its contribution to sustainable development. Int J Technol 9(8):1569–1578

Digital Environment of Entrepreneurial Structures as a Basis for Spatial Regional Development: Problems and Ways to Solve Them

Irina A. Morozova, Ekaterina V. Kuzmina, Svetlana A. Shevchenko, Mariya I. Kuzmina, and Valentina A. Dorzhdeeva

1 Introduction

In today's conditions of increasing international integration and global competition, developing the country's economy is possible only based on digital transformation. However, there are significant territorial disproportions and regional differences in the level of economic development in the Russian Federation. Digital entrepreneurship occupies a central place and can become a significant factor in the development of regions due to the peculiarities of its functioning.

The global development of digital technologies within the framework of Industry 4.0 directly affects the economic system, particularly entrepreneurship. Entrepreneurial structures actively respond to measurements of the external and internal environment, being the main driving force behind the development of all sectors of the economy. Business entities are producers and accelerators of digital innovations because they primarily carry out the production, testing and transfer of information technologies. Digital transformation accelerates the exchange of information, increases the competitiveness of business in Russia and the world market and makes it possible to develop and make a choice in favour of an entrepreneurship development strategy while reducing transaction costs and increasing value added. In this regard, the research topic is particularly relevant. Nowadays, the development of digital entrepreneurship is hindered by several significant problems, including the need to improve the regulatory and legal framework, the lack of IT specialists for developing and implementing digital innovations, the search for investments and sources of financing for projects to produce digital technologies, the creation of effective cybersecurity systems, the design of digital infrastructure and the expansion of technical and organisational capabilities.

I. A. Morozova (✉) · E. V. Kuzmina · S. A. Shevchenko · M. I. Kuzmina · V. A. Dorzhdeeva
Volgograd State Technical University, Volgograd, Russia

E. G. Popkova (ed.), *Smart Green Innovations in Industry 4.0*, Springer Climate,
https://doi.org/10.1007/978-3-031-45830-9_2

The research aims to analyse the main problems of digital entrepreneurship in the regional economy and find ways to solve them.

The digitalisation of the economy depends in many aspects on the architecture of entrepreneurial ecosystems. The development of digital entrepreneurship is possible only with the formation of a high-quality environmental ecosystem, which contributes to an increase in labour productivity in some regions and cities [1, 9, 10].

The digital entrepreneurial environment will change the landscape of the entire economy and region of the country through the creation of information technologies, the production of high-tech products and services and the transformation of business processes and business models in all industries and areas of activity. Analysts predict an increase in the global industry of digital goods and services to $26.5 trillion by 2022 [2]. This provides great potential for developing regions and enterprises planning to operate in the area of digital entrepreneurship.

The research novelty is as follows:

1. Grouping and identification of the problems of developing digital entrepreneurship: regulatory, human resources, investment, infrastructure, technical and organisational, which are based on the problems of economic, innovative and social development of regions.
2. Justification of the effectiveness of creating a digital eco-environment in business as a new format of territorial development.
3. Development of a conceptual scheme for the unification of digital ecosystems of business structures containing universal platforms and services and including three stages: (1) creation of innovative technologies in entrepreneurship, (2) innovation of all areas of business and creation of basic infrastructure platforms and (3) formation of a single digital business eco-environment, including compatible industry environments.

The approach proposed by the authors will ensure the interaction of digital business entities on an innovative basis.

2 Materials and Methods

Theoretical and methodological issues related to the development of entrepreneurship are in the focus of scientists, businesses and governments worldwide. Practical studies show that countries with developed entrepreneurial structures will increase economic and social effects within the country and strengthen the competitive position in the world economy [5, 6].

Approaches to regional development were considered in the works of foreign and Russian scientists, including Morozova, Kuzmina, Shevchenko and Simonov [8], Morozova, Kuzmina, Kuzmina, Shevchenko and Avdeyuk [7] and Stam [9]. The problems of the regional economy, spatial development, stimulation of economic growth of territories and search for innovative growth points were studied by Morozova, Kuzmina, Shevchenko and Simonov [8], as well as Stam and Spigel [10].

The development of digital entrepreneurship can ensure the innovative development of the country's economy and its individual territories. Business entities have the potential to become active technology developers or adapt technologies created in other countries to improve business processes [2]. Certain aspects of digitalisation of the economy and digital entrepreneurship are considered in the works of Alvedalen and Boschma [1], Kadatskaya and Lavrova [3] and Karelina [4].

Scientific and technological transformation in the area of engineering and information technology has intensified applied research. Science and practice develop and test technologies, such as blockchain, artificial intelligence and neural networks. Based on innovative technologies, new forms and methods for developing digital entrepreneurship are being identified. The problems of developing digital entrepreneurship and economic entities, territorial disproportions and improving infrastructure elements that conduct digital innovations are still relevant.

Based on the application of theoretical, empirical and general logical methods of scientific knowledge and systemic approach, the research highlights the legal, personnel, investment, infrastructure, technical and organisational problems of the development of digital entrepreneurship and suggests separate ways to solve them.

3 Results

3.1 Grouping and Identification of Problems in the Development of Digital Entrepreneurship and Ways to Solve Them

1. Problems of legal regulation and insufficient development of the legislative framework and ways to solve them. The most pressing issue from the point of view of legal regulation of digital entrepreneurship is the unpreparedness of the legislation for the emergence of innovative digital technologies, which are practically not reflected in the current regulatory framework. According to the authors, legal regulation in Russia is based on the application of foreign experience without considering the specifics of the national system. Technologies of the legislative process in the area of legal regulation of the use of digital innovations should be based on scientific, informational, technical, material, organisational and educational foundations. With the introduction of innovative technologies in the activities of entrepreneurship, the problem of different levels of readiness of regions for digitalisation arises. It is very important to provide for a federal law that gives regions the right to issue acts that simplify the decision-making process and contribute to the introduction of digital innovations. In total, a synergistic effect will be achieved in developing entrepreneurship and the digitalisation of all areas of the region's economy.
2. Problems of staffing and ways to solve them. The development of digital entrepreneurship is possible only if there are highly qualified and specially trained

personnel for information and communication activities in the area of digital technologies. The authors believe that it is especially important to achieve systemic unity of scientific, educational and methodological support for training personnel for activities with innovative digital technologies in entrepreneurship.

3. Problems of investment and financing and ways to solve them. The issue of financing development and the creation of innovative digital technologies is very acute. The solution to this problem is designed to ensure investment at all stages of the process: from initial investments at the stage of a start-up to loan financing at the final stages. It is possible to solve the problems of attracting investments with the help of research and fundamental science support funds, which finance innovative developments of high-tech and science-intensive products.
4. Problems of cyber security and ways to solve them. One of the main problems of digital entrepreneurship is information security, which must be addressed in conjunction with a continuously improved cybersecurity system. The security issue includes two aspects: the security of information and digital technologies and the reduction of the dependence on imported software products as part of the import substitution program. A required condition for solving the problem of cybersecurity is the centralised management of information processing. In this case, an information security policy should be developed in such a way that it can ensure the security of digital infrastructure facilities, maintaining the confidentiality and integrity of data, as well as the unified information space.
5. Infrastructural problems and ways to solve them. The digital infrastructure provides opportunities for access to databases and arrays of information resources on mutually beneficial terms for small and medium-sized businesses. The solution to the problem can be the creation of digital infrastructure that provides the opportunity for the subjects to use the formed unified information and analytical base at the federal and regional levels, the purpose of which is ensuring innovation and exchanging information communications in the area of digital technologies.
6. Technical problems and ways to solve them. Digitalisation of the economy of the country and regions is of a point nature; it affects mainly urban conglomerations. The biggest part of digital technologies and tools is not sufficiently accessible for most economic entities operating in regions. The elimination of disproportions in the socio-economic development of regions is possible through the digital innovation of entrepreneurship, which is a regional development driver.
7. Organisational problems and their solutions. The organisational aspect is the fact that there is no single digital environment and single ecosystem of digital entrepreneurship. Separate digital platforms and services are actively developed and used, mainly in public sector organisations, as a result of which the need for the use of digital innovations by business structures increases many times over. The exchange of information resources, the creation of relevant services and software products and the transfer of digital innovations by business entities in a single digital environment are currently impossible. The authors propose combining individual services containing platforms and services into a single digital

ecosystem to solve the problem. Diffusion of digital innovations can occur through the adaptation of platforms and technologies, on the basis of which general and specific competencies are formed for the development of entrepreneurship.

Nowadays, innovative digital technologies are recognised as drivers of economic development, which can be ensured through their active implementation and use. According to the authors, promising digital technologies can be used in all areas of business and increase the efficiency of the functioning of medium and small businesses and the region's economy. These technologies include the following:

- Cloud technologies provide large computing power to achieve high performance in business, production and science. In entrepreneurship, they allow scaling production processes, deploying and creating business models and bringing new products to the market faster, ahead of competitors while reducing corporate costs for storing and transmitting information without losing the level of information security.
- Big data allows processing structured and unstructured data in different systems to increase work efficiency. Big data business enterprises provide the highest level of generalisation and analysis of information, which is important for the development and adoption of management decisions, profit forecasting and building logistics supply chains of resources.
- Artificial intelligence and robotisation contribute to creating and implementing new models and business processes, which makes it possible to get great competitive advantages. In entrepreneurship, it tracks and analyses incremental changes in demand, making it easier to translate strategic opportunities into business model elements. Robots are used to increase the efficiency of routine and labour-intensive procedures in order to optimise and automate the internal processes of business enterprises.
- Machine learning is essential for entrepreneurship. For entrepreneurship, machine learning has a number of undeniable advantages. The Internet of Things (IoT) will combine many technologies and automation systems, providing unlimited possibilities for monitoring, controlling and managing technological and production processes in real-time. Blockchain technology is a revolutionary breakthrough in innovative digital technologies. The advantages of the blockchain methodology for business structures are effective defense mechanisms against external interference and achieving global network performance with a high degree of cyber security.

3.2 Conceptual Scheme for Creating a Single Digital Eco-environment That Ensures the Development of Digital Entrepreneurship

The need to transform business systems by business structures is dictated by the current trends in the digitalisation of the economy. The use of a contemporary approach to the formation and use of a business model of entrepreneurship makes it possible to achieve high-performance indicators and a high level of competitiveness and occupy a niche in the market.

Innovative technologies are an effective tool for transformation, the introduction of which solves the large-scale task of creating an ecosystem of digital entrepreneurship. The main task of creating a digital ecosystem of entrepreneurship is to develop the country's economy, increase the socio-economic indicators of territories' development and eliminate regional structural imbalances through the production of high-tech products.

Creating a single digital space within the digital eco-environment is the formation of a new business model that will provide business entities with a sustainable development strategy and the acquisition of additional competitive advantages. Technological integration of digital platforms and services is a promising trend in developing the digital business environment.

To solve the problems posed, the authors propose a conceptual scheme for creating a single digital environment that combines industry environments and platform technologies (Fig. 1).

The conceptual scheme for creating a digital eco-environment involves the consistent implementation of three basic blocks: (1) creation of innovative technologies in entrepreneurship, (2) innovation of all areas of business and creation of basic infrastructure platforms and (3) integration of single digital entrepreneurship eco-environment.

Stage I is the creation of innovative technologies in entrepreneurship. The first stage includes the creation and implementation of technologies that are joint products of industry innovations and digital products. The introduction of innovative technologies allows business entities the following:

- To search for promising technological solutions for the introduction of digital innovations to optimise current business processes and generate additional sources of income.
- To conduct an examination to predict possible risks and produce additional profit as a result of the digitalisation of business processes.

Stage II is an innovation of all areas of business and the creation of basic infrastructure platforms. The presence of platforms allows creating innovative service products. Integration of industry environments and infrastructure platforms accumulates innovative activity and provides an opportunity to realise the potential of business entities.

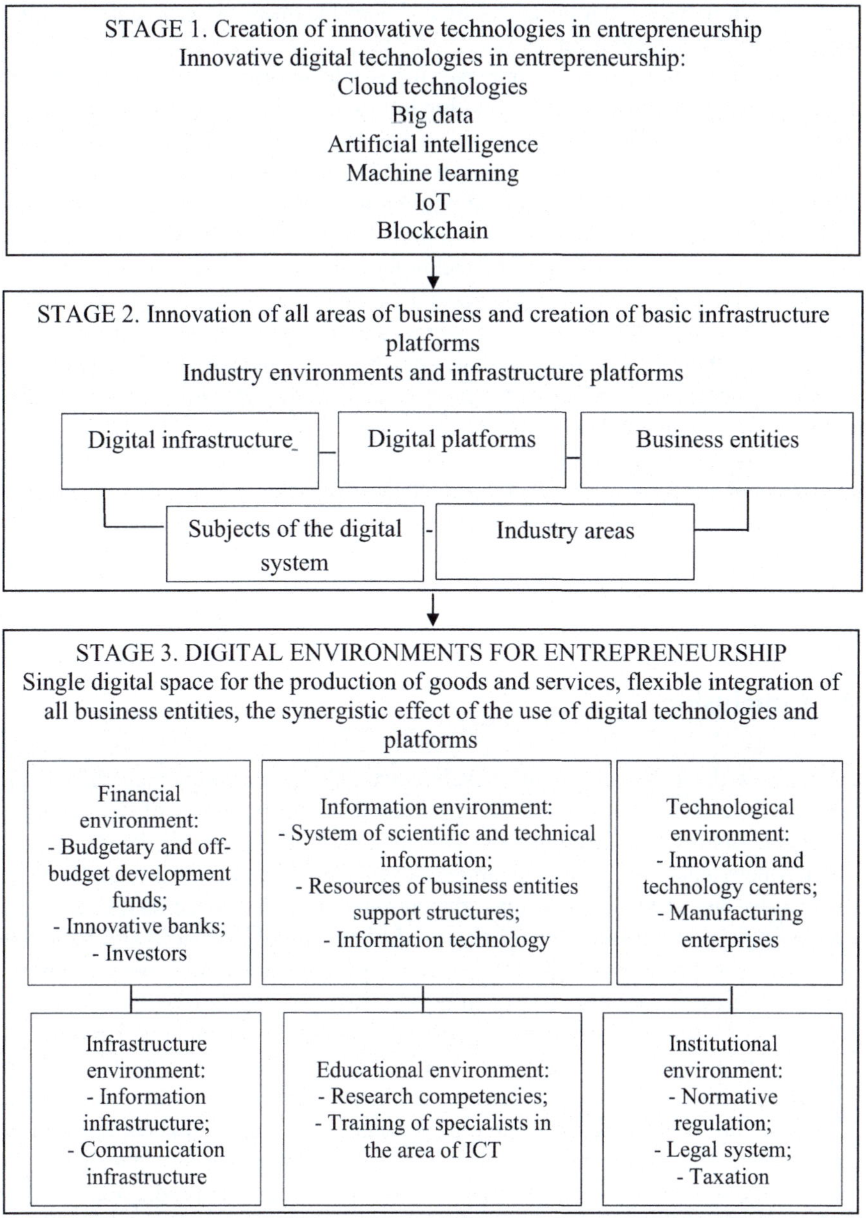

Fig. 1 Conceptual diagram of the process of creating a digital entrepreneurial environment. (*Source* Compiled by the authors)

Stage III is the formation of a single digital business environment, including compatible industry environments.

The infrastructure environment makes it possible to form a single digital network for processing, analysing and transmitting information, which contributes to the use of qualitatively new business processes in the digital ecosystem. The development of the infrastructure environment is extremely important for business entities and the region. The functioning of digital entrepreneurship in the ecosystem is impossible without regional information technologies, access to information services, knowledge-intensive production and technological telecommunications services.

The institutional environment ensures the regulatory and legal regulation of the digital environment, the system and the procedure for taxation. The improvement of these areas should be carried out as infrastructure platforms and technologies develop. The educational environment prepares personnel with industry-specific digital competencies and skills to carry out scientific and technical research, as well as developments in digital innovation and information technology. The information environment allows entrepreneurship to integrate into the architecture of the digital ecosystem. The technological environment provides opportunities for the production of high-tech products and unique services based on the use of innovation and technology centres and enterprises.

4 Conclusion

A study of the problems of the development of digital entrepreneurship at the regional level has demonstrated that it is based on the problems of the country's economic and political development. The main problems are of regulatory and legislative, personnel, investment, technical, organisational and infrastructural nature.

To solve the identified problems, the authors propose a conceptual scheme for combining ecosystems containing universal platforms and services and including three stages: (1) creation of innovative technologies in entrepreneurship, (2) innovation of all areas of business and creation of basic infrastructure platforms and (3) formation of a single digital entrepreneurial eco-environment, including compatible industry environments. The proposed approach to the systemic formation of a single digital eco-environment, presented in the developed scheme, will contribute to developing digital entrepreneurship in a regional aspect. The digital entrepreneurial environment will change the landscape of the entire economy and each region of the country through the creation of information technologies, the production of high-tech products and services and the transformation of business processes and business models in all industries and areas of activity.

Thus, the ways of solving problems proposed by the authors and the conceptual scheme for creating a digital business environment contribute to the progressive development of regions and the national economy.

Acknowledgements This study was supported by the Russian Foundation for Basic Research, project No. 20-010-00072 "Formation of creative centers of spatial development as a mechanism for improving the quality of life of the population of rural areas".

References

1. Alvedalen J, Boschma R (2017) A critical review of entrepreneurial ecosystems research: towards a future research agenda. Eur Plan Stud 25(6):887–903. https://doi.org/10.1080/09654313.2017.1299694
2. Broadband Commission (2018) Working group report on digital entrepreneurship. Broadband commission for sustainable development. ITU and UNESCO, Geneva. Retrieved from https://www.broadbandcommission.org/workinggroups/Pages/Digital-Entrepreneurship.aspx. Accessed 18 Oct 2021
3. Kadatskaya DV, Lavrova YS (2020) Trends in the development of innovative technological entrepreneurship in the digital economy. Russ J Innov Econ 10(2):985–992. https://doi.org/10.18334/vinec.10.2.100800
4. Karelina EA (2020) Opportunities for building effective digital platforms and ecosystems in developing countries. E-Management 3(1):59–67. https://doi.org/10.26425/2658-3445-2020-1-59-67
5. Kosmin AD, Kosmina EA, Malysheva VA (2018) Small and medium-sized entrepreneurship in The Russian Federation: comparative analysis of trends and performance. Creative Economy, Moscow
6. Kostin KB, Khomchenko EA (2021) Improving the efficiency of Russian entrepreneurship based on foreign experience in international technology transfer. Russ J Innov Econ 11(2):729–744. https://doi.org/10.18334/vinec.11.2.112161
7. Morozova IA, Kuzmina EV, Kuzmina MI, Shevchenko SA, Avdeyuk OA (2019) Economic development of the territory on the basis of formation of regional innovation system in conditions of digitalization. In: Popkova EG, Sergi BS (eds) Modern global economic system: evolutional development vs. revolutionary leap. Springer, Cham, pp 974–982. https://doi.org/10.1007/978-3-030-69415-9_108
8. Morozova IA, Kuzmina EV, Shevchenko SA, Simonov AB (2021) Digital and information technologies as a tool for creative spatial development of territories. Scientific works of the Free Economic Society of Russia 230(4):332–338. https://doi.org/10.38197/2072-2060-2021-230-4-332-338
9. Stam E (2015) Entrepreneurial ecosystems and regional policy: a sympathetic critique. Eur Plan Stud 23(9):1759–1769. https://doi.org/10.1080/09654313.2015.1061484
10. Stam E, Spigel B (2018) Entrepreneurial ecosystems. In: Blackburn R, De Clercq D, Heinonen J (eds) The SAGE handbook for entrepreneurship and small business. SAGE Publications, London, UK. https://doi.org/10.4135/9781473984080

Platform Approach as an Innovative Trend in the Development of Digital Entrepreneurship: Regional Aspect

Irina A. Morozova, Ekaterina V. Kuzmina, Svetlana A. Shevchenko, Mariya I. Kuzmina, and Valentina A. Dorzhdeeva

1 Introduction

In today's conditions of escalation of the processes of globalisation and integration into international spheres of activity, the digital version of the development of the country's economy presents unlimited opportunities for modernising all industries and areas of activity, developing innovations and improving socio-economic indicators. An innovative development path is possible only through implementing digital transformation processes, which implies the introduction of information and communication technologies in all areas of activity. The socio-economic development of regions is possible only with the use of digital technologies, the use of software products by business entities and the introduction of information services in the process of making managerial decisions.

Digital entrepreneurship is the institutional basis for the socio-economic, industrial and innovative development of regions. Due to their agility, susceptibility and instant adaptation to innovative technologies, business entities can become drivers for the creation and use of digital products. The use of digital tools and technologies in digital entrepreneurship contributes to improving the socio-economic indicators of the development of the territorial unit, increasing the investment attractiveness of the subject's industries and acquiring competitive advantages of the region's economy.

Digital entrepreneurship is based on doing business by using a leadership style with information and communication technologies to create competitive advantages, covering all industries and areas of activity: the production of high-tech products, the provision of unique services, innovation, business processes, digital technologies, marketing, interaction producers and consumers. In this regard, the

I. A. Morozova (✉) · E. V. Kuzmina · S. A. Shevchenko · M. I. Kuzmina · V. A. Dorzhdeeva
Volgograd State Technical University, Volgograd, Russia

E. G. Popkova (ed.), *Smart Green Innovations in Industry 4.0*, Springer Climate,
https://doi.org/10.1007/978-3-031-45830-9_3

manufactured products and services provided by business structures become digital and knowledge-intensive, which opens up unlimited opportunities for innovative business processes and business models due to the information environment of digital platforms. In total, this emphasises the particular relevance of the study.

The research aims to substantiate the effectiveness of the use of digital platforms in digital entrepreneurship to increase the sustainability of the spatial development of the regional economy.

The research novelty is as follows:

1. The effectiveness of the use of digital platforms in entrepreneurial activity as a new format in social, economic and innovative development of the region is substantiated.
2. The research formulated and proposed conceptual provisions for the use of digital platforms in entrepreneurship to increase the economic and innovative efficiency of digital entrepreneurship.
3. The authors developed an algorithm for managing the development of digital entrepreneurship through the use of digital platforms, which includes three stages: (1) studying business processes of business entities and assessing the possibility of modifying them; (2) digital transformation of the activities of business entities based on digital platforms built into the business model; (3) evaluation of the effectiveness of digital and information technology applications.

2 Materials and Methods

Theoretical and methodological issues of regional development have been widely studied by such scientists as Gelishanov, Yudina and Babkin [1], Morozova, Kuzmina, Shevchenko and Simonov [2] and Vorobyev and Drozdov [3]. Digital transformation characterises a new economic format that is emerging through opening up information and technological opportunities. Extended views on the problems of digitalisation of the economy are presented in the works of Ghazawneh and Henfridsson [4] and Parker, Van Alstyne and Choudary [5].

Practical aspects of the functioning of entrepreneurial activity and the possibility of business participation in the development of regions were studied by Kuzmina, Avdeyuk, Kuzmina, Tarasova and Rayushkina [6] and Solodilova, Malikov and Grishin [7]. The transformation of the economy caused by digital innovations was studied by Evans [8] and Parker, Van Alstyne and Choudary [5]. Entrepreneurship as the most flexible and receptive environment for technological innovations was the object of study by Polyanin, Soboleva and Tarnovskiy [9]. Of particular interest are the works of Evans [8], Gelishanov, Yudina and Babkin [1] and Thomas, Autio and Gann [10], devoted to contemporary trends in the development of digital platforms.

However, despite the rapid use of information and communication technologies and the active development of digitalisation processes in all areas of activity, the theoretical and methodological issues of applying innovative technologies in digital entrepreneurship are poorly understood. The general issues of entrepreneurial activity have been mainly studied. The application of technological innovations in digital entrepreneurship, particularly the impact of digital entrepreneurship on the regional economy, has not received sufficient attention.

The problems of introducing advanced digital technologies for economic entities, territorial issues of development and improving business models of business environment organisations that conduct digital innovations also remain relevant.

The methods used in the research make it possible to substantiate the impact of digital entrepreneurship on regional development. The paper uses forecasting and SWOT analysis to assess the possibility of using digital platforms in entrepreneurship.

3 Results

3.1 Problems and Difficulties in the Development of Digital Entrepreneurship in the Regional Economy and the Need for Digital Platforms

In the real sector of the economy, digital entrepreneurship activates the creation of innovations, the development of high-tech products, the production and launch of new goods and services, as well as the development of new digital technologies. Due to the rapid pace of technological development and the systemic impact on all sectors of the economy, the change in the technological formation as part of the deployment of the fourth industrial revolution entails fundamental changes in the management of regional development.

Industry 4.0 creates the potential for developing digital entrepreneurship by strengthening the interaction of cyber-physical systems. The integration of digital entrepreneurship in all areas of the economy contributes to an increase in gross value added and the gross regional product of territorial units. According to the McKinsey Global Institute, the economic effect of the digitalisation of the economy of the Russian Federation's GDP by 2025 will be 4.1–8.9 trillion rubles, i.e. will be from 19% to 34% of the total growth of the indicator [11].

The distinctive features of digital entrepreneurship are as follows:

1. Carrying out activities using information and communication technologies and software products.
2. Increasing exports by entering world markets.
3. Implementation of opportunities for activities in electronic markets, participation in international electronic trading processes and functioning on marketplaces.

4. The use of digital platforms to ensure the relationship between the manufacturer, seller and buyer (Grid technology, BioTech, NanoTech, BlockChain, Digital marketing, etc.)
5. Increase in labour productivity.
6. Reduction of the transaction, management and non-operating costs.
7. Increasing the consumer value of the services provided and goods produced due to their innovativeness and high technology.
8. Increasing the level of uncertainty and risks due to dynamic changes in technology.

The organisation and development of digital entrepreneurship are possible in external and internal conditions. The most important external condition is a favourable business environment, which is an ecosystem that can ensure effective interaction of participants and provide the necessary resources to generate innovation, produce high-tech products, create new industries and develop new markets. Internal conditions include scientific and technical developments, subdivisions of research laboratories, highly qualified personnel, application of high-quality standards and attraction of innovative business.

Digital platforms are an effective tool for the interaction of innovation, science and business. The digital platform in digital entrepreneurship forms an information platform that contributes to the activation of the process of creating new products and innovative technologies, which are promising for commercialisation, attracting resources for development and research work with the participation of science, business and government agencies and improving the regulatory framework in the area of research and innovation – investment activity.

The authors formulated and proposed conceptual provisions for the use of digital platforms in business activities to increase economic and innovative efficiency.

The processes of digital transformation in entrepreneurial activity necessitate changes in business models, the effective functioning of which is possible based on digital platforms. The use of digital platforms will ensure a significant increase in market volumes and increase the competitiveness of business entities. The effectiveness of the use of digital platforms by business entities is disclosed in the following conceptual provisions:

1. A digital platform, built into the business model and based on information technology, creates additional value, improves the production process and optimises the interaction between participants in the business process.
2. A digital platform includes a combination of high information and communication technologies, which provides competitive advantages.
3. A digital platform is a step-by-step algorithm of actions that makes the most optimal interaction of all participants in the business environment, as a result of which it is possible to increase labour productivity, reduce transaction costs and build efficient logistics by optimising business processes.

In accordance with the first provision, the authors note that the main goal of entrepreneurial activity is the creation of value-added. The increase in added value occurs as a result of the formation of the optimal ratio of cost and profit. It is possible to achieve such a ratio through the optimal organisation of technology in the production process of products and services. The classical business model builds the process of creating the value of a product or service linearly (i.e. the stage of production of goods is initially carried out, and then it is delivered to the consumer through marketing and logistics). Using digital platforms, business entities carry out these processes in parallel or in parallel-sequentially, thereby ensuring continuous value creation.

The second provision is about obtaining additional competitive advantages. Obtaining competitive advantages is achieved by entrepreneurial structures, when interacting participants have the opportunity to build activities on their own, gaining access to the appropriate platform services and resources. Nowadays, digital platforms create communication between producers and consumers of goods and services in compliance with information cyber security standards. Competitive advantages are achieved by expanding information capabilities, increasing resistance to high loads and increasing security against cyber-attacks. The architecture of digital platforms opens up the possibility for business entities to respond flexibly and dynamically to rapidly changing business conditions, allowing them to make and correct management decisions in a timely and efficient manner.

According to the third provision, the mutually beneficial relationship of business environment entities that carry out consistent activities in the form of a certain algorithm in the information environment based on digital platforms leads to a reduction in transaction costs. Digital platforms provide the opportunity to apply big data using cloud technologies, conduct transactions using blockchain and implement RFID to reduce the costs of entrepreneurial activity. In addition to transaction costs, the use of digital platforms allows reducing other types of costs, including non-operating and other expenses.

The authors believe that the digitalisation of the economy necessitates changes in business models. Business entities must use innovative technologies to improve performance, enhance competitive advantages and increase profits for this. In today's conditions, the use of digital platforms can become the most effective tool for entrepreneurship.

3.2 *Opportunities and Prospects for the Introduction of Digital Platforms in the Activities of Business Structures*

The digital platform is an information and communication system, the functionality of which includes ensuring multilateral interaction between manufacturers and buyers in the process of creating and exchanging information and selling goods and services. The implementation of the activities of organisations in the area of digital entrepreneurship based on digital platforms ensures the optimisation of business processes, the reduction of transaction costs and the increase in the efficiency of the cycle from production to marketing of products (goods and services).

Specialists create digital infrastructure and innovative environment, improving business models and changing hierarchical relationships in the structure of digital entrepreneurship platforms.

The authors apply SWOT analysis to assess the possibility of using digital platforms by business entities in terms of the state of the external and internal environment.

The authors also propose an algorithm for managing the development of digital entrepreneurship through the use of digital platforms (Fig. 1).

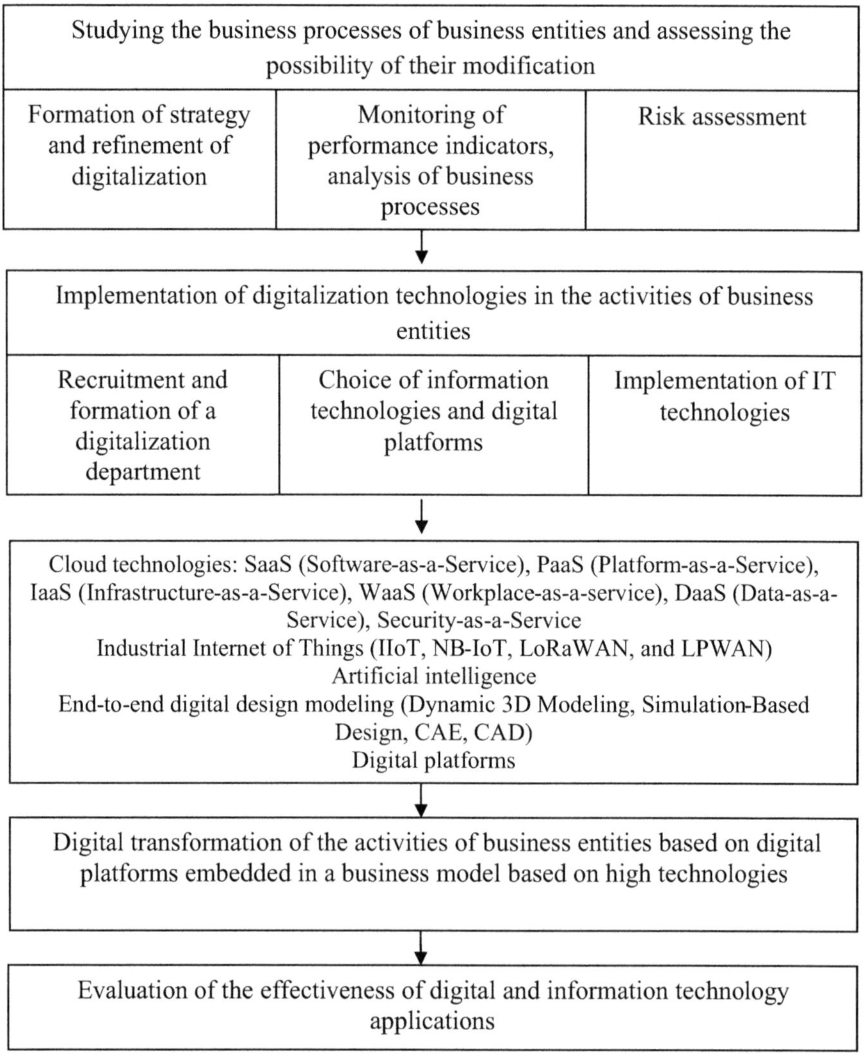

Fig. 1 Algorithm for managing the development of digital entrepreneurship through the use of digital platforms. (Source: Compiled by the authors)

The digitalisation of entrepreneurial activity must be started with a thorough analysis of the results of its functioning to identify existing problems and risks. Depending on the developed strategy and the tasks set, it is necessary to choose information technologies and digital platforms. After that, it becomes possible to carry out digital transformations by changing the business model based on digital platforms. The results of monitoring and evaluating the use of digital and information technologies will present the level of effectiveness of the changes made.

The results of the SWOT analysis conducted in Table 1 demonstrate that threats arise from the external environment: the emergence of new competitors, an increase in cyber threats and a decrease in information security while using digital platforms in the business. Along with this, digital platforms provide opportunities for obtaining information about consumer preferences, expanding new markets and integrating digital platforms into the ecosystem.

The implementation of entrepreneurial activities based on the use of digital platforms opens up such strengths as a high potential for transmission and storage of information, creation of innovations, development and production of high-tech products, reduction of time and money costs of information transfer, reduction of transaction costs, elimination of intermediary organisations and services and expansion of sales markets.

The external environment for the use of digital platforms for digital entrepreneurship enterprises is characterised by weaknesses associated with commission

Table 1 SWOT analysis of the assessment of the possibility of carrying out activities by business entities based on digital platforms

External environment	
Threats	Opportunities
Decrease in information security	Studying demand and identifying customer preferences
Increase in cyber-attack capabilities	Integration of digital platforms into a single digital environment
The emergence of new competitors	Entering new markets
Internal environment	
Weaknesses	Strengths
Commission costs	High potential for transmission and storage of information
	Creation of innovations, development and production of high-tech products
Weak infrastructural and technological accessibility of network forms of interaction	Reduction of time and money costs of information transfer
Professional incompetence of personnel	Establishment of direct sales channels
The need to restructure corporate processes	Reduction of transaction costs
	Elimination of intermediary organisations and services
	Expansion and increase in activity volumes

Source: Compiled by the authors

costs, underdeveloped infrastructure and technological accessibility of network forms of interaction, professional incompetence of personnel and the need to restructure corporate processes.

Thus, strengths prevail over weaknesses. This factor makes it possible to talk about the effectiveness and prospects of using digital platforms in entrepreneurship. Digital platforms reveal the high potential of opportunities for processing and transmitting information, contributing to the growth of added value, increasing profitability, reducing transaction costs and optimising supply chains.

4 Conclusion

Digital platforms have a huge potential for developing and modernising all areas and sectors of the economy. Globalisation and integration dictate the need to change business models of entrepreneurial activity using digital platforms. The authors formulated the conceptual provisions for the effectiveness of the introduction of digital platforms in business activities. The use of platforms will increase competitiveness and create conditions for the generation of technological, innovative and information resources of digital entrepreneurship and areas of the regional economy that are closely integrated with it.

Digital platforms in business activities can enhance competitive advantages based on emerging effects. In the future, digital platform technology will provide additional value-added growth factors, optimise information flows, reduce transaction costs, increase labour productivity and improve the exchange of information and technologies between participants in the business environment.

Thus, in today's conditions, digital platforms and digital ecosystems develop entrepreneurship and economic sectors, become drivers of innovative activity and improve the socio-economic condition of regions.

Acknowledgments This study was supported by the Russian Foundation for Basic Research, project No. 20-010-00072 "Formation of creative centers of spatial development as a mechanism for improving the quality of life of the population of rural areas."

References

1. Geliskhanov IZ, Yudina TN, Babkin AV (2018) Digital platforms in economics: essence, models, development trends. π-Economy [St Petersburg State Polytechnical University Journal Economics] 11(6):22–36. https://doi.org/10.18721/JE.11602
2. Morozova IA, Kuzmina EV, Shevchenko SA, Simonov AB (2021) Digital and information technologies as a tool for creative spatial development of territories. Sci Work Free Econ Soc Russ 230(4):332–338. https://doi.org/10.38197/2072-2060-2021-230-4-332-338

3. Vorobyev AS, Drozdov OA (2017) Theoretical approaches and practical aspects concerning the analysis of factors for the social and economic development of regions. Theory Pract Soc Dev 3:42–45. https://doi.org/10.24158/tipor.2017.3.8
4. Ghazawneh A, Henfridsson O (2015) A paradigmatic analysis of digital application marketplaces. J Inf Technol 30(3):198–208. https://doi.org/10.1057/jit.2015.16
5. Parker GG, Van Alstyne MW, Choudary SP (2016) Platform revolution: how networked markets are transforming the economy – and how to make them work for you. W.W. Norton and Company, New York/London
6. Kuzmina M, Avdeyuk O, Kuzmina C, Tarasova I, Rayushkina A (2020) Strategy of development of regional infrastructure for creating innovative production complexes in the digital economy. In: Kolmykova T, Kharchenko E (eds) Digital future economic growth, social adaptation, and technological perspectives. Springer, Cham, pp 31–37. https://doi.org/10.1007/978-3-030-39797-5_4
7. Solodilova NZ, Malikov RI, Grishin KE (2018) Methodological tools to measure the state of regional entrepreneurial ecosystem. Econ Reg 14(4):1256–1269. https://doi.org/10.17059/2018-4-16
8. Evans DS (2003) Some empirical aspects of multi sided platform industries. Rev Netw Econ 2(3):191–209. https://doi.org/10.2139/ssrn.447981
9. Polyanin AV, Soboleva YP, Tarnovskiy VV (2020) Digitalization of processes of small and average business. Adm Consult 4:80–96. https://doi.org/10.22394/1726-1139-2020-4-80-96
10. Thomas LD, Autio E, Gann DM (2014) Architectural leverage: putting platforms in context. Acad Manag Perspect 28(2):198–219. https://doi.org/10.5465/amp.2011.0105
11. McKinsey (2017) Digital Russia: a new reality. Retrieved from https://www.mckinsey.com/~/media/mckinsey/locations/europe%20and%20middle%20east/russia/our%20insights/digital%20russia/digital-russia-report.ashx. Accessed 15 Apr 2021

High-Tech Export as a Vector of Development of Climate-Responsible Entrepreneurship in the Markets of the Digital Economy in Developed Countries

Saida M. Ibraimova, Azisbek A. Beksultanov, Larisa V. Shabaltina, and Tatiana A. Dugina

1 Introduction

Modern technological solutions allow the creation of new types of products in various sectors of the economy. Digitalisation is a driving force behind their competitiveness. Society and economy in developed and quickly developing countries are oriented toward high technologies and high-tech products to such a high extent that failures in their delivery might influence the work of many processes in business activities (banks, insurance companies, etc.), government sector, life activities of people and functioning of the critical infrastructure (activities of communal structures, municipal and national medical establishments). Companies that manufacture high-tech products are mostly located in developed countries or China, which is the largest developing country in the world. The manufacture and export of new types of high-tech products stimulate the growth of further consumer expectations, which is the basis for the competitive struggle between manufacturers. There are many characteristics of high-tech products, which parameters are compared before the purchase. Fewer materials are required for the manufacture of high-tech products, but often they are more expensive. For example, the production of ultra-thin

S. M. Ibraimova
Kyrgyz Economic University Named After M. Ryskulbekov, Bishkek, Kyrgyzstan

A. A. Beksultanov
Kyrgyz National University Named After Jusup Balasagyn, Bishkek, Kyrgyzstan

L. V. Shabaltina
Plekhanov Russian University of Economics, Moscow, Russia

T. A. Dugina (✉)
State-Funded Budget Educational Institution of Higher Education, Volzhsky Institute of Economics, Pedagogy and Law, Volzhsky, Russia
e-mail: deisi79@mail.ru

E. G. Popkova (ed.), *Smart Green Innovations in Industry 4.0*, Springer Climate,
https://doi.org/10.1007/978-3-031-45830-9_4

smartphones and laptops involves the use of less plastic and synthetic materials compared to earlier models of this equipment. Accordingly, after the end of their useful life, these products require less effort for utilising. The leading manufacturers of high-tech digital technologies try to bring down to the minimum the use of environmentally hazardous substances. Most of the world's leading manufacturers of high-tech products focus on the modern tendencies of climate preservation. With time, parameters of ecologisation become important, equally to technical characteristics. Growth of production and export of these products leads to an increase in GDP and an eco-oriented image of countries in which the world's largest technological giants are based.

The goal of this paper is to study the impact of high-tech export on the growth of the indicator of climate responsibility of the entrepreneurial sector in developed countries. This goal involves the following tasks: determination of the level of export of high-tech products in developed countries and assessment of the connection between the indicator of export and climate-responsible entrepreneurship in developed countries.

The hypothesis of this research consists in the existence of a dependence of the two variables and opportunities for stimulation of high-tech production for climate conservation.

2 Materials and Method

The influence of high-tech production and export on ecologisation of entrepreneurship in the context of digital technologies was studied by [5, 10–12, 19, 21]. Special attention should be paid to [5], in which the authors use the regression method to assess the influence of the factors of growth of high-tech industry on the reduction of CO_2 level and substantiate the positive results of decarbonisation in the regional context in China, which took place due to the implementation of low-carbon technologies. Miller and Wunsch-Vincent [12] is devoted to the statistical analysis of high-tech export and the evaluation of the factors of its growth during the 2020 pandemic. Shi [19] dwell on China's participation in the resolution of the problems of CO_2 emissions reduction, including through an upgrade of industrial installations on CO_2 emissions capture. The existing works demonstrate a positive influence of high-tech export on the state of the climate. In this research, we attempt at presenting a systemic picture of this influence in the selected developed countries.

The methods used in this paper include the following: the method of induction – to unify separate facts and provisions into conclusions on the connection between the development of high-tech export and growth of climate responsibility; the statistical and economic method – to determine the parametric indicators of economic growth of high-tech export; the systems method – to describe the complex development of the influence of high-tech export on ecologisation of companies that use ICT.

The research was performed based on the data on the main developed countries that have achievements in the sphere of high-tech export, responsible production and climate-responsible entrepreneurship.

3 Results

The choice of the course at technological is determined by the necessity for constant investing in the improvement and creation of new products based on innovative digital technologies. Apart from the financial components, the market element is very important. It is connected with the creating and offering to consumers (individuals, companies, government sector) products that are most relevant, i.e. products which parameters conform to the level of society's formation. Thus, it is possible to state that the export of high-tech products that are aimed at mass demand should have innovative parameters and features, but, at the same time, should not be very complex in the technical aspect for the use by consumers in specific conditions and territories. For example, the export of smartphones with 5G function will not be an advantage for regions without 5G coverage, and, thus, consumers will not seek this function. Apart from the importance of innovative technological characteristics, the products should comply with ecological standards. This parameter is supported by consumers and governments of importers and exporters.

To study the dynamics of the high-tech export volume in developed countries, we selected the following countries: China, Hong Kong, Germany, South Korea, Singapore, the USA, Japan, the Netherlands, France, the United Kingdom and Ireland.

Volumes of high-tech export in the selected developed countries are shown in Fig. 1.

As is shown in Fig. 1, China demonstrates a stable growth of the indicator. The main structure of high-tech export in China is as follows: phones (14.7%), processing machines (13%), routers (5.7%), integrated circuits (6.6%), technology and phone parts (5.3%) and other products [12]. The effectiveness of R&D, including in the high-tech sphere, implies the commercialisation of knowledge, which is conducted at the level of the entrepreneurial sector [21]. This direction is actively developed in China. In many parameters, China equals or even exceeds the USA [10].

Among the high-tech companies of China, we may note the following [16]:

- Apple Inc., which activities were stopped on the territory of China in 2020 due to the pandemic;
- Beijing Xiaomi Science & Technology Co. Ltd. – a world leader in the export of smartphones and other digital gadgets. R&D expenditures in 2021 reached CNY 13 billion ($1.43 billion US$) [20].
- Huawei Technologies Ltd., which works in 170 countries and has 300,000 personnel. R&D expenditures in 2021 equalled CNY 142 billion ($15.61 billion) [6].

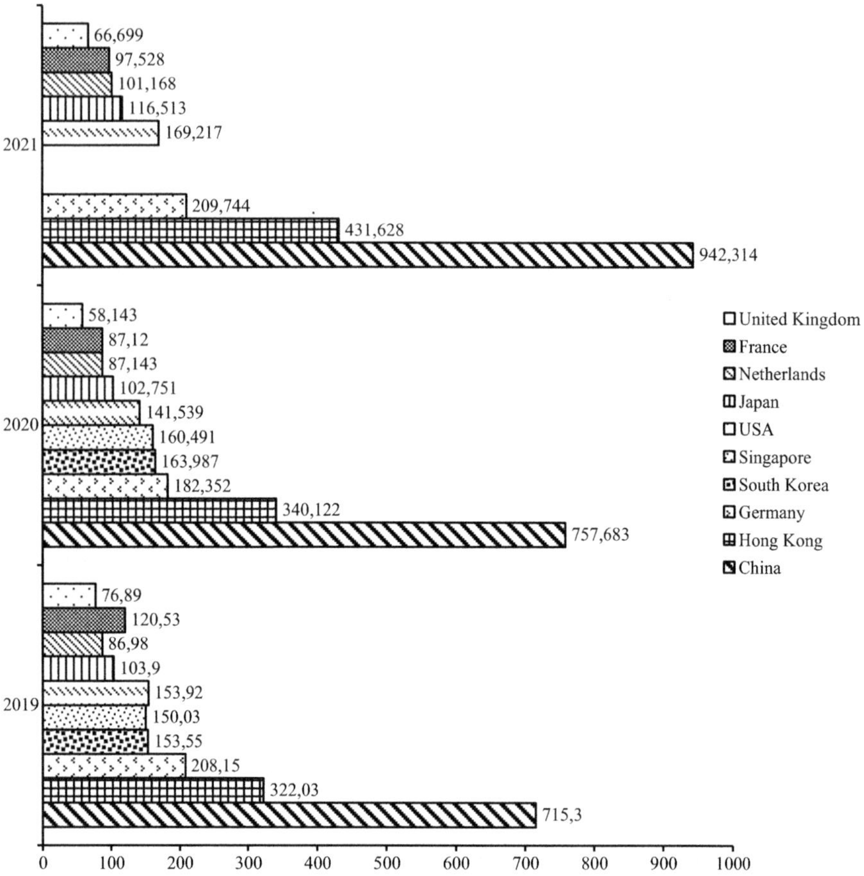

Fig. 1 Dynamics of high-tech export in developed countries, billion \$. (Source: Compiled by the authors using [3, 7])

- Guangdong Oppo Mobile Telecommunications Corp. Ltd. – manufacturer and exporter of smartphones, tablets and audio equipment accessories; it works in 50 countries.
- Vivo Communication Technology Co. Ltd. – a global manufacturer of smartphones, which works in 60 countries and provides services to 400 million users.

A study of climate change that took place in China in 2013–2020 demonstrated that the country reached positive effects. Since the main influence of the manufacture of exported high-tech products is connected with CO_2 emissions, it is necessary to determine its effect on the environment. Over the given period, there was a reduction of CO_2 emissions by 2.43 Gt CO_2, achieved with the help of special installations that ensured low-carbon production; the general growth of emission was 2.03 Gt CO_2 [19].

According to [5], the growth of high-tech productions and export facilitates entrepreneurship's implementing low-carbon technologies, which leads to climate protection. Equal distribution of such production in the main regions will allow ensuring investment in purification installations, which prevent CO_2 from entering the environment. Emphasis is made on the fact that an increase in high-tech export, which depends on the growth of production scales, will allow implementing the programmes of improvement the quality of labour resources for the achievement of the SDGs in this sphere [5]. Since high-tech companies of the world focus on the achievement of the SDGs, the social component is of top priority in this direction. To successfully integrate into the markets of other countries, Chinese high-tech companies demonstrate the indicators of sustainability management in the sphere of ecologisation and social sphere at the level of similar companies in developed countries.

One of the reasons for the growth of high-tech export was the pandemic, which led to remote work, and the need to buy electronic equipment – all of which raised the demand [12].

Entrepreneurship of Hong Kong that functions in the sphere of precision digital technologies achieved high indicators of export: an increase of 26.9% in 2021 compared to 2020. The main direction of high-tech export is the re-export of computer equipment, telecommunication equipment and semiconductors [15]. The main production facilities are located in continental China, while promotion, advertising and sales offices are located in Hong Kong. Accordingly, these companies do not raise CO_2 emissions in Hong Kong. Such an approach could be considered effective from the position of environmental protection in its own territory.

Germany has an important place among countries dealing with high-tech export (3rd position after China and Hong Kong). However, in 2020, the level of export was reduced by 12.4% (Fig. 1), which was mainly due to the temporary disruptions of logistics in the delivery of components and materials used in production [11]. In 2021, this indicator grew by 15% (Fig. 1).

Germany, France and the Netherlands form a structure of high-tech export of the EU that demonstrated changes due to the change in demand, which was caused also by socio-economic phenomena. A particular effect was due to the COVID-19 pandemic. In 2018, in the structure of high-tech export, there was a following situation: Pharmacy and Electronics-telecommunications (22%), Aerospace 23%, Scientific instruments – 16%, Computers and office machines – 8%, Non-electrical machinery – 4%, Chemistry – 3%, Electrical machinery – 2% and Armament – 1% [4]. In 2019, this structure changed in favour of Pharmacy, the export of which products accounted for 24%. In 2020, the share of high-tech export of the Pharmacy category in the EU was 29%, which was due to the growth of demand for new diagnostics products (influence of the pandemic), and 2021 saw an increase in this indicator up to 33% [4]. Over 2018–2021, the level of export of Aerospace reduced from 23% to 15%, which was a result of Brexit, since up until 2020 the UK exported 30% of Aerospace products in the entire EU [8].

German companies that manufacture high-tech products include the following:

- Novartis – a transnational company with the main office in Switzerland, with production facilities and twelve offices in various states of Germany. It is an active participant in international programmes on climate change and invests in countries, where its production capacities are located: in the sphere of energy decarbonisation and various renewable energy sources. The company takes part in ambitious projects to achieve carbon neutrality of production and logistics processes [13].
- Janssen (production and export of Johnson & Johnson products). Production capacities and offices of the company are located in various countries. The main office is in Belgium, with production capacities and branches in Germany. The company focuses on energy efficiency, which allows for the reduction of the level of CO_2 emissions [9].
- Bayer – a large German manufacturer and exporter of innovative, high-tech pharmaceutical products in the spheres of haematology, cardiology and oncology. The climate responsibility of the company is manifested in the use of renewable energy. In 2021, its share in the company's energy use was 24.7%. By 2030, Bayer plans to perform a transition to 100% of renewable energy in its activities [2].

Climate-responsible companies of France that deal with high-tech export include the following [14]:

- Servier – the world's leading pharmaceutical company with headquarters and production capacities in France. In manufactures and exports innovative products to more than 150 countries. The contribution of this company to environment protection includes the implementation of projects on CO_2 capture at production and energy efficiency. Over 2016–2019, the level of CO_2 emissions was reduced by 13% per one unit [18].
- Sanofi. The company manufactures and exports pharmaceutical products to many countries. The key directions of climate-responsible activities include the use of eco-friendly packaging, and measures to reduce carbon footprint (technological installations to capture CO_2 at production lines, use of renewable energy) [17].

European high-tech are the leaders in climate-responsible activities. The analysis showed that the growth of production and export of their products supports the positive tendencies in climate protection.

The USA is the world's leading manufacturer and exporter of high-tech products. The structure of high-tech export includes Pharmacy (50%), Scientific instruments (16%), Electronics-telecommunications (12%), Aerospace (12%) and other high-tech products (10%) [4]. Before the pandemic, the level of high-tech exports in the sector of Pharmacy was twice as low, with the priority belonging to Electronics-telecommunications and Scientific instruments. Change in demand, connected with the growth of consumption of pharmaceutical products, allowed an increase in their export. Such high-tech companies as Johnson & Johnson, Pfizer Inc., F. Hoffmann-La

Roche Ltd. and Amgen Inc. demonstrate high indicators of energy efficiency. Apple is carbon neutral and constantly improves the level of equipment's energy efficiency [1].

4 Discussion

In this research, we provided proof of the direct influence of the growth of high-tech production and export in developed countries on climate protection and prospects for motivation for the growth of high technologies implementation as the basis of environmental security.

The considered activities of the world's leading exporters of high-tech products demonstrated that the organisation and expansion of production and logistics chains of such companies allow reducing CO_2 emissions due to the implemented optimisation measures. These measures include the use of energy-efficient systems and equipment, the use of renewable energy (including investments in this sphere in the countries where production capacities are located) and the installation of equipment for CO_2 capture.

It is important to study the practice of the entrepreneurial sector's expanding high-tech export as the basis for ecologisation in developed countries. These companies have production capacities and offices in developed and developing countries. Their approaches to the work based on the use of modern digital technologies, which support economic growth and climate responsibility, are an example for companies in developing countries that aim at quick growth and international integration. The practice of implementing innovations in the sphere of carbon footprint reduction allows achieving the decrease in harmful emissions into the environment and, in certain cases, becomes a reason for the development of new types of products that are manufactured with fewer materials spent, compared to analogues. Therefore, economic and environmental goals are achieved, and companies demonstrate high climate responsibility before society.

5 Conclusion

Based on the results obtained, it is possible to state the necessity to implement strategic programmes for raising the technological level of modern productions in the main sectors of industry. Strategies can be stimulated by the government, various international organisations and R&D establishments that are aimed at the commercialisation and promotion of their developments (including the ones concerned with the environmental component).

The described effect of the influence of each of the leading high-tech companies on climate showed that these subjects facilitate the fight against the manifestations of global warming and support tendencies regarding the environment at the national level (in countries where production and logistics centres are situated). Given this, countries and supra-national structures can work on various motivation programmes for the liberalisation of conditions for doing business in case of compliance with or achievement of positive climate indicators.

The positive experience of developed countries is a model that could be used by companies in developing countries. The priority of companies' manufacturing products with a certain positive history of consumption could be increased by high technologies and consideration of modern climate requirements. Such an approach may open the way for innovative developing countries to international markets.

References

1. Apple (2022) The products you love also love the planet. https://www.apple.com/environment/. Accessed 24 Dec 2022
2. Bayer (2022) Protecting the environment. https://www.bayer.com/en/sustainability/environmental-protection. Accessed 24 Dec 2022
3. Data.worldbank (2022) High-technology exports (current US$). https://data.worldbank.org/indicator/TX.VAL.TECH.CD. Accessed 24 Dec 2022
4. Eurostat (2022) International trade and production of high-tech products. https://ec.europa.eu/eurostat/statistics-explained/index.php?title=International_trade_and_production_of_high-tech_products#EU_exports_of_trade_in_high-tech_products. Accessed 24 Dec 2022
5. Han M, Zhou Y (2022) The impact of high-tech product export trade on regional carbon performance in China: the mediating roles of industrial structure supererogation, low-carbon technological innovation, and human capital accumulation. Environ Sci Pollut Res 29:31148–31163
6. Huawei (2022) Our company. https://www.huawei.com/en/corporate-information. Accessed 24 Dec 2022
7. Indexmundi (2022) High-technology exports (current US$) – country ranking. https://www.indexmundi.com/facts/indicators/TX.VAL.TECH.CD/rankings. Accessed 24 Dec 2022
8. Irp.cdn (2022) The UK-EU trade and cooperation agreement: impact on the aerospace and defense sector. https://irp.cdn-website.com/d1c52dda/files/uploaded/BREXIT%20-%20impact%20on%20aerospace.pdf. Accessed 24 Dec 2022
9. Janssen (2022) Our responsibility to the planet. https://www.janssen.com/emea/our-responsibility/society/society-planet. Accessed 24 Dec 2022
10. Mahoney JG (2022) China's rise as an advanced technological society and the rise of digital orientalism. J Chin Polit Sci. https://doi.org/10.1007/s11366-022-09817-z
11. Maihold G, Mühlhöfer F (2021) Supply chain instability threatens security of supplies options for industry and policymakers. SWP Comment 60. https://www.swp-berlin.org/en/publication/supply-chain-instability-threatens-security-of-supplies. Accessed 24 Dec 2022
12. Miller J, Wunsch-Vincent S (2021) High-tech trade rebounded strongly in the second half of 2020, with new Asian exporters benefiting. https://www.wipo.int/pressroom/en/news/2021/news_0001.html. Accessed 24 Dec 2022
13. Novartis (2022) Environmental sustainability. Climate. https://www.novartis.com/esg/environmental-sustainability/climate. Accessed 24 Dec 2022
14. Pharmaboardroom (2022) Top 10 pharma companies in France. https://pharmaboardroom.com/facts/top-10-pharma-companies-in-france/. Accessed 24 Dec 2022
15. Research.hktdc (2022) Electronics industry in Hong Kong. https://research.hktdc.com/en/article/MzExMzM1NDk1. Accessed 24 Dec 2022

16. Researchandmarkets (2022) Smart phone manufacturing in China – industry market research report. https://www.researchandmarkets.com/reports/4062438/smart-phone-manufacturing-in-china-industry?gclid=Cj0KCQiA45qdBhD-ARIsAOHbVdFLGTdHgtjiHQoO0G8mZcil8idi7Nb9FB3NSafygAFSvJjn2mVn3KEaAuaMEALw_wcB. Accessed 24 Dec 2022
17. Sanofi (2022) Our responsibility. Environment. https://www.sanofi.com/en/our-responsibility/documents-center/environment. Accessed 24 Dec 2022
18. Servier (2022) Taking tangible action to protect the climate. https://servier.com/en/our-commitments/for-the-planet/. Accessed 24 Dec 2022
19. Shi Q, Zheng B, Zheng Y et al (2022) Co-benefits of CO_2 emission reduction from China's clean air actions between 2013-2020. Nat Commun 13:5061. https://www.nature.com/articles/s41467-022-32656-8#citeas. Accessed 24 Dec 2022
20. Statista (2022) Xiaomi's research and development (R&D) expenses worldwide from 2017 to 2021. https://www.statista.com/statistics/1125785/china-xiaomi-randd-expenses/. Accessed 24 Dec 2022
21. Zarea H, Esmaeelzadeh I, Jafariyeh HS, Buitek E, Aliei M (2021) Knowledge commercialization framework: factors affecting developing countries. AD-minister 39(2021):217–236. https://doi.org/10.17230/Ad-minister.39.10

Effects of the Economic Crisis on the Development of Climate-Responsible Entrepreneurship in the Markets of the Digital Economy in Developed and Developing Countries

Timur M. Israilov, Elena A. Sergodeeva, Valentina I. Rodionova, and Milyausha K. Khalilova

1 Introduction

Economic crises are caused by the totality of the factors of external and internal character. Their emergence and growth are connected with various negative consequences for the socioeconomic, technological and political spheres. The consequences of these phenomena are dealt with at the level of governments and entrepreneurship. Insufficiently effective management of anti-crisis measures is connected with the level of national governments' readiness for the support for the protection of their countries from the growth of negative indicators in all spheres of the economy, in the social sector and in the sphere of environmental protection from industrial pollution, which may take place due to the absence of purification works and control over their functioning. On the contrary, the successful use of modern capabilities and attraction of the innovative entrepreneurial sector allows countries to reduce the negative influence of certain crisis phenomena in the economy. A clear understanding of the causes of crises and capabilities of the internal potential of the main active sectors of the economy and their subjects, which can form and

T. M. Israilov (✉)
Osh State University, Osh, Kyrgyzstan
e-mail: israilov@oshsu.kg

E. A. Sergodeeva
North-Caucasus Federal University, Stavropol, Russia
e-mail: esergodeeva@ncfu.ru

V. I. Rodionova
Institute of Service and Entrepreneurship (Branch) of Don State Technical University, Shakhty, Russia

M. K. Khalilova
Finance University Under the Government of the Russian Federation, Moscow, Russia

E. G. Popkova (ed.), *Smart Green Innovations in Industry 4.0*, Springer Climate,
https://doi.org/10.1007/978-3-031-45830-9_5

implement the measures of struggle, linked to their economic and technological growth and the focus on the foundations of climate protection, is very important. Successful reactions to crisis-related challenges can be more effective in developed countries, which have better financial markets and a digital economy. Developing countries can also implement successful strategies, which ensure the overcoming of crisis phenomena (fighting the decrease in GDP, growth of unemployment rate, etc.). Given the above, it is very topical to study the specifics of the positive practice in overcoming crises under the conditions of various economic systems. It should be also analysed in the context of the participation of the innovative climate-responsible entrepreneurial sector, which can provide a stabilisation effect.

The purpose of this work is to reveal the main aspects of climate-responsible entrepreneurship that uses digital technologies as a result of the emergence of economic crises in developing and developed countries. For this, we selected countries that ensured effective responses to the effects of economic crises due to the successful activity of the entrepreneurial sector and determined the directions for the formation of climate-responsible entrepreneurship which facilitated the fight against economic crises in developing and developed countries.

2 Materials and Method

Analysis of the main features of the formation of a climate-responsible entrepreneurial sector during periods of economic crises and the issues of its using digital tools that facilitate the growth and fight against climate change was conducted based on various scientific materials, namely [5, 12, 13, 18, 20]. Yu et al. [20] conducted a systemic analysis of the effectiveness of the functioning of Amazon (USA) in the sphere of achievement of specific UN SDGs in the context of climate problems and corporate social responsibility. It was shown that the company implemented a range of positive environmental practices, which may be used by other companies. Cook et al. [5] characterised the features of sustainable energy development of Iceland under the conditions of the energy transition to renewable sources. There are also works in the sphere of the historiography of economic risks' emergence. Thus, [15] presented a complex analysis of the dot-com bubble, which took place in the USA in 2000–2002 and was connected with the collapse of the ICT companies' stock. Paluteder [15] classified the reasons for the emergence of this crisis. We should also note the evaluation of companies that suffered during the COVID-19 pandemic in the USA [16].

Though at the current state, there is a certain array of studies in the given direction, there is still a need for a system approach to determining the characteristics of the formation of an environmentally oriented business sector, which functions with the use of modern ICT under the influence of crises.

We studied the indicator of the emergence and development of economic crises with the use of statistics, which allowed identifying the level of these phenomena in the parametric measuring (level of GDP change, level of unemployment). We also used the comparative method, which allowed comparing transformations of the studied indicators. The method of systematisation allows establishing the specifics of the effects of crises on business and the main aspects of its work for economic growth and achievement of climate goals.

3 Results

Let us elaborate on the characteristics of the emergence of economic crises and measures to fight them in the activities of climate-responsible entrepreneurship in the markets of the digital economy in 2000–2022 (Tables 1 and 2).

We considered the USA and Iceland among developed countries. The USA is a high-tech country, which entrepreneurship is focused on the support for climate-responsible parameters, which are demonstrated in periods of complex economic crises. Analysis of the features of the activity of climate-responsible entrepreneurship in Iceland was made because Iceland is a unique example of an energy-autonomous country of Europe [5].

The described ecologisation is formed within the following directions:

1. Balance of economic and environmental goals at all stages of product management (from the selection of resources and raw materials by partner companies (suppliers) and production to logistics of final products – example of Amazon). Here we speak of the use of eco-friendly materials, which will be safe for the final consumer and for further circulation after the end of life, during regular recycling. Not only does this direction of development influence the level of climate protection, it is also a precondition for raising image, competitive positions and volumes of product sales, which are less safe compared to similar products from rivals. The focus on environmental safety in the considered companies is not just a declared component, for the level of quality and safety of products is subject to certification. For example, all products that are sold on Amazon are subject to certification [20]. On the whole, all considered companies in the USA and Iceland achieved a high level of control over the safety and quality of products, despite the emergence of the economic crises.
2. High dynamism in the creation of unique offers as a response to changes in consumer needs. This allows individual companies to achieve high results of functioning and to be effective during economic crises. An example is the application of innovative solutions in the creation of new products, which conform to the current expectations of consumers, including during crises.
3. World leading positions on the commercialisation of new digital technologies in the sphere of climate protection. The world's leading manufacturers and online retailers achieved improvement in the UN SDGs in the sphere of climate protection at the national level and in other countries and regions. This is largely due to the fact that such companies have more capabilities due to a more stable entrepreneurial environment and financial establishments that give access to finances.

Table 1 Indicators and characteristics of dealing with the effects of economic crises in selected developed countries in 2000–2022

No.	Country/economic crisis (period, name)	Indicators	Companies damaged by the crisis	Companies that were able to deal with the effects of the crisis
1	USA			
1.1	Dot-com bubble (2000, 2001, 2002)	The collapse of the start-ups and new ICT companies stock exchange led to bankruptcy, losses and the end of attraction of venture investment. Decrease in the stock market index Nasdaq Composite by 78% compared to the growth of 400% in 1995–2000. Losses: $5 trillion of investment capital. Mass dismissal of programmers.	Webvan, Pets.com, NorthPint, Boo.com, Worldcom, Global Crossing	Cisco Systems, *Inc.*, Amazon
1.2	Economic decline (2006, 2007, 2008) (beginning of the Great Recession)	GDP growth in 2005–2006 was 2.8%, in 2006–2007 – 2%, in 2007–2008 – 0.1% and in 2008–2009 there was GDP reduction of 2.6%	Most American companies had an economic decline.	Despite the decrease in revenue, Amazon implemented innovative projects (AWS, Amazon Kindle, Amazon Prime). These projects attracted consumers, were dynamic innovations and allowed Amazon to become the leader in the e-market trade. Against the background of reduction in the activity, LEGO implemented the strategy of financial revival (growth of financial turnover, reduction of time for bringing new products to the market, reduction of assortment), which allows stabilising the situation. Over 2007–2011, the company's net profit grew by four times.

1.3	COVID-19 pandemic (2020) and the period of overcoming its consequences (2021). Global energy crisis (beginning – 2021)	Reduction of GDP by 3.4% in 2020 compared to 2019. Growth of energy prices, companies' conducting energy transition to renewable energy. Energy transition to the use of renewable energy (in 2019 and 2020, the level of its use was 7.9% of the total volume of consumption, and in 2021 – 12.4–12.5%)	Large, medium and small companies. Large companies: J. Crew (shopping malls); Tailored Brands (production and sales of brand clothes); Hertz (car rental and sales); JC Penney (shopping malls); Whiting Petroleum (oil business); Chesapeake Energy (oil and gas production); GNC (health food products). Deficit of energy resources, growth of energy prices.	Amazon (growth of annual income in 2018–2019 was 20.5%, in 2019–2020 – 37.6%, and in 2020–2021 – 21.7%). Use of renewable energy – 85%. Reduction of weight of packaging by 38%, which ensured the reduction of effect on climate. The focus is on strict control over the ecological characteristics of products. In 2018–2021, LEGO used eco-friendly materials produced from renewable resources and renewable energy in the main production processes and logistics processes.

(continued)

Table 1 (continued)

No.	Country/economic crisis (period, name)	Indicators	Companies damaged by the crisis	Companies that were able to deal with the effects of the crisis
2	Iceland			
	Icelandic financial crisis (period: 2008, 2009, 2010, 2011). Global energy crisis (beginning – early 2021)	2008–2011: bankruptcy of the main financial companies in Iceland, reduction of access to capital for entrepreneurship. Default on foreign debt ($62 million). 2021: growth of energy resources prices, search for a replacement for traditional energy resources, transition to renewable energy sources.	2008–2011: losses in the main sectors of the economy at the initial stages of the crisis. Growth of energy prices and deficit of energy resources.	Arctic Green Energy (development, installation and maintenance of equipment for renewable energy (geothermal heating of buildings, production of electric energy from wind and solar energy)). Effect on the reduction of CO_2 in Iceland, Scandinavian countries and China (many large companies-consumers). The largest consumer is the Chinese company Sinopec Green Energy. During the 2008–2011 crisis, Arctic Green Energy ensured stable functioning due to partnerships with Chinese companies, which were regular customers. During the energy transition in 2021, the company continued demonstrating growth and stability. IceWind (manufacturer of wind turbines for residential buildings and telecommunication towers) started working in 2008. It implements six new digital technologies each year. Since 2012, it has been supporting the transition of telecommunication companies in Europe to autonomous renewable wind energy.

Source: Made by the authors based on [1–3, 6, 7, 9–11, 13, 15–17]

Table 2 Specific features and indicators of overcoming economic crises in selected developing countries in 2000–2022

No.	Country/economic crisis	Characteristics of indicators	Environmentally active companies and characteristics of anti-crisis results
1	China		
	World energy crisis (since early 2021)	Growth of energy prices for final consumers and industry; search for alternative energy sources; country's participation in the energy transfer to alternative energy.	Sinopec Green Energy (national operator on the production and sales of electric and heat energy based on renewable energy sources to individual consumers and businesses). Cooperation with Arctic Green Energy (Iceland) ensured the reduction of CO_2 by 16 million tons in the main regions of the country. The company invested in digitalisation and technologies for wastewater treatment. This facilitated the growth of support from the government and consumers and enabled partner companies to use purified water resources for communal purposes.
2	Brazil		
	Brazilian economic crisis (2014 – early 2017)	Fall in GDP: \$2456.04 billion in 2014, \$1802.2 in 2015, (decrease of 26.6%), \$1795.7 in 2016 (decrease of 0.36%) and \$2063.5 in 2017 (growth by 14.9%). The fall in GDP was caused by ineffective management of the economy, and a reduction in export prices (a large share of products were exported to China, so the reduction of the Chinese economy led to such a decline). Growth of unemployment, social divide, mass emigration of intellectual and skilled personnel. The economy stabilised in early 2017.	Climate-responsible companies used the "win-win situation" approach, which involves the design and implementation of such business models that allow for positive influence on climate and economic effect given the impact of the crisis. Companies that were able to oppose the crisis phenomena include the following: Naturalle Tratamento de Residuos Salvador. A national company dealing with the cleaning of urban territories and environmental management of waste (processing with the use of innovative digital technologies and smart processing into secondary raw materials, which are used in various productions). Projects on the training of environmental agents who support the philosophy of climate protection; Kaizen servicos de construcao civil Curitiba. A national construction company that deals with the control of the quality and safety of construction materials, uses energy-efficient technologies (construction of energy-efficient buildings and other types of construction products). The main focus in the management of sustainable environmental development: climate protection through the reduction of energy consumption, which facilitates the reduction of CO_2 emissions.

Source: Made by the authors based on [3, 4, 10–12, 14, 18]

According to [8], the success of the commercialisation of innovations in developed countries (shown by the example of countries of Northern Europe) depends on support from financial organisations.

Let us consider the indicators and characteristics of crisis phenomena in the selected developing countries under the conditions of development of the environmentally responsible entrepreneurial sector (Table 2).

The analysis (Table 2) demonstrated the key features of the formation of climate-responsible entrepreneurship, which functions with the use of innovative digital technologies under the conditions of crises in modern developing countries. These features include the following:

1. Entrepreneurial sector's focus on climate and technological indicators as equal goals of achieving effectiveness of functioning under the conditions of economic instability (level of income and profit and market share). Determination of companies' real commitment to the balanced approach is connected with the indicators of the companies' achievements in reducing the effects on climate.
2. Partnership with world leaders in the management of climate-responsible goals. The interaction with manufacturers of new digital technologies and equipment is implemented. An example of such a partnership was shown in Tables 1 and 2 – the Icelandic manufacturer of equipment for renewable energy (Arctic Green Energy) and the Chinese operator of the energy market (Sinopec Green Energy) [19]. In this case, climate-responsible entrepreneurship in developing countries raises its technological advantages (equipment and technologies) within the partnership interaction with companies that are world leaders in a certain sphere (energy, industry, etc.).

4 Discussion

In this research, we formulated the key features of the functioning of environmentally oriented entrepreneurship under the influence of crisis phenomena in developing and developed countries. It was shown that the advantages and stable functioning of companies of the two types of economic systems are achieved at the modern stage due to the use of digital technologies. This condition is the key basis of the effectiveness of the leading companies of the world, aimed at the achievement of the UN SDGs in the sphere of ecologisation. Innovative leading companies, which support environmental indicators, were able to oppose the effects of the crisis due to the successful identification of market needs and the creation of offers that a modern consumer cannot refuse even in a complicated economic situation (examples of American and Icelandic companies).

The considered features of successful functioning, demonstrated by the modern leading climate-responsible companies in developing and developed countries, differ by vectors that are connected with technologies and goals. Companies in developed countries create branches and productions and implement technologies and innovative equipment (including in the climate sphere) in developed and developing countries. That is, the entrepreneurial sector facilitates technological and environmental development in developed and developing countries. As for the leading companies in developing countries, we should note their significance and decisive contribution to digitalisation and climate protection at the national level. Such tendencies could be preserved up until the moment of possible economic and technological growth of countries. This concerns the conditions of the functioning of China, which somewhat lacks stabilisation of the social component for it to be assigned to the category of developed countries.

5 Conclusion

Analysis of the influence of crises on acceleration and transition to a new level of activity of climate-responsible entrepreneurship demonstrated the existence of similar and different characteristics. Similarities include the balance of goals in the sphere of ecologisation, digitalisation and economic growth, which is peculiar to entrepreneurship under the conditions of economies of various types and which allows – thanks to a range of factors – achieving the effectiveness of activity even during economic crises; partnership, which allows implementing the goals and tasks of each interested party (it is effective for the models of interaction between companies in developed countries and between companies from developed and developing countries). The specific features of entrepreneurship in developed countries include leadership in the sphere of commercialisation of R&D and high activity in the sphere of creation of actual innovative products (services) that can influence the attraction of wide groups of consumers even during crises.

References

1. Amadeo K (2021) Iceland's economy, its bankruptcy, and the financial crisis. https://www.thebalancemoney.com/iceland-financial-crisis-bankruptcy-and-economy-3306347. Accessed 29 Dec 2022
2. Amazon (2022) Environment. https://sustainability.aboutamazon.com/environment. Accessed 29 Dec 2022
3. Arcticgreen (2022) Who we are. https://arcticgreen.com/about/. Accessed 29 Dec 2022
4. Bolddata (2022) List of environmental companies Brazil. https://bolddata.nl/en/companies/brazil/environmental-companies/. Accessed 29 Dec 2022
5. Cook D, Davíðsdóttir B, Gunnarsdóttir I (2022) A conceptual exploration of how the pursuit of sustainable energy development is implicit in the genuine progress indicator. Energies 15:2129. https://www.mdpi.com/1996-1073/15/6/2129. Accessed 29 Dec 2022
6. Eia (2022) How much of U.S. energy consumption and electricity generation comes from renewable energy sources? https://www.eia.gov/tools/faqs/faq.php?id=92&t=4. Accessed 29 Dec 2022
7. IceWind (2022) Who we are. https://icewind.is/about-us/. Accessed 29 Dec 2022

8. Ilina I, Zharova E, Turginbayeva A, Agamirova E, Kamenskiy A (2019) Network platform of commercializing the results of R and D. Int J Civ Eng Technol 10(1):2647–2657. https://iaeme.com/MasterAdmin/Journal_uploads/IJCIET/VOLUME_10_ISSUE_1/IJCIET_10_01_237.pdf
9. LOOP (2022) 5 Icelandic startups to keep an eye on. https://circulareconomyloop.com/5-icelandic-startups-to-keep-an-eye-on/. Accessed 29 Dec 2022
10. Macrotrends (2022) Amazon revenue 2010–2022. https://www.macrotrends.net/stocks/charts/AMZN/amazon/revenue. Accessed 29 Dec 2022
11. Macrotrends (2022) Brazil GDP. https://www.macrotrends.net/countries/BRA/brazil/gdp-gross-domestic-product. Accessed 29 Dec 2022
12. Marquetti AA, Hoff C, Miebach A (2020) Profitability and distribution the origin of the Brazilian economic and political crisis. Lat Am Perspect 47(1):115–133
13. McLaughlin K, Bird L (2021) The US set a record for renewables in 2020, but more is needed. World Resources Institute. https://www.wri.org/insights/renewable-energy-2020-record-us. Accessed 29 Dec 2022
14. Naturalleresiduos (2022) Conservação da fauna e da flora. https://naturalleresiduos.com.br/. Accessed 29 Dec 2022
15. Paluteder D (2022) Dot-com bubble explained. The true story of 1995–2000 stock market. https://finbold.com/guide/dot-com-bubble/. Accessed 29 Dec 2022
16. Reeth M (2021) 7 companies that went bankrupt due to COVID. https://money.usnews.com/investing/stock-market-news/slideshows/covid-bankrupt-companies. Accessed 29 Dec 2022
17. Sdgindex (2019) Sustainable development report 2019 transformations to achieve the Sustainable Development Goals. https://www.sdgindex.org/reports/sustainable-development-report-2019/. Accessed 29 Dec 2022
18. Seles BMRP, Lopes de Sousa Jabbour AB, Jabbour CJC, Latan H, Roubaud D (2019) Do environmental practices improve business performance even in an economic crisis? Extending the win-win perspective. Ecol Econ 163:189–204
19. Sinopecgroup (2022) Green & low-carbon development. http://www.sinopecgroup.com/group/en/socialresponsibility/greenlow/. Accessed 29 Dec 2022
20. Yu W, Hassan A, Adhikariparajuli M (2022) How did Amazon achieve CSR and Some Sustainable Development Goals (SDGs)-climate change, circular economy, water resources and employee rights during COVID-19? J Risk Financ Manag 15:364. https://www.mdpi.com/1911-8074/15/8/364. Accessed 29 Dec 2022

Conditions of Doing Business for the Development of Climate-Responsible Entrepreneurship in the Markets of the Digital Economy on the Example of Developed and Developing Countries

Ainura A. Adieva, Kanikei T. Samieva, Milyausha K. Khalilova, and Olga B. Tokareva

1 Introduction

Business is one of the driving forces of the market economy. Under the conditions of domination of the neo-liberal paradigm in economic science, the maximum support for the development of business is treated as a factor of the socio-economic development of a country. It covers the processes of an increase in the level of investment attractiveness, improvement of the quality of the business environment and level of the population's economic activity. However, together with the priorities of economic progress, the Sustainable Development Goals, including the fight against climate change, become more topical.

Under the conditions of the formation of the global economic space, global ratings or so-called global performance indicator (GPI) [8] played an important role. In particular, the ease of doing business index (DB) allowed the World Bank to create a global regulatory environment [6]. This index inspired more than 3500 reforms in 190 economies. In 2017–2018, 128 economies implemented record 314 reforms [4]. For many countries, it became the basis for the improvement of investment climate and attraction of foreign investments.

A. A. Adieva (✉)
International University of the Kyrgyz Republic, Bishkek, Kyrgyzstan

K. T. Samieva
Osh Technological University Named After M.M. Adyshev, Osh, Kyrgyzstan

M. K. Khalilova
Finance university Under the Government of the Russian Federation, Moscow, Russia

O. B. Tokareva
Sebryakovskiy Branch of Volgograd State Technical University, Mikhailovka, Russia

E. G. Popkova (ed.), *Smart Green Innovations in Industry 4.0*, Springer Climate,
https://doi.org/10.1007/978-3-031-45830-9_6

An increase in risks connected to climate change and the quick development of the digital economy predetermine the need to supplement the indicators of the success of business activities. Business and financial goals are not considered without a focus on their influence on the SDGs and without the assessment of using the tools of the digital economy. According to this, assessment of the role of doing business for the development of climate-responsible entrepreneurship in the digital economy is a current scientific task, which is equally peculiar for countries with different levels of economic development.

2 Materials and Method

This research is based on different approaches to business. They cover its classic treatment as activities aimed at profit-making; responsible activities, which, within the main direction, also ensure the achievement of the SDGs; and activities that use digital technologies to make a profit. Within each approach, we evaluate developed countries and developing countries according to relevant indices. Based on the global ratings, which characterise these approaches, we perform the analysis of the correlation between the values of indices or rank of countries in the ratings. According to this, we determine a mutual connection between the conditions for doing business and successfulness of a country's policy on climate change and the development of the digital economy. To generalise the results obtained, we use a systems approach, which allowed coordinating the positions of each approach and establishing the directions for improvement of the policy on the strengthening of climate-responsible entrepreneurship in the digital economy.

In this research, we use a large array of theoretical materials, which allowed ensuring reasoning for the results obtained. The key materials that were the basis of the research include the ranking of developed countries and developing countries by Doing Business [13] and explanations of the change in the methodological approach to the formation of this ranking in the future [14]; assessment of the influence of the Doing Business Index on economic processes [5, 8, 11, 12] and economic behaviour [6]; ranking of the studied groups of countries by the indicators of the digital economy development [1, 4, 7] and the indicators of successfulness of the policy of fighting climate change [3].

The important provisions, used in this work, are based also on the identification of the drawbacks of ranking of countries by indicators, which characterise ecology, social responsibility and government policy (ESG) [2] and the emphasis on the fact that the ease of doing business is often achieved through the aggravation of the environmental situation in the country through reduction of ecological control over the activities of entrepreneurial structures [9, 10].

The main goal of this research is to identify the conditions for doing business for the development of climate-responsible entrepreneurship under the conditions of the digital economy with differentiation by developed and developing countries.

3 Results

The climate crisis is treated as one of the largest threats not only to the economy but also to life on Earth. To solve it, all countries agreed on climate-responsible behaviour, which will allow limiting global warming up to 1.5 °C, according to the Paris Agreement [3]. Also, the issue of achieving economic efficiency and attractiveness of investment climate are top-priority tasks, which are implemented in national policies through national and global institutions. Against this background, the acceleration of the development of the digital economy is among the factors that facilitate the achievement of environmental and economic goals. An important task today is ensuring the rational interaction of all the above processes for common benefit.

Studying the conditions for doing business from the position of climate responsibility in the digital economy combines the research of the three main approaches to business, which cover the following aspects:

- From the position of the Doing Business Index – maximum elimination of barriers and cheapening and acceleration of all operations connected with the starting, doing and closing business.
- From the position of the fight against climate change – restraints upon the actions that negatively influence or may negatively influence the state of the environment.
- From the position of the digital economy – formation of digital infrastructure and simplification of procedures for starting, doing and closing online business.

As we see, the positions of Doing Business and the digital economy have similarities, which are aimed at eliminating the barriers and creating conditions for doing business, including its digital form. From the position of fighting climate change, the focus is on limitations for business. Based on this, conditions for doing business, as factors of the maximum elimination of restrictions for business, poorly correlate with the concept of climate responsibility in the digital economy. In many cases, ease of doing business implies cancelling permits to accelerate the procedure of registration of a business structure.

To determine the degree of interconnection between the three above approaches to business, we performed a correlation analysis of global rankings, which characterise countries' position by the ease of doing business (Doing Business Index – DB), the successfulness of fighting climate change (Climate Change Performance Index – CCPI) and the development of the digital economy (Digital Platform Economy Index – DPE).

The Doing Business Index is based on the idea that the maximum elimination of bureaucratic and technical obstacles to the creation, functioning and liquidation of business is an important factor in ensuring the attractiveness of the business environment. This index has been used for quite a long time by national governments to improve the policy of internal regulation of conditions for doing business. However, in September 2021, the World Bank stopped calculating and publishing the Doing Business report, announcing that they were working on a new approach to assess business and investment climate [7].

The main indicators that were included in the Doing Business Index contained the assessment of the conditions on the following aspects:

- Speed and simplicity of registration of business
- Absence of bureaucratic and other obstacles to obtaining construction permits
- Ease of obtaining access to a stable energy supply
- Possibility for quick, unhindered registration of property
- Access to crediting
- Tax burden and tax administration costs
- Protection of investors' rights
- Simplicity and inclusiveness of international trade
- Degree of adherence to provisions of contracts
- Possibility for quick liquidation of a company [14]

The above indicators' focus is closely connected with the simplification and speed of actions on the organisation, doing and liquidation of the business. However, the index does not take into account the current problems with climate change or the potential of the digital economy. To remove some of the calculations, it was offered to use the Digital Platform Economy Index, which would take into account not only the basic factors of doing business but also the development of digital capabilities. The structure of such an index includes the level of development of e-commerce, digital media, the economy of joint use digital products, activity and qualification of freelancers (digital block), as well as the accessibility of data, development of digital and analogue platforms and generalised assessment of Doing Business [4].

Neither of the above indices reflects the state of the development of climate-responsible entrepreneurship. To assess this direction, we used the Climate Change Performance Index. It considers four categories of 14 indicators; they include CO_2 emissions, use of renewable energy, use of fossil fuel and climate policy. The unique division of climate policy of CCPI assesses the progress of countries in the implementation of policy that is aimed at the achievement of the goals of the Paris Agreement [3].

The sample of countries by the designated list of indices covers the 2020 data (the last available ranking by the Doing Business Index) and 41 countries (15 developing countries (Algeria, Argentina, Brazil, Egypt, India, Indonesia, China, Colombia, Malaysia, Mexico, Morocco, Thailand, Turkey, South Africa and Chile) and 26 developed countries (Australia, Austria, Belgium, UK, Denmark, Estonia, Ireland, Spain, Italy, Canada, Latvia, Germany, New Zealand, Norway, South Korea, Poland, Portugal, Slovakia, Slovenia, USA, Hungary, France, Czech Republic, Sweden and Japan)) (list of countries by the CCPI). The results of the correlation analysis of the indicators of countries by these indices are presented in Table 1.

For the better objectivity of results, we performed a correlation analysis of indices for the indicators of countries, determined by the value of each index (a) and the country's position in the ranking (b). Differences between the results of analysis for each sample were insignificant, which allowed using the country's position in rankings and values of indices for the treatment of the results obtained. All samples are similar in the following: the index of successfulness of fighting climate change

Table 1 Results of the correlation analysis of the indices CCPI, DB, DPE for developed and developing countries

Correlation matrix for the indices of developed countries, calculated based on the values of indices (a) and country's position in the ranking (b)							
(a)				(b)			
	CCPI	DB	DPE		CCPI	DB	DPE
CCPI	1			CCPI	1		
DB	0.177	1		DB	0.228	1	
DPE	0.196	0.634	1	DPE	0.271	0.677	1
Correlation matrix for the indices of developing countries, calculated based on the values of indices (a) and the country's position in the ranking (b)							
(a)				(b)			
	CCPI	DB	DPE		CCPI	DB	DPE
CCPI	1			CCPI	1		
DB	0.026	1		DB	−0.046	1	
DPE	−0.105	0.599	1	DPE	−0.225	0.550	1
Correlation matrix for indices of the entire sample of countries, calculated based on the values of indices (a) and country's position in the ranking (b)							
(a)				(b)			
	CCPI	DB	DPE		CCPI	DB	DPE
CCPI	1			CCPI	1		
DB	0.038	1		DB	0.080	1	
DPE	0.018	0.765	1	DPE	0.098	0.720	1

Source: Compiled by the authors based on the data from [3, 7, 13]

(CCPI) poorly correlates with the indices of ease of doing business (DB) and development of the digital economy (DPE). Among all results, the highest value of the correlation coefficients was found during the assessment of the position of developed countries in the corresponding rankings ($r = 0.271$ and $r = 0.228$). The received results point to the weak connection between the given positions of countries in the rankings. Taking into account the absence of the connection between CCPI and DB and DPE in all other calculations, this demonstrates that the conditions for doing business, expressed through the index (DB) and conditions for doing digital business (DPE), do not influence or influence slightly the development of climate-responsible entrepreneurship.

Unlike this, in all presented matrices (Table 1), we observe a medium or close direct connection between the values of DB and DPE. At that, for developed countries, this connection is higher ($r = 0.634$ and $r = 0.677$) than for developing countries ($r = 0.599$ and $r = 0.550$). However, for the entire totality of countries, these values are the highest ($r = 0.765$ and $r = 0.720$). This shows that despite the fact that in developed countries, ease of doing business is better imposed on the conditions for the development of digital business, this tendency is common in the global economy.

The results obtained show that the influence of doing business on the development of climate-responsible entrepreneurship in the digital economy is insignificant. At the same time, the perception of this index as one of the factors of aggravation of the environment, including the impact on climate change [9, 10], was not confirmed either. Therefore, an important task of all three described concepts of the development of business is the coordination of economic and digital goals with the SDGs, which includes supplementing doing business with the criteria of support for sustainability.

4 Discussion

Based on the received results, we can state that the interaction of approaches that characterise business from the position of economic results, fight against climate change and development of the digital economy is based on the search for the ways for optimal combination of the conditions for regulating business from the position of ecology and digitalisation. At present, business regulation has many indicated and institutionalised provisions, which are the object of public authorities' activity. A much lower level of institutionalisation is peculiar to environmental and digital conditions of doing business. Thus, the main task of government, international organisations and the public sector is the search for the means to combine all three approaches based on a "win-win" strategy.

According to this, further research on the influence of conditions for doing business in the development of climate-responsible entrepreneurship in the digital economy should include a discussion on the integration of global indices, which rank countries by different approaches, within one ranking. This will allow supplementing the concept of the maximum simplification of conditions for doing business with value-based provisions in the fight against climate change and additional capabilities based on digital technologies. The expected results from such actions, based on the extrapolation of the impact of doing business on the reformation of the government policy of business development in previous years, will allow for better integration of the Sustainable Development Goals into the system of registering, starting and liquidating the business and will create additional opportunities for simplifying, accelerating or improving these processes. On the other hand, the achieved results, integrated into the digital economy, will allow creating additional value of digital products and facilitating the formation of new effects.

A separate direction of further discussion should be the analysis of the assessment of the impact of liberalisation of conditions for doing business on the achievement of the SDGs. Refusing the existence of such a connection will allow avoiding the potential implementation of the restrictions aimed at the necessity to adhere to the SDGs during the registration, functioning or liquidation of entrepreneurial structures.

5 Conclusion

Climate change and the development of the digital economy become more topical and acquired the role of inseparable parts of strategic planning. At the same time, the evaluation of the attractiveness of the business environment according to the Doing Business Index, which is important for ensuring the most favourable conditions for the development of business, does not take into account climate factors or factors of the digital economy development. Given this, as for the development of climate-responsible entrepreneurship in the markets of the digital economy, there is no common approach to its assessment. It is performed within three approaches, which include economic attractiveness (Doing Business Index), level of the development of the digital economy (Digital Platform Economy Index) and assessment of the successfulness of fighting climate change (Climate Change Performance Index). Analysis of the indicators that characterise each of these approaches showed the absence of the connection between the position of countries in the ranking of Doing Business and indicators that reflect the reaction to climate change, though the connection between the index of the ease of doing business and digital ease of doing business is rather significant. The outlined results are equally peculiar for developing countries and developed countries. The value of the coefficients of correlation of the Doing Business Index and the Digital Platform Economy Index in developed countries is a bit higher, but the highest degree of connection is peculiar for the common sample, without the division of countries into groups.

References

1. Acs ZJ, Szerb L, Song A, Komlosi E, Lafuente E (2021) The digital platform economy index 2020. Global Entrepreneurship and Development Institute. https://thegedi.org/wp-content/uploads/2020/12/DPE-2020-Report-Final.pdf. Accessed 25 Jan 2023
2. Berg F, Kölbel JF, Rigobon R (2022) Aggregate confusion: the divergence of ESG ratings. Rev Financ 26(6):1315–1344. https://doi.org/10.1093/rof/rfac033. https://academic.oup.com/rof/article/26/6/1315/6590670/. Accessed 25 Jan 2023
3. Burck J, Hagen U, Höhne N, Nascimento L, Bals C (2021) Climate Change Performance Index – CCPI. Results 2020. Germanwatch, NewClimate Institute & Climate Action Network. https://newclimate.org/sites/default/files/2019/12/CCPI-2020-Results_Web_Version.pdf. Accessed 25 Jan 2023
4. Chakravorti B, Chaturvedi RS (2019) Ranking 42 countries by ease of doing digital business. Harvard Business Review. https://hbr.org/2019/09/ranking-42-countries-by-ease-of-doing-digital-businessE. Accessed 25 Jan 2023
5. Dinuk J (2021) Improvements in the World Bank's ease of doing business rankings: do they translate into greater foreign direct investment inflows? Policy Research Working Papers, September. https://doi.org/10.1596/1813-9450-5787
6. Doshi R, Kelley J, Simmons B (2019) The power of ranking: the ease of doing business indicator and global regulatory behavior. Int Organ 73(3):611–643. https://doi.org/10.1017/S0020818319000158

7. EoDDB (2020) Ease of doing digital business 2019. Which countries help expedite entry, growth, and exit of technology-based businesses? Digital Planet. https://sites.tufts.edu/digitalplanet/ease-of-doing-digital-business-2019/. Accessed 25 Jan 2023
8. Kelley J, Simmons B (2019) Introduction: the power of global performance indicators. Int Organ 73(3):491–510. https://doi.org/10.1017/S0020818319000146
9. Kukreti I (2017) Ease of doing business comes at an environmental cost. Down to Earth. https://www.downtoearth.org.in/news/governance/the-environmental-cost-of-making-business-easy-59001. Accessed 25 Jan 2023
10. Rieger A (2019) Doing business and increasing emissions? An exploratory analysis of the impact of business regulation on CO_2 emissions. Hum Ecol Rev 25:69–85. https://doi.org/10.22459/HER.25.01.2019.04
11. Scalet S, Kelly T (2010) CSR rating agencies: what is their global impact? J Bus Ethics 94:69–88. https://doi.org/10.1007/s10551-009-0250-6. https://www.researchgate.net/publication/227016775_CSR_Rating_Agencies_What_is_Their_Global_Impact. Accessed 25 Jan 2023
12. Skagerlind H (2020) The power of indicators in global development policy: the millennium development goals. https://doi.org/10.1017/9781108763493.005
13. WBG (2021) Doing business 2020. Comparing business regulation in 190 economies. https://documents1.worldbank.org/curated/en/688761571934946384/pdf/Doing-Business-2020-Comparing-Business-Regulation-in-190-Economies.pdf. Accessed 25 Jan 2023
14. WBG (2022) Business Enabling Environment (BEE). https://www.worldbank.org/en/programs/business-enabling-environment. Accessed 25 Jan 2023

The Contribution of Climate-Responsible Entrepreneurship in the Digital Economy Markets to Green Growth in Developed and Developing Countries

Vladimir S. Osipov, Uran N. Busurmankulova, Tatiana V. Popova, and Tatiana I. Barsukova

1 Introduction

Modern innovative companies in developing and developed countries strive toward integration in the international economic space. To create a successful image of a company and to conquer the market environment, it is not enough just to create and move to the market a quality product (service) – it is necessary to take into account the problems connected with climate protection of territories in which production and services capacities are located. This is required for demonstrating to society that the company is a part of society and functions for its well-being and support for life activities. At the modern stage, the climate is protected due to the efforts of the green economy, which is an economic system – part of the environment in which it functions; it cannot destroy it, but should support it and ensure green growth. The main parameters of green growth include clean water; biodiversity underwater and on land, which influences human health and the production of safe food; clean air [13]. These parameters could be achieved due to the reduction of polluting emissions in the environment, which is implemented thanks to energy efficiency

V. S. Osipov (✉)
Institute of Public Administration and Civil Service RANEPA, Moscow, Russia

Lomonosov Moscow State University, Moscow, Russia

U. N. Busurmankulova
Kyrgyz National University Named After Jusup Balasagyn, Bishkek, Kyrgyzstan

T. V. Popova
Academy of Management of the MIA of Russia, Moscow, Russia
e-mail: lapopova@ncfu.ru

T. I. Barsukova
North Caucasus Federal University, Stavropol, Russia
e-mail: tbarsukova@ncfu.ru

E. G. Popkova (ed.), *Smart Green Innovations in Industry 4.0*, Springer Climate,
https://doi.org/10.1007/978-3-031-45830-9_7

(transition to renewable energy sources, which use does not negatively influence the ecology, the use of eco-friendly approaches to waste management). Activities on the management of green growth in most countries are implemented due to the active role of national companies, which, in their turn, promote eco-friendly products (services) due to the effective management of the demand of climate-responsible consumers; for them, the top-priority role belongs to green growth, not the pricing component [22]. In the case of a large participation of climate-responsible entrepreneurship in green growth, countries can achieve the goals of this strategy. Good results in this direction were achieved in recent years when the digital economy offered a large range of opportunities for the implementation of climate-oriented improvements which are preconditions for the effective economic development of entrepreneurship. Developed and developing countries take part in the implementation of the strategic concept of green growth. Thus, it is very topical to study this process in developed and developing countries.

The goal of this work is to study the character of the contribution of climate-responsible entrepreneurship in the digital economy markets to green growth in developed and developing countries. Achievement of this goal involves the following tasks: determining countries with a high level of digitalisation of entrepreneurship and a high level of green growth; elaborating on the directions of the selected countries' companies' influence on green growth.

2 Materials and Method

The issues of entrepreneurship's participation in the formation of the green economy and the parameters of the green growth of countries were studied in [1, 11, 15–17, 20, 22, 23]. It is worth mentioning the provisions of [11], which allows for a comprehensive assessment of the effect of the formation of the wind energy sector of the Netherlands on the support for the country's employment rate. Attention should be also paid to [1], which characterises the influence of agricultural companies in South Korea on climate protection, biodiversity, economic growth and employment. The authors of [16] determine the key characteristics of energy transition in the UAE. The above studies focused on the specific features of the considered problems at the level of isolated countries. In this paper, we analyse the issues of the impact of entrepreneurship on green growth in developing and developed countries.

As for the methods used, we applied comparative analysis of the formation of economic systems of the two categories. We also used the method of main factors, which allowed discovering the reasons for support for green growth. The statistical method was used to find indicators that characterise the parameters of the green economy, digitalisation and isolated estimate indicators.

3 Results

The list of developed and developing countries in the context of the indicator of digitalisation of entrepreneurship is given in Fig. 1. This indicator was studied with the help of the index of Business agility (an element of the World Digital Competitiveness Index).

We selected five developed countries that achieved key positions by digital support for entrepreneurial activities: Denmark (5th position in 2020, 7th in 2021 and 1st in 2022), South Korea (3rd, 5th and 2nd positions, accordingly); the USA (2nd position in 2020, 1st in 2021 and 4th in 2022); the Netherlands (7th position in 2020, 8th in 2021 and 2022); Switzerland (6th position in 2020, 4th in 2021 and 7th in 2022). The most progressive developing countries in the digitalisation of business include China (1st position in 2020, 3rd position in 2021 and 2022); Kazakhstan (13th position in 2020, 6th position in 2021 and 2022); Indonesia (24th position in 2020, 26th in 2021, 22nd in 2022); India (52nd position in 2020, 36th in 2021 and 26th in 2022); the UAE (12th position in 2020, 10th in 2021 and 26th in 2022).

Let us analyse the state of green growth under the conditions of specific economic situations in the considered countries (Table 1).

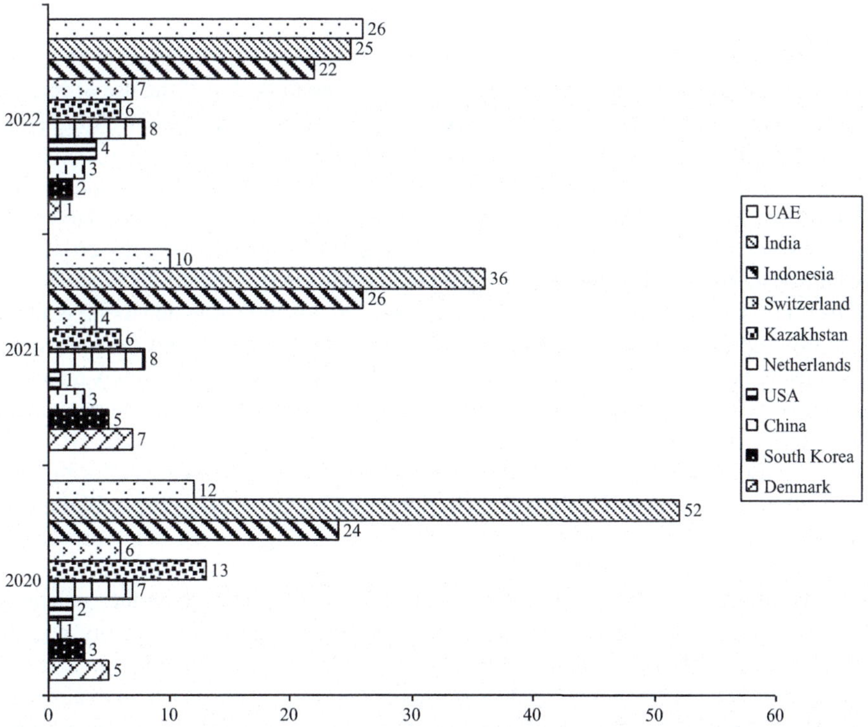

Fig. 1 Level of digitalisation of entrepreneurship in the selected countries. (*Source:* Created by authors based on [6–8])

Table 1 Level of green growth of countries that implement the digitalisation of entrepreneurship

		% of the population with access to drinking water		% of protected areas on land to preserve biodiversity		% of renewable energy in the total volume of energy consumption	
No.	Country	2017	2020	2019	2020	2019	2020
1	Denmark	100	100	86.2	88.82	37.02	31.681
2	South Korea	99.79	99.93	37.48	37.57	2.36	2.251
3	China	92.85	94.26	9.94	10.06	9.673	15.9
4	USA	99.27	99.88	51.21	34.93	7.92	8.496
5	Netherlands	100	100	97.9	79.59	8.886	13.999
6	Kazakhstan	95.63	95.44	11.15	13.05	1.464	3
7	Switzerland	100	100	35.48	37.03	22.45	23.833
8	Indonesia	89.34	92.42	26.11	25.91	23.88	30.18
9	India	92.68	90.49	21.02	20.72	23.135	40 (2022)
10	UAE	98.05	99.97	57.9 (2018)	51.55	0.777	19.63 (2021)

Source: Created by the authors based on [2–5, 9, 12, 14, 18, 19]
Eurostat (2022), OECD (2021), Sdgindex (2022), Carbonbrief (2022), Energycentral (2021), BP (2021), Investindia (2022), Pudovkina (2023), Sdgindex (2021)

Analysis of the results (Table 1) shows that the lowest effectiveness is observed in certain developed and developing countries by the indicator of the use of renewable energy (including at the level of entrepreneurship in the main sectors of the economy).

UAE has the lowest value of this indicator (0.777% in 2019). This is mainly due to the fact that the UAE has approximately 6% of the world's oil reserves. However, these reserves are expected to be depleted by 2100. To avoid an energy crisis, the country started implementing projects on renewable energy, including solar energy. Over the 2 years, there was a significant increase in this indicator, due to the gradual energy transition by entrepreneurship. The main sectors that started implementing and using renewable energy are the service sphere, commercial buildings management, as well as industry, transport and communal sphere. The latter are the largest consumers of energy, with the highest negative influence on climate [16]. An increase in the level of energy transition (by 18.853%) in 2020 allowed for the stabilisation of GDP in 2021 after its reduction in 2020. Over 2019–2021, GDP grew by 0.6% [21], which is a positive result for a developing country in the process of overcoming the consequences of the COVID-19 crisis.

The influence of the level of implementing renewable energy on the growth of GDP could be connected with the fact that the use of clean solar energy was an advantage in attracting demand from consumers. There was an increase in demand for commercial property in which clean energy was used. Consumers were aimed at ensuring their ecological safety within certain territories where commercial buildings were located and, accordingly, cooperated with companies that started implementing such projects. Here we see the existence of green growth under the conditions of support for the improvement of ecological parameters.

Given the low, but improving, level of renewable energy in the UAE, we should note a large percentage of protected land with supported biodiversity (more than 50%). This is a large part of the territory, especially for developing countries in which entrepreneurship is aimed at quick economic growth and the use of resources. We should also note the high level of the population's access to drinking water, which is improving. The organisation of works on the provision of drinking water is performed within public-private partnership projects. Entrepreneurship plays an important role in the projects on installation and maintenance of purification works. Entrepreneurship also makes investments in modern equipment with digital tools. We should also note the contribution of national companies to the reduction of the carbon footprint in the oil refining industry, which includes the use of equipment for CO_2 capture. This helps the efforts on water purification. At the current stage, the technologies of Porous Alpha and Badia Farms are used [20].

1. Similarly to the UAE, Kazakhstan has a low level of renewable energy use (Table 1). Analysis of the materials of [17] showed that despite the reserves of fossil energy resources, Kazakhstan may face an energy crisis in the near future. Another problem is the negative effect of the use of non-renewable energy on climate, which is especially topical for the economy of a country focused on a large energy intensity. The indicator of Kazakhstan's economy energy intensity is 6 MJ per dollar of GDP [10]. This indicator was gradually reducing since 2000, but it is still high – which shows the necessity to continue the implementation of measures on energy efficiency. The main sectors of the economy that feature such measures include science-driven innovative sectors (mainly IT and service sphere), which use electric energy produced by individual sources (solar and wind stations). The highest consumption of non-renewable energy resources accounts for the processing and mining industries (more than 98%) [15].

Transitioning to renewable energy sources is not a reason for an economic downturn; it may ensure the same values of GDP and, in some cases, it may stimulate its growth. According to [15], the influence of the transition to renewable sources on the economy is the same; also, there is a reduction of the climate footprint, which reduces proportionally to the decrease in the use of non-renewable energy.

The innovations-oriented entrepreneurial sector in Kazakhstan is the main participant of the economy that demonstrates participation in the reduction of the negative influence on climate and ensures green growth with simultaneous economic growth.

It is worth noting the rather effective management of services in the provision of drinking water in Kazakhstan (95% in 2019–2020) (Table 1). The management of the water purification processes is performed by the entrepreneurial sector with support from local authorities.

2. South Korea also does not demonstrate a high level of the use of renewable energy (Table 1). However, the country has positive indicators in two other con-

sidered parameters of green growth: access to drinking water (more than 99%); % of protected land that ensures biodiversity. On the whole, despite the relatively low value of renewable energy use, the parameters of green growth in South Korea demonstrate positive dynamics.

Since entrepreneurship in South Korea demonstrates world leadership in the sphere of digitalisation business processes (Fig. 1), such process as the provision of drinking water to the population is performed with the use of innovative digital technologies. We should also note the achievements of the agrarian sector in the creation of farms that do not violate biodiversity on land. Analysis shows that South Korea has centuries-old traditions of agricultural production without the violation of the eco-system, connected with the following: focus on the specifics of the landscape; growing cultures that are most adaptable to certain territories; care for forests, as the natural component of the protection of agricultural plants from damage; change of cultures that are planted on the same territory, which ensures the preservation of soil's water balance [1]. The approach of agrarian entrepreneurship of South Korea to climate protection improved in the twenty-first century due to a wide application of digital technologies connected with the monitoring of climate characteristics. All these advantages influence productiveness and green growth in this sphere; thus, economically developed South Korea demonstrates good indicators of employment and decent wages in the agricultural sector (10–15% of the population) [1].

3. Netherlands demonstrated a growth in the level of renewable energy resources use in 2020 (Table 1). According to the analysis, the Netherlands implements a sustainable energy transition. Stable growth is predetermined by business and population's support for climate protection and CO_2 reduction. The able-bodied population of the Netherlands has a positive attitude to retraining at a company's expense: teaching new speciality in the case of dismissals in the sphere of electric power plants maintenance (ones that work on non-renewable sources); personnel are trained to work at wind power plants [11]. On the whole, the basis for human resources' supporting these changes is the desire to retain usual incomes and decent work given the need to master new skills and knowledge. Activation of business correlates with the activation of labour resources, which is a sign of the adoption of the green growth concept and improvement of climate indicators in the long term.

The Dutch business sector's results in providing drinking water to the population are very good (100% of satisfaction of the needs). Such a result was achieved in the course of the evolution of social norms to the standards of drinking water quality, which took place over the course of various stages of resolution of this problem in 1850–1990 [23]. Digitalisation allows companies involved in the sector of drinking water purification and supply to achieve improvement of the quality of water resources purification. These companies do not increase the burden on the traditional energy system but use renewable energy sources in their work.

The USA retained a stable level of renewable energy implementation in 2019–2020 (Table 1). As in the case of the Netherlands, the USA demonstrated a decrease in % of land protection ensuring biodiversity preservation. The latter was caused mainly by the expansion of territories used for agricultural production. Due to the Dutch and American agrarian sectors' focus on the growing of monocultures and insufficient focus of agricultural companies on the change of cultures, there is a threat of drying up of soil and further reduction of productiveness and biodiversity. As we see, the entrepreneurial sector can have a positive effect on green growth; at the same time, if there is no focus on support for climate characteristics, companies' activities may lead to aggravation of the green growth parameters.

4 Discussion

Based on the above results, we can state that climate-responsible entrepreneurship in the markets of the digital economy can contribute to green growth in developed and developing countries. Many countries strive toward an increase in economic growth and toward international integration. Innovations-active companies implement ICT, which allows for the more effective application of eco-oriented projects, which influence the main parameters of green growth. The difference in the level of digitalisation and environmental activity between rapidly developing countries and developed countries is not large. This facilitates the growth of social standards and the reduction of the carbon footprint. Accordingly, the goal of green growth is achieved. This goal is connected with the economy's integration into the surrounding ecosystem, which provides resources for the stable and sustainable development of the current and future generations.

Assessment of the activities of the entrepreneurial sector showed that its participation in the realisation of the green growth projects is driven by different motives, which include the following: improvement of image related to ecologisation, which facilitates the promotion of products (services); government's support for green initiatives, which may include certain subsidies, guarantees, etc.; possibility of involvement in large projects on climate protection at the national and international levels; possibility to attract investments for prospective growth and increase in market positions.

5 Conclusion

In this work, we analysed possible directions for the participation of climate-responsible entrepreneurship, which uses digitalisation, in the achievement of green growth in developing and developed countries. We also identified the pros and cons of the entrepreneurial environment's functioning in the formation of the green economy. Based on empirical and statistical data, we proved the assumption that green growth can be achieved under the conditions of developing and developed countries. Thus, it is possible to conclude that the main driving force of climate protection is entrepreneurship, while national governments and supra-national organisations sometimes limit themselves by only declaring possible changes.

Therefore, perspective directions for green growth of countries could be determined at the level of factual and expected (potential) capabilities of entrepreneurship, within benchmarks and acute problems of the environment of territories. Accordingly, the involvement of innovative entrepreneurship may facilitate a better and more precise determination of the goals of green growth of regions and countries.

References

1. Abulu L (2020) In South Korea, centuries of farming point to the future for sustainable agriculture. https://news.mongabay.com/2020/05/in-south-korea-centuries-of-farming-point-to-the-future-for-sustainable-agriculture/. Accessed 01 Jan 2023
2. BP (2021) Statistical review of world energy – 2021. Indonesia's energy market in 2020. https://www.bp.com/content/dam/bp/business-sites/en/global/corporate/pdfs/energy-economics/statistical-review/bp-stats-review-2021-indonesia-insights.pdf. Accessed 01 Jan 2023
3. Carbonbrief (2022) Guest post: will China's new renewable energy plan lead to an early emissions peak? https://www.carbonbrief.org/guest-post-will-chinas-new-renewable-energy-plan-lead-to-an-early-emissions-peak/. Accessed 01 Jan 2023
4. Energycentral (2021) Kazakhstan: 3% of all energy generated by renewable sources in 2020. https://energycentral.com/news/kazakhstan-3-all-energy-generated-renewable-sources-2020. Accessed 01 Jan 2023
5. Eurostat (2022) Share of energy from renewable sources. https://ec.europa.eu/eurostat/databrowser/view/NRG_IND_REN/default/table?lang=en&category=nrg.nrg_quant.nrg_quanta.nrg_ind.nrg_ind_share. Accessed 01 Jan 2023
6. IMD (2020) IMD world digital competitiveness ranking 2020. https://www.imd.org/wcc/docs/release-2020. Accessed 01 Jan 2023
7. IMD (2021) IMD world digital competitiveness ranking 2021. https://www.imd.org/wcc/docs/release-2021. Accessed 01 Jan 2023
8. IMD (2022) IMD world digital competitiveness ranking 2022. https://www.imd.org/centers/world-competitiveness-center/rankings/world-digital-competitiveness/. Accessed 01 Jan 2023
9. Investindia (2022) Renewable energy. India is 3rd most attractive for Renewable energy investments and deployments. https://www.investindia.gov.in/sector/renewable-energy. Accessed 01 Jan 2023
10. Knoema (2020) Kazakhstan – energy intensity level of primary energy. https://knoema.com/atlas/Kazakhstan/Energy-intensity. Accessed 01 Jan 2023
11. Knol E, Coolen E (2019) Employment analysis (2019–2023) of various fields of activities in the Dutch offshore wind sector. https://www.researchgate.net/publication/336922754_Employment_analysis_2019-2023_of_various_fields_of_activities_in_the_Dutch_offshore_wind_sector/link/5dbae9364585151435d6e9d4/download. Accessed 01 Jan 2023
12. OECD (2021) Renewable energy. https://data.oecd.org/energy/renewable-energy.htm. Accessed 01 Jan 2023
13. OECD (2022) What is green growth and how can it help deliver sustainable development? https://www.oecd.org/greengrowth/whatisgreengrowthandhowcanithelpdeliversustainabledevelopment.htm. Accessed 01 Jan 2023

14. Pudovkina K (2023) The UAE announced the implementation of 11 projects in the sphere of green energy at the cost of $43.2. https://profile.ru/news/economy/v-oae-zayavili-o-realizacii-11-proektov-v-sfere-zelenoj-energetiki-na-43-2-milliarda-1243533/. Accessed 15 Jan 2023
15. Raihan A, Tuspekova A (2022) Role of economic growth, renewable energy, and technological innovation to achieve environmental sustainability in Kazakhstan. Curr Res Environ Sustain 4. https://www.sciencedirect.com/science/article/pii/S2666049022000433. Accessed 01 Jan 2023
16. Salimi M, Hosseinpour M, Borhani TN (2022) Analysis of Solar energy development strategies for a successful energy transition in the UAE. Processes 10(7):1338. https://www.mdpi.com/2227-9717/10/7/1338. Accessed 01 Jan 2023
17. Sarkhanov T, Huseynli N (2022) Econometric analysis of renewable energy consumption and economic growth: the case of Kazakhstan and Kyrgyzstan. Int J Energy Econ Policy 12(6):163–167
18. Sdgindex (2021) Sustainable development report 2021. The decade of action for the Sustainable Development Goals. https://2021.dashboards.sdgindex.org/. Accessed 01 Jan 2023
19. Sdgindex (2022) Sustainable development report 2022. From crisis to sustainable development, the SDGs as roadmap to 2030 and beyond. https://www.sdgindex.org/reports/sustainable-development-report-2022/. Accessed 01 Jan 2023
20. Shawish AA, Nabhan T, Almadidy A (2019) Potable water in UAE: an overview of water characteristics and sources of contamination. J Environ Toxicol Stud 3(2). https://doi.org/10.16966/2576-6430.120. https://www.researchgate.net/publication/338695772_Potable_Water_in_UAE_An_Overview_of_Water_Characteristics_and_Sources_of_Contamination. Accessed 01 Jan 2023
21. Statista (2022) United Arab Emirates: gross domestic product (GDP) from 1987 to 2027. https://www.statista.com/statistics/297605/uae-gross-domestic-product/. Accessed 01 Jan 2023
22. Turginbayeva A, Shaikh, Aijaz A (2022) How price sensitivity influences green consumer purchase intention? Sustainable business concepts and practices. EuroMed Academy of Business 15th annual conference book of proceedings (Book series). ISSN 2547-8516. September 21st–23rd, pp 1388–1390
23. Zetland D, Colenbrander B (2018) Water civilization: the evolution of the Dutch drinking water sector. Water Econ Policy 04(03). https://www.worldscientific.com/doi/epdf/10.1142/S2382624X18500121. Accessed 01 Jan 2023

Sustainable Development of Climate-Responsible Entrepreneurship of Central Asia and Eastern Europe in the Digital Economy Markets Under the Crisis Conditions

Ainura A. Adieva, Erkegul M. Biimyrsaeva, Tariel Ch. Tashibekov, and Platon A. Lifanov

1 Introduction

The use of market methods of management, creation of attractive conditions for business and participation in the liquidation of barriers in various spheres characterise market-oriented state management of the economy. In such an environment, entrepreneurship is aimed at support for the government's initiatives in the sphere of sustainable development, focusing also on the implementation of own interests and the protection of the ecosystem. Different regions and countries demonstrate different levels of entrepreneurial climate, which influences the climate-responsible behaviour of the business environment's subjects. On the whole, supporting such interaction at the level of the mentioned participants of an economic system (government and entrepreneurship) ensures the balance under the conditions of crisis phenomena. Despite the differences in the level of economy and sectoral structure, countries of Eastern Europe and Central Asia demonstrate certain results in the functioning of climate-responsible entrepreneurship, which is connected with the digitalization of climate-responsible entrepreneurship and the creation of a favourable climate by the government. Though crisis phenomena are a serious barrier to development, they do not lead to the business environment's refusal of the adopted

A. A. Adieva (✉)
International University of the Kyrgyz Republic, Bishkek, Kyrgyzstan

E. M. Biimyrsaeva
Russian-Kyrgyz Institute of Business Management Automation of the International University of Innovative Technologies, Bishkek, Kyrgyzstan

T. C. Tashibekov
Osh Technological University Named After M.M. Adyshev, Osh, Kyrgyzstan

P. A. Lifanov
Volgograd State Technical University, Volgograd, Russia

E. G. Popkova (ed.), *Smart Green Innovations in Industry 4.0*, Springer Climate,
https://doi.org/10.1007/978-3-031-45830-9_8

environmental responsibilities. This ensures the achievement of the UN SDGs in the sphere of climate-responsible production.

The purpose of this paper is to describe the features of sustainable development of climate-responsible entrepreneurship of Central Asia and Eastern Europe in the digital economy markets under the conditions of a crisis. To achieve this purpose, we determine the level of stability of the functioning of entrepreneurship in the selected regions' countries under the conditions of a crisis and describe the key features of the impact of entrepreneurship on the parameters of sustainable development with the help of digital tools.

2 Materials and Method

Provisions of the theoretical and empirical studies in the sphere of climate responsibility of the entrepreneurial sector in countries of Eastern Europe and Central Asia allowed discovering the features of their impact on the economy in the digital economy markets under the conditions of crises. These works include [1–3, 6–9]. Audonin et al. [1] considered the directions and prospects for intensive development of the system of transport communications in Kazakhstan, which might be competitive in the context of prices and quality with other participants of the Chinese project "One belt, one way." Harizanova-Bartos and Stoyanova [3] demonstrate the contribution of Bulgaria's agricultural companies to the implementation of digital climate technologies. Nazare [6] demonstrated the participation of the entrepreneurial sector of Romania in the realisation of the project of renewable (solar) energy and determine the financial problems of business and the government in this sphere.

The methodology of this research comprises parametric and combined methods based on the statistical and analytical materials of the formation of the studied countries' entrepreneurial sector. The statistical method was used to consider such indicators as national GDP and the share of the activities of entrepreneurial subjects of all sectors in GDP. The index method was used to identify the ranking of countries by the selected directions and indicators. The method of comparison was used to compare the qualitative and quantitative indicators with quantitative measuring (index or value).

3 Results

We shall consider the state of the activities of the entrepreneurial sector of the selected countries under the conditions of crises and the influence of the digital economy through the assessment of change in national GDP (Table 1) and we shall determine the level of their digitalization (including digital support for the entrepreneurial sector) and elaborate on the dependence of GDP on the level of digital support for business.

Table 1 GDP change in the selected countries of Central Asia and Eastern Europe in 2019–2022

	Region/country	GDP, $ billion				GDP change, +/−, %		
		2019	2020	2021	2022	2019–2020	2020–2021	2021–2022
I	Eastern Europe							
1	Moldova	11.97	11.86	13.68	14.048	−0.92	15.35	2.69
2	Romania	251.02	251.36	284.09	299.885	0.14	13.02	5.56
3	Bulgaria	68.92	70.24	84.06	86.66	1.92	19.68	3.09
4	Hungary	163.99	157.18	181.85	184.65	−4.15	15.70	1.54
5	Czech Republic	252.55	245.97	281.78	295.62	−2.61	14.56	4.91
6	Slovakia	105.72	106.7	116.53	127	0.93	9.21	8.98
7	Poland	596.05	599.45	679.44	716	0.57	13.34	5.38
II	Central Asia							
8	Kazakhstan	181.67	171.08	197.11	224.34	−5.83	15.22	13.81
9	Kyrgyzstan	8.87	7.78	8.54	8.94	−12.29	9.77	4.68
10	Tajikistan	8.3	8.13	8.75	9.39	−2.05	7.63	7.31
11	Uzbekistan	59.91	59.89	69.24	74.2	−0.03	15.61	7.16

Source: Compiled by the authors based on [5, 12, 15]

The timeframe of the research is 2019–2022. This period covers the energy crisis of 2021–2022 [4] and the COVID-19 crisis of 2020.

As is shown in Table 1, all selected countries were able to increase their GDP under the conditions of crisis phenomena in the economy (energy crisis and the COVID-19 crisis).

Table 2 shows the level of information and communication technologies, through the index of Information and communication technologies [14].

Based on the determined dependencies, we can describe the existing specifics of the influence of the entrepreneurial sector on climate protection under crisis conditions.

As shown in Table 2, the use of digital tools in Poland allowed entrepreneurship to ensure stable production and sales of products (services), which led to the growth of GDP.

Poland's GDP is the largest in the considered sample of countries ($596.05 billion in 2019, $599.45 billion in 2020, $679.44 in 2021 and $716 billion in 2022). According to the IMF [5], Poland is ranked 22nd in the world by GDP, which is a sign of the high economic potential of the country and its entrepreneurship.

Based on the data of [11], it is possible to state that Polish companies adopted certain responsibilities in the sphere of climate protection. In 2022, Poland was ranked 12th in the world by the indicator of sustainable development, including in the context of SDG 12 (reduction of Production-based SO_2 emissions down to 14.2 kg/capita in 2018, compared to the 2012 value (30.73 kg/capita) and 2019 value (32.1 kg/capita)) [10, 11].

As for the factors ensuring the reduction of SO_2 emissions, we should mention entrepreneurship's investments in the upgrade of technologies and equipment in the processing and mining industries, and electric energy production (transition to renewable energy sources in certain large industrial clusters). The main

Table 2 Level of information and communication technologies in the selected countries

	Region/ country	Information and communication technologies (Rank)				Change, +/−, rank		
		2019	2020	2021	2022	2019–2020	2020–2021	2021–2022
I	Eastern Europe							
1	Moldova	52	61	62	68	9	1	6
2	Romania	63	60	52	50	−3	−8	−2
3	Bulgaria	45	44	42	37	−1	−2	−5
4	Hungary	54	54	55	59	0	1	4
5	Czech Republic	64	63	53	54	−1	−10	1
6	Slovakia	47	48	54	56	1	6	2
7	Poland	28	30	24	22	2	−6	−2
II	Central Asia							
8	Kazakhstan	40	42	29	25	2	−13	−4
9	Kyrgyzstan	85	86	82	77	1	−4	−5
10	Tajikistan	112	119	123	123	7	4	0
11	Uzbekistan	73	72	65	55	−1	−7	−10

Source: Compiled by the authors based on [14]

technologies that ensure the reduction of SO_2 emissions include the following: innovative systems of monitoring and assessment of emissions, systems of capture and utilisation of emissions, technologies of transformation of waste into energy. The problems connected with the financing of projects, which are implemented by Polish entrepreneurship in the context of technologies and equipment upgrades, are solved with the help of support and subsidies from the EU aimed at climate programmes [13].

However, there is a range of problems in the sphere of climate protection, connected with the reduction of CO_2 emissions from the activities of industry in Poland.

A positive aspect is the entrepreneurial sector‘s participation in the support of infrastructure for the provision of drinking water to the population. In 2017, 99.73% of the Polish population had access to clean drinking water, and in 2020 – 99.97% [10, 11]. According to [8], Poland’s large business deals with the purification of water with the help of technological installations with filters, which ensure the continuous turnover of water resources. The infrastructure of the provision of drinking water to the population and municipal services involves also small companies. Pawełek and Bergel [8] provided an example of economically and environmentally effective technology for the purification of water in Mszana Dolna (Małopolska region). The authors considered the advantages of ozone technologies, which allow keeping standards of water quality in different seasons. The requirements of the focus on seasonal parameters in the assessment of the quality of water were adapted by Poland within the practices of the EU member states [7].

The achieved level of information and communication technologies in Kazakhstan facilitated economic growth and became a good result for a developing country. It is particularly important that such indicators were reached by a country of Central Asia, which has been actively integrating into the global economic system over the last 5 years.

According to [11], Kazakhstan improved its position by sustainable development in 2019–2022. This was achieved due to the entrepreneurship's implementing their responsibilities in the environmental sphere. The transition to renewable energy is not large-scale, but it is supported by a business environment that is aimed at the continuous functioning of entrepreneurship, reduction of payments for the use of autonomous sources of energy and support for the reduction of CO_2 emissions from the burning of non-renewable sources of energy. The modern export-oriented textile industry of Kazakhstan may become more competitive in the world market thanks to the reduction of product prices – this could be achieved through the use of a combined energy supply (traditional sources and solar energy) [2]. The projects of companies' transition to such form are supposed to be financed partially by the government. There are also large perspectives in the implementation of the processes of energy transition in the transport sphere, which competitiveness is very important due to international trade and passenger communications with China and Russia. There is a need to reduce the cost of services in this sphere, which could be achieved through the use of renewable energy sources [1].

Despite certain efforts of entrepreneurship in the implementation of the projects on energy transition to renewable energy, there are still problems with regulation and reduction of Production-based SO emissions (68.16 kg/capita in 2018, compared to 14.2 kg/capita in Poland) [11].

Romania demonstrates significant results in information and communication technologies (Table 2). Due to the focus on digital innovations, entrepreneurship improved sales volumes, which positively influenced the national GDP (growth by 5.56% in 2022 – $299.885) (Table 1).

The entrepreneurial sector of the main economic spheres in Romania demonstrates a serious impact on the achievement of the SDGs. The largest contribution of national business to climate protection is the implementation of renewable energy, which production and use ensure the reduction of CO_2 emissions. Despite certain problems with the financing of renewable energy in Romania [6], there are positive results. The problems in this sphere include the toughening of the government's regulation of the electric energy market. In 2012, there was adopted a law that prohibited selling electric energy (including electric energy produced from renewable sources) without the participation of the regulatory body; if a company (individual) produced electric energy, it was not able to sell excessive volumes of energy to other consumers without applying to the regulator. This coincided with an energy crisis.

Bulgaria experienced certain problems with investments in renewable sources of energy since the government's participation in the creation of favourable conditions for business that implements projects on the installation of solar stations are not sufficient.

A positive fact is the implementation of innovative technologies in agriculture, which allowed reducing the negative influence on the biodiversity of territories. In 2019, the indicator "Mean area that is protected in marine sites important to biodiversity" was 87.5% [10], and in 2020 – 96.61% [11]. On the whole, agricultural companies strive to protect soils from drying up and pesticide contamination [3].

Uzbekistan's entrepreneurship made a significant contribution to the implementation of digital technologies in the main export-oriented sectors. The main burden in investments in innovative technologies in the sphere of textile, chemical and food industries was put on entrepreneurship. The government develops plans on implementing projects to support the entrepreneurial sector in online trade and other directions [9].

Uzbekistan does not demonstrate serious positive results in the sphere of sustainable development. A significant contribution of business in this sphere is the use of renewable energy, the level of which in the total energy consumption was 1.2% in 2022 [11].

4 Discussion

Analysis of the features of sustainable development of climate-responsible entrepreneurship of Central Asia and Eastern Europe in the digital economy markets under the crisis conditions allowed discovering the entrepreneurship's serious influence on the level of the countries' digitalization, which ensures the competitiveness of products (services), and the achievement of the environmental SDGs.

The main features in countries of Eastern Europe include the implementation of climate-oriented technologies to improve competitive positions, sales and economic results due to price reduction through the use of less expensive renewable energy sources; guarantees of compensation or support from the governments (EU funds).

The development of climate-responsible entrepreneurship amid crises in the considered regions is characterised by certain specific features.

An important aspect of the influence of Polish entrepreneurship on climate growth during the energy crisis was the realisation of effective technologies of the transformation of industrial SO_2 emissions into energy. This allowed reducing the environmental pollution. The role of Romanian business in the implementation of renewable energy became particularly actual during the COVID-19 crisis. The climate-responsible agricultural sector of Bulgaria was able to implement effective measures in to protect soils, despite the COVID-19 crisis.

The general key features of climate-responsible activities of entrepreneurship in countries of Central Asia include the leadership (compared to the government) in the sphere of digital environmental technologies implementation (renewable energy projects and technologies of soil management in agriculture). Countries of Central Asia, similarly to certain countries of Eastern Europe, have natural potential for satisfying the entrepreneurship and population's needs for renewable energy, but the problems of the legislative and organisational character do not allow for rapid growth in this sphere. Thus, we can state the existence of barriers for the achievement of climate-related Sustainable Development Goals. Elimination of these barriers depends on the change in economic and energy strategies in the considered countries.

5 Conclusion

In this work, we analysed and systematised the specific features of the sustainable development of climate-responsible entrepreneurship in countries of Central Asia and Eastern Europe in the markets of the digital economy in the context of crisis conditions. Companies that adopted climate-responsible behaviour have more opportunities in Eastern Europe. This particularly applies to the members of the European Union, which have support from the EU. Also, the accessibility of credit funds allows entrepreneurship in this region to start ambitious technological projects, which envisages the achievement of many goals, including satisfaction of consumer demands and reduction of final prices.

The positive influence of digitalization of the entrepreneurial sector on the economic growth of the considered countries is connected with the introduction of eco-oriented digital technologies.

It is possible to state that entrepreneurship in countries of Eastern Europe and Central Asia demonstrates high readiness for the countries' achievement of climate-related Sustainable Development Goals. A very important aspect here is the increase in the government's role in the provision of support for the initiatives of climate-responsible entrepreneurship.

References

1. Audonin A, Turginbayeva A, Askerov A, Yergobek D (2020) Modern economic and logistical trends in Eurasia: how do new trans-Eurasian mega-projects influent to national economic growth. E3S Web Conf 159:06009. https://doi.org/10.1051/e3sconf/202015906009
2. Durru O, Yessengeldina A, Kosherbayeva A, Shakeyev S, Shaikin D (2021) State policy of Kazakhstan on implementing of renewable energy sources in textile industry companies. Int J Energy Econ Policy 11(3):51–56. https://www.econjournals.com/index.php/ijeep/article/view/10659. Accessed 08 Jan 2023
3. Harizanova-Bartos H, Stoyanova Z (2019) Impact of agriculture on soil pollution in Bulgaria. Econ Agric 66(2):375–387. https://doi.org/10.5937/ekoPolj1902375H
4. IEA (2022) Global energy crisis. https://www.iea.org/topics/global-energy-crisis. Accessed 08 Jan 2023
5. IMF (2022) World economic outlook database: October 2022. https://www.imf.org/en/Publications/WEO/weo-database/2022/October/weo-report. Accessed 08 Jan 2023
6. Nazare L (2021) The Romanian renewable energy sector: a potential still untapped: report. https://bankwatch.org/wp-content/uploads/2021/04/Romanian-renewable-energy-sector-bankwatch.pdf. Accessed 08 Jan 2023
7. Ober J, Karwot J (2021) Tap water quality: seasonal user surveys in Poland. Energies 14(13):3841. https://doi.org/10.3390/en14133841

8. Pawełek J, Bergel T (2019) Ozone water treatment in small water purification plants in Poland – Mszana Dolna case study. E3S Web Conf Ecol Environ Eng 86:00029. https://www.researchgate.net/publication/331281844_Ozone_water_treatment_in_small_water_purification_plants_in_Poland_-_Mszana_Dolna_case_study. Accessed 08 Jan 2023
9. Raimjanova MA, Shadiyeva DK, Zoyirov LS, Saidov RB, Askarov MT (2021) Digitalization of the economy in the geography of the Republic of Uzbekistan. E3S Web Conf 244:10044. https://www.e3s-conferences.org/articles/e3sconf/pdf/2021/20/e3sconf_emmft2020_10044.pdf. Accessed 08 Jan 2023
10. Sdgindex (2021) Sustainable development report 2021 the decade of action for the Sustainable Development Goals. https://www.sdgindex.org/reports/sustainable-development-report-2021/. Accessed 08 Jan 2023
11. Sdgindex (2022) Sustainable development report 2022. From crisis to sustainable development, the SDGs as roadmap to 2030 and beyond. https://www.sdgindex.org/reports/sustainable-development-report-2022/. Accessed 08 Jan 2023
12. Statista (2022) Turkmenistan: gross domestic product (GDP) in current prices from 1997 to 2027. https://www.statista.com/statistics/1034327/gross-domestic-product-gdp-in-turkmenistan/. Accessed 08 Jan 2023
13. Trade.gov (2023) Environmental technologies. https://www.trade.gov/country-commercial-guides/poland-environmental-technologies. Accessed 08 Jan 2023
14. WIPO (2022) Global Innovation Index 2022. https://www.wipo.int/global_innovation_index/en/2022/. Accessed 08 Jan 2023
15. Worldbank (2022) GDP current US$. https://data.worldbank.org/indicator/NY.GDP.MKTP.CD. Accessed 08 Jan 2023

Practical Experience of Climate-Responsible Business in the Digital Economy Markets in the Eurasian Economic Union (EAEU)

Aidarbek T. Gyiazov, Nurgazy Zh. Isakov, Zhypara Asilbek Kyzy, and Tanirbergen S. Darkenbaev

1 Introduction

The key global processes of the twenty-first century are common – to a large extent – for all countries and regions of the world. They are directly connected with the use of resources and the protection of the environment in which mankind exists. According to this, the fourth energy transition to the wide use of renewable energy sources [15] is closely connected with the institutionalisation of the processes of fighting climate change and is set on the quick development of the digital economy.

The Eurasian Economic Union was created in 2014 with the purpose of coordinating the issues of economic and customs & tariff policy of a group of countries with historically close economic ties. The EAEU members are Russia, Belarus, Kazakhstan, Armenia and Kyrgyzstan. The main goals of the Union are "to ensure the freedom of the movement of goods, services, workforces and capital and to conduct coordinated, agreed or common economic policy for comprehensive modernisation, cooperation and increase in the competitiveness of national economies and population's living standards" [13].

The territory of the EAEU covers approximately 34% of the area of Eurasia and has different natural and climate zones. Thus, it has the same climate risks that other countries of the continent face: an increase in the average annual temperature and the related risks of drying out of large agricultural areas and protected areas, growth

A. T. Gyiazov (✉)
Batken State University, Batken, Kyrgyzstan

N. Zh. Isakov · Z. A. Kyzy
Osh State University, Osh, Kyrgyzstan

T. S. Darkenbaev
International University of Kyrgyzstan, Bishkek, Kyrgyzstan

E. G. Popkova (ed.), *Smart Green Innovations in Industry 4.0*, Springer Climate,
https://doi.org/10.1007/978-3-031-45830-9_9

of the number of natural disasters, including hurricanes, anomalous temperatures, storms, etc., as well as threats to the biodiversity of the region.

The Global Climate Risk Index for countries of the EAEU is different. While for Belarus, Kazakhstan and Kyrgyzstan, climate risks are minimal (118 points), for Armenia it is 98.2 points and for the Russian Federation – 50.67. The global average value is 78.68 points [9].

In the context of fighting climate change, the indicator of CO_2 emissions is used most often. CO_2 emissions per capita characterise the country's influence on climate change. Among countries of the EAEU, only Kyrgyzstan and Armenia demonstrate a low level of CO_2 emissions per capita – 1.56 tons and 2.19 tons per capita, accordingly, with a global average level of 4.13 tons per capita. The indicators of other countries of the EAEU are much higher: 6.12 tons per capita in Belarus, 11.5 tons per capita in Kazakhstan and 44.8 tons per capita in the Russian Federation [16].

Thus, the EAEU – together with other countries and blocs – is responsible for climate change and must take measures to solve the related problems. Assessment of the effectiveness of the actions of each country of the EAEU from the position of climate risks management is shown with the help of the Environmental Performance Index (EPI) The value of this index for all countries of the sample is lower than the global average value (48.9), which is a sign of the insufficient effectiveness of government measures aimed at the reduction of the negative influence of the economic activity on climate situation. The highest environmental effectiveness among the EAEU countries is observed in Armenia and Belarus (48.3 and 48.5 points, accordingly), and the lowest – in the Russian Federation and Kyrgyzstan (37.5 and 35.7 points, accordingly); for Kazakhstan, the value of the indicator is 40.9 points [3].

2 Materials and Methods

The methodology of studying the issue of discovering, systematising and implementing the successful practical experience of climate-responsible business implies the use of the process approach. According to this, we identify climate problems, assess the level of development of the digital economy and dwell on the practical experience of development and introduction of digital solutions to solve the problems of climate change and adaptation to it. The tools of evaluation of climate problems and the level of the EAEU countries' reaction to the current climate challenges and the level of development of the digital economy include the use of global indices and rankings, which allow comparing the results obtained with other countries and determining the potential on the improvement of the EAEU countries' position in the context of the studied problem.

The basis for studying, generalising and interpreting the practical experience of doing climate-responsible business in the digital economy markets is the Green Tech concept, which implies the development and implementation of digital solutions that are aimed at the resolution or alleviation of problems connected with climate change.

The information base of this research is comprised of data from global institutes and expert portals with the assessment of the level of climate risks [9] and effectiveness of environmental measures to fight climate change [3]; characteristics of the activity of the EAEU (EEC 2022), the volume of CO_2 emissions per capita [16], level of the development of the digital economy [4], applied examples of functioning of climate-responsible entrepreneurial entities in the countries of the EAEU [5–8, 14].

The practical experience of climate-responsible business in the digital economy markets in the EAEU has not been widely studied in the scientific literature. Certain aspects of this problem were described in works devoted to the activity of the EAEU countries from the position of fighting climate change [12, 15], the issue of interaction of the digital economy and green technologies in the EAEU [11], implementation of GreenTech in the industry of the Russian Federation [10], development of digital markets of the EAEU [17] and economic processes in countries of the EAEU [1].

The goal of this research consists in the search and generalisation of the applied examples of climate-responsible business in the digital economy markets of the EAEU, which envisages the assessment of the level of the EAEU members' development level from the position of climate risks and fight against them, formation of the digital economy and work on digital solutions aimed at the resolution or alleviation of the problem of climate change.

3 Results

All countries of the EAEU signed the Paris Agreement on fighting climate change, adopting the responsibility to seek the declared goals connected with the reduction of greenhouse emissions, decrease in the anthropogenic burden on the environment, etc. Participants of the Paris Agreement adopted national programmes aimed at an increase in the level of energy saving and energy efficiency, development of renewable energy sources and improvement of the environmental situation. Apart from the national measures of such character, the EAEU also has a coordinated policy aimed at the common actions of the participating countries. According to the main goal of the Paris Agreement – not allowing the excess of the global average annual temperature by more than 2 °C by 2100 as to the level of the pre-industrial age – each country of the EAEU adopted the responsibility to reduce CO_2 emissions, increase the forested areas and ensure the balanced socioeconomic development of the country [15].

Countries of the EAEU are considered a group of the largest suppliers of raw materials – hydrocarbons, agricultural products, etc. However, one of the goals of the creation of the EAEU was the acceleration of innovations and increase in value added of products. According to this, the digital economy is among the top-priority spheres of the EAEU's development. The basis for the coordinated development of the digital economy in the EAEU countries is the EAEU Digital Agenda [2]. It

Table 1 Ranking of countries of the EAEU by the level of the digital economy development

Country	UNCTAD B2C E-commerce Index, Rank	ICT Development Index, Rank	WEF Networked Readiness Index
Armenia	84	75	55
Belarus	35	32	65
Kazakhstan	60	52	56
Kyrgyzstan	97	109	94
Russian Federation	41	45	48

Source: Formed by the authors based on [4]

declares the main goals for development of the digital society and digital markets within the EAEU, which include stimulation and support for new digital initiatives and projects, which cover different spheres, including climate change, environmental protection, etc.

The level of digital economy development in the EAEU countries is shown with the help of three indices: the B2C UNCTAD E-commerce index, the ICT Development Index and the Networked Readiness Index by the WEF (Table 1).

As is shown in Table 1, Belarus and the Russian Federation have the highest level of digital economy development in the EAEU. A slightly worse situation is observed in Kazakhstan and Armenia. Within the sample, the lowest level of digital economy development is observed in Kyrgyzstan. Thus, the average level of development of the digital economy shows the need for improvements in digital capabilities, level of digital readiness, etc.

An important direction of the digital economy, which combines the potential and capabilities of the EAEU and facilitates the resolution of climate-related problems, is the development of Green Tech start-ups. Given the diversity of ecological and climate problems, as well as the large potential of digital technologies, climate-responsible business in the EAEU member states deals with different tasks connected with the development of alternative energy, optimisation of the use of resources, waste management, etc. Such projects reflect practical solutions and practical experience of the complex or partial resolution of climate and ecological problems. The most well-known start-ups in the EAEU in this sphere are shown in Table 2.

Digital economy is considered a tool for fighting climate change or adapting to climate change. According to this, as is shown in Table 2, entrepreneurship in the digital markets helps track carbon footprint, provides tools for environmental or climate-responsible crowdfunding, ensures the monitoring of human-nature interaction from the position of effect on climate change and offers models and solutions to solve climate problems.

Table 2 Characteristics of the best-known Green Tech start-ups in the countries of the EAEU

Problem that is solved	Start-up	Country
Development of alternative energy solutions	Biohydrogen from Waste	Armenia
Production of raw materials for bioenergy and production of biofuel	GreenWayTechnology	Armenia
Accessible digital agricultural consulting	Garoon Tech	Armenia
Smart lighting of agricultural plants	FSLS Grow Lights	Armenia
Monitoring and rational use of land resources and food production	OneSoil	Belarus
Energy for IoT autonomous device	SolGate	Belarus
Air purification and disinfection around gadget users	CleanUV	Belarus
Transformation of an old smartphone into an IoT device for a smart home and social activity	SolGate mobi	Belarus
Monitoring of energy production and emissions from this production	Dereknet	Kazakhstan
Use of wind energy at electric car charging stations	RETEV	Kazakhstan
Manufacture and installation of a wind & sun gree – a smart device that generates alternative energy	Smart Energy Tree	Kazakhstan
Development of innovations in the sphere of wind energy technologies	Ecowatt LLP	Kazakhstan
Mobile device and platform for improvement of the waste management system	Tazar	Kyrgyzstan
Teaching and implementation of the practice of waste sorting and use of eco-friendly garbage cans	Begreen	Kyrgyzstan
Effective use of resources, including water, land, air and forest resources	TION, AirTron, Lintech, Prof-IT group, EnWatec	Russian Federation
Waste management and recovery of resources	Wasteout, QTech, Fineplastic, Ubirator, Binology	Russian Federation
Development of alternative energy sources	Solartrek, IFO, Altechnologies, NeoSun Energy, Green Energy	Russian Federation
Energy storage	Watts Battery, Volts Energy Storage, BM Power, NeoSun Energy, ENRU	Russian Federation
Development of smart networks of energy supply	Volts Energy Storage, ChermoGlass, Watts Battery, Insyte Electronics	Russian Federation

Source: Formed by the authors based on [5–8, 14]

4 Discussion

Researching the practical experience of climate-responsible business in the digital economy markets in the EAEU countries is an important scientific task, with many directions. Depending on the goal set, the perspectives of studying the practical experience within the considered problem can allow assessing the real impact of each project on the processes of the fight against climate change or adaptation to it. The obtained results might be tools for optimising the projects and the system of projects management.

An in-depth study of the algorithm of development and implementation of Green Tech start-ups in the EAEU countries will allow creating a structured guide for potential entrepreneurs and projects. The creation of such a guide and the development of a corresponding algorithm are an important factor for the stimulation of the development of climate-responsible business, since the key problems of the activity in this sphere are most often connected with a lack of information on the sources of investments, order of development, promotion and implementation of start-ups and assessment of their effectiveness from the position of the fight against climate change.

It is also important to systematise and generalise not only the successful experience of solving climate problems with the help of digital solutions but also unsuccessful practices. Ignoring problems during the implementation of start-ups often leads to the idealisation of the process of their development and leads to underestimation of the seriousness of this direction in the system of business. Thus, significant scientific and practical aspects of the resolution of the studied problem lie in the research, generalisation and systematisation of the unsuccessful practice of development and implementation of start-ups in the sphere of fighting climate change.

On the whole, the described perspectives of the research form important directions of the scientific discussions, which will help improve the information and investment support for climate-responsible business.

5 Conclusions

Countries of the EAEU are an inseparable part of the global economic system, with common economic and climate challenges. In this context, countries of the EAEU have adopted – separately and as a union – obligations on the decarbonisation of the economy, increase of green areas, development of alternative energy, etc. Despite this, the effectiveness of the current measures in this sphere should be improved. This requires more attention to government programmes and real examples of the development and introduction of projects at the level of entrepreneurship.

An important direction for solving the problems of low effectiveness of measures to fight climate change is the development of digital solutions by the climate-responsible business. Studying the practice of climate-responsible business allowed discovering and systematising information on the existence of successful Green Tech start-ups in the EAEU, which deal with the problems of development of alternative energy, waste management, optimal use of resources, smart agriculture, etc.

References

1. Amrin AK, Kaliyeva SA, Alzhanova FG (2020) Migration processes in Kazakhstan in the context of the formation of the silk road economic belt. J Popul Soc Stud 28(2):011, 156–011, 174. https://doi.org/10.25133/JPSSV28N2.011
2. EEC (2022) Czifrovaya povestka EAES [Digital agenda of the EAEU]. Evrazvijskaya ekonomicheskaya komissiya [Eurasian Economic Commission]. http://digital.eaeunion.org/extranet/ (in Russian). Accessed 18 Jan 2023
3. EPI (2022) 2022 EPI results. Environmental Performance Index. https://epi.yale.edu/epi-results/2022/component/epi. Accessed 18 Jan 2023
4. eTrade for all (2023) Country profiles. https://etradeforall.org/country-profiles/. Accessed 18 Jan 2023
5. F6S (2023a) 10 top startups and companies in Kyrgyzstan. https://www.f6s.com/companies/kyrgyzstan/lo. Accessed 18 Jan 2023
6. F6S (2023b) 5 top energy startups and companies in Yerevan. https://www.f6s.com/companies/energy/armenia/yerevan/co. Accessed 18 Jan 2023
7. F6S (2023c) 6 top energy & cleantech startups and companies in minsk. https://www.f6s.com/companies/energy-cleantech/belarus/minsk/co. Accessed 18 Jan 2023
8. F6S (2023d) 6 top energy startups and companies in Kazakhstan. https://www.f6s.com/companies/energy/kazakhstan/co. Accessed 18 Jan 2023
9. GermanWatch (2020) Global Climate Risk Index 2021. https://www.germanwatch.org/sites/default/files/Global%20Climate%20Risk%20Index%202021_2.pdf. Accessed 18 Jan 2023
10. Lola IS, Bakeev MB (2022) «Zelenye» Tekhnologii v promyshlennosti: trendy GREENTECH-napravlenij v 2021–2022 gg. [«Green» technologies in industry: trends in GREENTECH areas in 2021–2022]. NIU VShE, Moscow, 19 P. https://www.hse.ru/data/2022/06/20/1860252879/Digital_industry_2_2022.pdf (in Russian). Accessed 18 Jan 2023
11. Makarov IN, Drobot E, Levchegov ON (2020) Zelenaya ekonomika, czifrovye tekhnologii i nanoinstrumentarij: osnovnye bazisy transformaczii proizvodstvennykh sistem v Evrazijskom ekonomicheskom soyuze [Green economy, digital technologies and nanotools: the main bases for the transformation of production systems in the Eurasian Economic Union]. J Int Econ Aff 10:719–742. https://doi.org/10.18334/eo.10.3.110822 (in Russian)
12. Migaleva TE, Podbiralina GV, Razumnova LL (2022) Sblizhenie poziczij gosudarstv – chlenov EAES po voprosam klimaticheskoj povestki i uglerodnogo regulirovaniya [Rapprochement of the positions of the EAEU member states on the issues of the climate agenda and carbon regulation]. Evrazijskaya integracziya: e'konomika, pravo, politika [Eurasian Integration: Economics, Law, Politics] 16(2):41–50 (in Russian)
13. RosStat (2022) Evrazijskij ekonomicheskij soyuz [Eurasian Economic Union]. Federalnaya sluzhba gosudarstvennoj statistiki [Federal State Statistics Service]. https://rosstat.gov.ru/folder/12880 (in Russian). Accessed 18 Jan 2023
14. Rusbase (2023) Karta rossijskogo rynka GreenTech [Map of Russian GreenTech market]. https://rb.ru/greentech/ (in Russian). Accessed 18 Jan 2023
15. Solovianov A, Saparov M, Solovianova A (2021) Vklad stran Evrazijskogo ekonomicheskogo soyuza v deyatelnost po predotvrashheniyu izmenenij klimata [Contribution of the countries of the Eurasian Economic Union to the prevention of climate change]. Okhrana okruzhayushhej sredy i zapovednoe delo [Environmental Protection and Reservation] 1(2):112–123. https://vniiecology.ru/wp-content/uploads/2021/08/ОХРАНА-ОКРУЖАЮЩЕЙ-СРЕДЫ_2_2021.pdf#page=112 (in Russian). Accessed 18 Jan 2023
16. WorldBank (2022) World Development Indicators (WDI). https://databank.worldbank.org/source/world-development-indicators. Accessed 18 Jan 2023
17. Yanik AA (2021) Czifrovoe prostranstvo EAES: aktualnoe sostoyanie i perspektivy [Digital space of the EAEU: current state and prospects]. Mezhdunarodnoe pravo i mezhdunarodny'e organizaczii [International Law and International Organizations] 3:42–61. https://doi.org/10.7256/2454-0633.2021.3.36454 (in Russian)

Case Experience of Adaptation of Climate-Responsible Entrepreneurship in the Russian Markets of the Digital Economy to the Conditions of the COVID-19 Pandemic and Crisis

Ainura K. Isagalieva, Marina L. Nechaeva, Fedor P. Potapov, and Tatiana A. Dugina

1 Introduction

The climate-responsible aspect of the activities of Russian companies in modern conditions demonstrates these subjects' support for the national course in this sphere and the focus on stable work in the long term, which can be implemented due to supporting and protecting the climate. The period of the COVID-19 pandemic and the following crisis led to the reduction of financial stability and companies' ability to realise socially important projects, including ones connected with the issues of emissions, transition to renewable energy and preservation of biodiversity.

The digital economy and its tools play a decisive role in the management of entrepreneurial strategies. Digitalisation allows achieving positive results in business that is focused on zero climate footprint or improvement of environmental indicators. Effective management of activities during a crisis requires from climate-responsible entrepreneurship, which uses innovative digital tools, and the introduction of the optimal methods of strategic management. The case (or situational) method, which becomes more popular, deserves attention and further assessment. A

A. K. Isagalieva (✉)
Kyrgyz State Technical University Named After I.Razzakov, Bishkek, Kyrgyzstan

M. L. Nechaeva
Vyatka State University, Kirov, Russia

F. P. Potapov
Sebryakovskiy Branch of Volgograd State Technical University, Mikhailovka, Russia

T. A. Dugina
State-Funded Budget Educational Institution of Higher Education, Volzhsky Institute of Economics, Pedagogy and Law, Volzhsky, Russia
e-mail: deisi79@mail.ru

E. G. Popkova (ed.), *Smart Green Innovations in Industry 4.0*, Springer Climate,
https://doi.org/10.1007/978-3-031-45830-9_10

special role within the case method belongs to the strategy of adaptation, which facilitates the integration of the entrepreneurial subjects' potential into the factual conditions of business environment formation.

This paper is aimed at determining the list of effective directions of the case experience of the strategy of adaptation, which allow for the successful implementation of climate-responsible projects by the Russian entrepreneurial sector under the conditions of the COVID-19 pandemic and crisis. Further in this research, a list of Russian companies that implement climate-responsible technologies in development strategies management is compiled, and effective directions for the use of the situational strategy of adaptation with a focus on climate responsibility and automation are described.

2 Materials and Method

To identify the main directions for the use of the strategy of adaptation, a range of scientific works were analysed. We should mention the work by [1, 6, 9]. In [9], the authors present the experience of the use of the strategy of adaptation by Spanish companies of various categories under the conditions of the COVID-19 pandemic and crisis. DiBella [1] dwells on the problems and possible solutions in the sphere of adaptation of business models to the consideration of climate aspects in the interaction with the environment. Gong and Janssen [6] formulate the directions for the use of big data analytics technology in the assessment of value-added creation from the introduction of various strategies, including the case strategy of adaptation.

The above scientific materials contain fragmentary provisions and conclusions, which require systematisation and assessment of implementation at the practical level, which predetermines the importance of this research.

The methods used in this research include the method of systematisation – to determine the characteristics of the case strategy of adaptation and identify it based on the practice of the considered Russian companies. We also analysed the provisions of scientific works that describe the characteristics of this strategy and determine its specific features, which allows distinguishing them in the companies' activities.

We also used statistical analysis, which enabled us to evaluate the financial indicators of the companies, which can be found in their financial statements, which are available in open sources.

The trends method was used to identify the tendencies of implementing digital technologies that facilitate the realisation of climate-responsible strategies of the considered Russian companies at the modern state. This method also allowed discovering dynamic changes, which took place in the activities of the analysed companies over 2019–2021.

3 Results

According to [1, 6, 9], the strategy of adaptation belongs to the category of case business strategies. It is aimed at the modification of the business model of functioning according to the challenges faced by the subject of entrepreneurship under the conditions of a crisis. Implementation of such a strategy has its specifics in the functioning of climate-responsible Russian business.

Let us consider a list of Russian companies which contributed to the environmental component of sustainable development by implementing certain digital technologies and adapting to the conditions of the COVID-19 pandemic and crisis. The key role in this sphere belongs to companies dealing with the production of renewable energy and technologies and equipment for production for industrial and household purposes. This is an important direction of climate-responsible entrepreneurship, which contributes to the reduction of CO_2 emissions.

The specifics of the work of PJSC Novatek – one of Russia's leaders in green energy production – are presented in Table 1.

Analysing the results of the activity management of PJSC Novatek (Table 1), it is possible to say that the company achieved significant results in the sphere of climate-responsible management. PJSC Novatek is Russia's second-largest natural gas production and sales company, but the potential depletion of natural gas reserves in the future and the need for energy transition led to the company's starting to produce renewable ("green") hydrogen. Demand for this product was high, compared to fossil energy resources, in the domestic and external markets. In 2020, despite the pandemic, the company worked on the energy transition, which continued in 2021. PJSC Novatek started producing and promoting an eco-friendly form of renewable energy – "green" hydrogen. This facilitated the economic growth (revenues and profit) and achievement of environmental indicators. Wind turbines and solar panels for domestic consumption were installed in 2020; this allowed reducing fossil energy resources consumption by 30%. This is a sign of the company's impact on climate protection on the territory where production and consumers are located.

Table 2 shows the indicators of the activities of TOK Arsenal – one of Russia's largest suppliers of equipment and technologies for renewable energy production.

Focusing on the situational necessity of adaptation to the needs of the market environment and the COVID-19 pandemic and crisis, TOK Arsenal changes its assortment of products in 2020 (Table 2). New types of equipment and technologies, which are used in the sphere of renewable (solar) energy, were introduced. In 2021, the assortment was expanded thanks to the stable growth of demand for these products.

The analysis has shown that energy transition in the context of the change in the product assortment (accompanying services) ensured economic effectiveness for TOK Arsenal (Table 2). The considered measures on energy transition in the assortment were implemented during the period of the Covid-19 pandemic (2020) and the following period of overcoming its consequences (2021). The measures taken involve the minimisation of the market and economic risks, which was achieved

Table 1 Characteristics of energy transition of PJSC Novatek

	Indicators of activities	Value of the indicators			Change, +/–	
		2019	2020	2021	2019–2020	2020–2021
1	Type of activities: production and sales (*including export*), including:	Natural gas, oil	Natural gas, oil	Natural gas, oil, renewable hydrogen	–	Renewable hydrogen
2	Product sales, RUB million, including:	862,803	711,812	1,156,724	−150,991	444,912
2.1	Sales structure:					
2.1.1	Natural gas, oil, %	100	100	61.6	0	−38.4
2.1.2	Renewable hydrogen, %	0	0	38.4	0	38.4
3	Profit from sales, RUB million, including:	222,340	159,750	281,565	−62,590	121,815
3.1	Natural gas, oil, RUB million	222,340	159,750	173,445	−62,590	13,695
3.2	Renewable hydrogen, RUB million	0	0	108,120	0	108,120
4	Use of renewable energy sources in the activities, in the total volume of energy used, %	0	30% of the total consumption of energy, due to the installation of 148 sets of wind generators and solar panels at gas condensate field areas and main pipelines telemechanics spots	30% of the total consumption of energy	30	0

5	Digital technologies that contribute to the energy transition	–	Technologies that capture CO_2 during "green" hydrogen production. Win generators, solar panels	Indicators that are similar to 2020 values	Use of innovative climate-responsible digital technologies	Continuation of the 2020 trends
6	Strategy of greening	Not specified, with a general focus on the activities that do not create threats to the environment	Energy transition	Energy transition	Changes in favour of renewable energy use	Trends of continuation of energy transition

Source: Created by the authors using the materials of [7, 8, 10, 11]

Table 2 Indicators of the energy transition by TOK Arsenal

	Indicator	Value of indicators			Change, +/–	
		2019	2020	2021	2019–2020	2020–2021
1	Type of activities: sales, maintenance and repairs of equipment and technologies for energy (industrial sector, service sector) products:	1. Equipment and technologies for use in the sphere of non-renewable energy (controllers, inverters, accessories, power units, batteries, etc.).	1. Equipment and technologies for use in the sphere of non-renewable energy (products presented in 2019) 2. Solar power stations and batteries.	1. Equipment and technologies for use in the sphere of non-renewable and renewable energy (controllers, inverters, accessories, power units, batteries, etc.). 2. Solar batteries, electric power stations, accessories, chargers, controllers, etc.	Implementation of solar energy products	Continuation of trends on the implementation of a range of products that allow producing and transmitting electric energy (from solar energy).Reduction of sales volume in the context of installations and technologies used in traditional energy
2	Sales volumes, RUB million, including:	51	76	178	21	102
2.1	Structure of sales:					
2.1.1	Products for producing and transmitting non-renewable energy, accompanying technologies and maintenance	100	85	76	–15	–9
2.1.2	Products for producing and transmitting solar energy, accompanying technologies and maintenance	0	15	24	15	9

3	Revenue, RUB million, including from:	0.25	0.875	1.5	0.655	0.325
3.1	Sales, installation, and maintenance of products used in non-renewable energy, RUB million	0.25	0.744	1.14	0.494	0.396
3.2	Sales, installation and maintenance in the sphere of green energy, RUB million	0	0.131	0.36	0.131	0.229
4	Use of digital technologies	Big data analytics in the assessment of market needs	Big data analytics in the assessment of market needs, imitation modelling of demand	Big data analytics in the assessment of market needs, imitation modelling of consumer behaviour and change in demand	Implementation of the element of big data analytics, imitation modelling	Expansion of the element of imitation modelling through forecasting of consumer behaviour
5	Strategies in the sphere of climate protection	General declaration of achievement of the UN SDGs in the sphere of climate protection	Declaration of climate-responsible goals in the sphere of clean energy	Strategy that is similar to the 2020 strategy	Energy transition	Continuation of energy transition trends

Source: Created by the authors using the materials of [5, 10–12]

with the help of a comprehensive approach to forecasting of the measures of implementing climate-responsible assortment. This involves big data analytics for assessing the market needs and imitation modelling of demand.

Table 3 presents the aspects of the activity of a climate-responsible company Fortum PJSC, which deals with the production and realisation of solar, geothermal and wind energy in Russia (Table 3).

As is shown in Table 3, Fortum PJSC focused on the implementation of renewable energy in the industrial sector and communal services sector of different regions of Russia and the gradual reduction of production and sale of electric energy from non-renewable fossil energy sources.

The company also works actively in the direction of reducing the climate footprint, through the use of AI technologies in the systems of control over CO_2 capture during the production of electric energy from non-renewable sources and the use of robotized technologies in various processes.

The company was able to increase revenues and sales profit during the crisis period of 2020–2021 when many similar large companies suffered losses and were not able to recover their pre-crisis positions. Such results were achieved due to consideration of possible risks and reliable partnership. We should also note clear reactions to market tendencies and tendencies of the energy market, in particular, the price factor (cheaper electric energy from renewable sources) and energy factor (problems connected with the issues of budget financing of the needs of the communal sector).

We also support the provisions of [13], where the authors prove the link between the innovativeness of the product range with high and low profitability in companies that function in modern market conditions. The analysis has shown that the studied Russian companies, which are among the leaders in the implementation of climate-responsible products (services), were able to reach their financial goals due to the orientation toward the changing market needs and the use of the new technologies of the digital economy.

4 Discussion

A study of the participation of climate-responsible Russian entrepreneurship in the achievement of national environmental goals showed that digitalisation is an important foundation in this context. Measures of climate protection, taken by business that is aimed at the introduction of digital technologies and environmental responsibility in consumption and production, are retained even in crisis periods (including the COVID-19 pandemic and the following period of economic recovery).

An example of the activities in the sphere of energy transition of PJSC Novatek allows distinguishing such direction for the use of the case experience of the

Table 3 Aspects of the course at energy transition of Fortum PJSC

	Indicators of activities	Value of indicators			Change, +/–	
		2019	2020	2021	2019–2020	2020–2021
1	Type of activities: production and sale of electric energy, including:	From non-renewable fossil energy sources, the production and use of which leads to the growth of CO_2 emissions. From renewable energy sources (using purchased solar power plants). Design of construction of solar, wind and geothermal power stations	From non-renewable fossil energy sources and renewable energy sources.	From non-renewable fossil energy sources and renewable energy sources (solar, wind and geothermal energy). New capacities are launched in regions of Russia that have the required climate conditions for their effective use. Supply of green energy to the leading industrial companies (e.g., Uralkali) and the housing and utilities sector in Kalmykia and Bashkiria	Expansion of electric energy production from renewable energy	Growth of the focus on green energy production
2	Sales of electric energy, RUB million, including:	77,000	79,000	82,500	2000	3500
2.1	Sales structure:					
2.1.1	Produced from non-renewable energy sources, %	60	58	55	−2	−3
2.1.2	Produced from renewable energy sources, %	40	42	45	2	3
3	Profit from sale of electric energy, RUB million, including from sale of energy:	16,250	17,000	18,600	750	1600
3.1	Produced from non-renewable energy sources, RUB million	9750	9860	10,230	110	370

(continued)

Table 3 (continued)

	Indicators of activities	Value of indicators			Change, +/−	
		2019	2020	2021	2019–2020	2020–2021
3.2	Produced from renewable energy sources, RUB million	6500	7140	8370	640	1230
4	Digitalisation of activities	Digital technologies of artificial intelligence in the systems of control over CO2 capture during the production of electric energy from non-renewable sources. Robotisation in the control over processes	Digital technologies that are similar to the ones used in 2019	New digital technologies, similar to the ones used in 2019. New functional solutions in technological support	No changes	Improvement of digital support for production processes
5	Strategy of greening	Strategy of energy transition and further implementation of the renewable energy objects with gradual reduction of the non-renewable energy segment	Strategy that was announced in 2019	Strategy that was announced in 2019	Course toward energy transition	Continuation of energy transition

Source: Created by the authors using the materials of [2–4, 10, 11, 14]

strategy of adaptation as a reaction to domestic economic and external economic and environmental needs. The market trend toward CO_2 reduction led to the necessity to implement measures of energy transition in the activities of producers and suppliers of energy resources that deal with green energy. The focus on the economic component led to the use of renewable energy sources for domestic consumers.

As for the case experience of adaptation of a climate-responsible company TOK Arsenal to the conditions of the COVID-19 pandemic and crisis, we should note the focus on vulnerable market segments of energy consumers, who can use their potential to deal with energy and economic problems. Consumers' ability to satisfy their own needs due to their financial potential is very important. An important aspect in the considered strategy is TOK Arsenal's using an optimal set of digital tools, which allow for precise forecasting of consumer behaviour and financial indicators. This allowed the company to avoid wrong managerial, financial, economic, organisational and personnel decisions over the crisis period.

PJSC Fortum's activity is characterised by the case adaptation strategy with a focus on interaction with reliable partners, which facilitates growth during economic crises. This strategy also includes links with climate growth in the sphere of green energy and the implementation of consumer goals regarding stable energy supply and price advantages.

5 Conclusion

The described aspects of the case experience of adaptation of Russian climate-responsible entrepreneurship to the conditions of the COVID-19 pandemic and crisis demonstrated effective measures in the case of the absence of companies' financial dependence, which would threaten their stable functioning and growth. If there are sufficient financial capabilities and access to financing (cheap credits), companies will be able to implement relevant climate projects, including the ones connected with the transition to renewable energy.

Entrepreneurship's influence on Russia's achieving the Sustainable Development Goals could be more significant with the support from the government, through the stimulation of programmes for the given projects' support. It would be expedient to stimulate the energy transition of large companies and SMEs in Russia. This will allow supporting the prospects for infrastructural development and growth of employment. The creation of stimuli for climate-responsible entrepreneurship, aimed at the adaptation to the crisis conditions, will allow ensuring a foundation for the life activities of future generations in the conditions of a favourable environment.

References

1. DiBella J (2019) The spatial representation of business models for climate adaptation: an approach for business model innovation and adaptation strategies in the private sector. Bus Strategy Dev 3(2):245–260
2. Eastcham (2021) Fortum invests in the construction of Russia's largest solar power station. https://www.eastcham.ru/novosti/fortum-investiruet-v-stroitelstvo-krupnejshej-solnechnoj-jelektrostancii-v-rossii/. Accessed 14 Jan 2023
3. Eprussia (2021) Fortum started supplying electric energy from solar power station to Uralkali. https://www.m.eprussia.ru/news/base/2021/7066934.htm. Accessed 14 Jan 2023
4. Excheck (2023) Financial statements of FORTUM PJSC according to the data from the Federal Tax Service and Rosstat for 2011–2021. https://excheck.pro/company/7203162698-fortum. Accessed 14 Jan 2023
5. Excheck (2023) Financial statements of TOK ARSENAL LLC according to the data from the Federal Tax Service and Rosstat for 2013–2021. https://excheck.pro/company/6670418524-ibp-ural. Accessed 14 Jan 2023
6. Gong Y, Janssen M (2021) Roles and capabilities of enterprise architecture in big data analytics technology adoption and implementation. J Theor Appl Electron Commer Res 16(1):37–51
7. Novatek (2022) Financial results. https://www.novatek.ru/en/investors/results/. Accessed 14 Jan 2023
8. Novatek (2022) Measures to protect climate. https://www.novatek.ru/ru/development/environmental/technologies/. Accessed 14 Jan 2023
9. Peñarroya-Farell M, Miralles F (2022) Business model adaptation to the COVID-19 crisis: strategic response of the Spanish cultural and creative firms. J Open Innov Technol Mark Complex 8:39. https://doi.org/10.3390/joitmc8010039
10. Tadviser (2022) Alternative energy in Russia. https://www.tadviser.ru/index.php/Статья:Альтернативная_энергетика_в_России. Accessed 14 Jan 2023
11. Tadviser (2022) Solar energy (Russian market). https://www.tadviser.ru/index.php/Статья:Солнечная_энергетика_(рынок_России). Accessed 14 Jan 2023
12. Tokarsenal (2023) Alternative energy. https://tokarsenal.ru/alternativnaya-energetika/. Accessed 14 Jan 2023
13. Turginbayeva A, Ustemorov A, Akhmetova G, Kose Z, Imashev A, Gimranova G (2018) Financing aspects of an effective strategy for innovative enterprise development. J Adv Res Law Econ 9(2):714–720. https://journals.aserspublishing.eu/jarle/article/view/2499. Accessed 14 Jan 2023
14. Vesti.ru (2021) The Russian Direct Investment Fund and Fortum to build Russia's largest solar power station. https://www.vesti.ru/finance/article/2530200. Accessed 14 Jan 2023

Case Experience of Conducting Climate-Responsible Entrepreneurship in the Digital Economy Markets in Russia

Aidarbek T. Gyiazov, Nadezda V. Gamulinskaya, Sergei A. Markeev, and Zoya V. Popkova

1 Introduction

The course towards maintaining and growing market positions in the new conditions of tough competition, the energy crisis, under the influence of a focus on environmental requirements, and social responsibility implies a review by enterprises of their own functioning strategies. The choice of new strategies and the adaptation of existing ones depend on the actual potential of enterprises, a common vision of further development and a focus on participation in certain regional, national, and global programs. Such participation in programs related to various areas of life, in turn, contributes to ensuring territorial competitiveness [14]. Climate-responsible activities of enterprises are connected both with the general course of the world community towards achieving the UN environmental SDGs and with the main goals in this area, formulated in the provisions of related strategies. Namely, these goals are presented within the framework of the concept of "green" – maintaining social justice, economic goals, and environmental benchmarks (reducing the level of CO_2 in the atmosphere) [7, 15], "blue" – overcoming the global problem of the water crisis, achieving the effective functioning of entities in the maritime industry, applying innovations in the field of water management [8], the

A. T. Gyiazov
Batken State University, Batken, Kyrgyzstan

N. V. Gamulinskaya (✉)
Vyatka State University, Kirov, Russia
e-mail: nv_gamulinsky@vyatsu.ru

S. A. Markeev
MGIMO University, Moscow, Russia

Z. V. Popkova
Volgograd State Pedagogical University, Volgograd, Russia

E. G. Popkova (ed.), *Smart Green Innovations in Industry 4.0*, Springer Climate,
https://doi.org/10.1007/978-3-031-45830-9_11

"circular" economy – a focus on production waste, life activity as an additional resource, the management of which requires resource and environmental needs of the territories [12]. The orientation of enterprises towards the adoption of standards, norms, and obligations of climate responsibility can be formed both under the influence of the desire to preserve the environment, which is initiated by the owners (management) and under the influence of the need to adopt new situational trends. These trends can be taken as a framework for the implementation of the main economic activity within the existing range of products (services) or with a focus on a new one. Accordingly, in this case, environmentally oriented entrepreneurship refers to case strategies that allow effectively responding to emerging situational challenges. If companies also actually adhere to positions regarding the need to protect nature, then focusing on this aspect can contribute to the current situational challenges that are acquiring persistent trends associated with climate change and the depletion of natural resources. Case experience in the implementation of entrepreneurial environmentally oriented initiatives related to the introduction of digital technologies has a certain distribution in Russia, which requires research and evaluation, and this determines the relevance of this article.

The purpose of the article is to characterise the features of the case experience in the implementation of climate entrepreneurial initiatives related to the focus on the digital economy in Russia. To achieve this goal, a list of tasks has been identified, including an assessment of the dynamics of climatic parameters demonstrated by Russia, the level of which also depends on the enterprises of the main infrastructures; identification of elements of case experience (strategies) of climate-responsible Russian enterprises that affect climate maintenance within the main territories of the state.

2 Materials and Method

The disclosure of the research problem of this article is implemented based on the provisions of the materials of theoretical, empirical data, including the works of [7, 8, 11, 12].

The considered articles [7, 8, 12] are devoted to the conceptualisation of concepts and categories of strategies and the course of states, the business sector for climate goals belong to research.

The study of [11] formulated possible directions and reserves for the stable sustainable functioning of entrepreneurship, including a focus on maintaining climate parameters. The article considers the practice and experience of businesses focused on the UN SDGs in a pandemic, and explores some of the possibilities for implementing a case adaptation strategy, the characteristics and application of which are studied in our study.

The research goals were realised with a focus on the following methods. The method of collecting facts made it possible to establish the main characteristics and performance indicators, the formation of the studied Russian enterprises, and determine their key case strategies. The focus on the differentiation method made it possible to categorise the ways of becoming business entities depending on the type of strategies. Statistical analysis was used to select statistical data on the formation of enterprises, entrepreneurship, and the state in the context of a benchmark for environmental indicators.

3 Results

Let us analyse the change in the main climatic indicators of Russia, the state of which is also associated with the activities of the business sector.

Firstly, an important aspect characterising the climatic state is CO_2 emissions from fuel combustion (Fig. 1). This is an indicator of the result of the energy transition at the level of entrepreneurship, private consumers and the public sector to renewable energy resources, the use of which does not affect the production of CO_2.

Data analysis from Fig. 1 shows that the studied indicator fluctuated during 2016–2021. It was determined that Russia (1795 $MtCO_2$) in 2021 was in 4th place, after China (10,398 $MtCO_2$), the USA (4632 $MtCO_2$), and India (2251 $MtCO_2$) by the value of this indicator. The indicated positions of the countries in terms of CO_2 emissions were maintained during the study period. Research materials of [2] link the dynamics of changes in this indicator with the dynamics of economic activity. Namely, it is noted that the period of the pandemic (2020), associated with a significant decrease in economic activity, led to a decrease in emissions of this category and that the latter increased in 2021 due to economic growth.

To determine the relationship between the change in this category of emissions and entrepreneurship activities, it is necessary to identify the relationship between the dynamics of GDP and the change in this environmental indicator in Russia (Fig. 2).

According to the results in Fig. 2, the Russian economy as a whole demonstrates the same trends in the dependence of CO_2 emissions on changes in economic activity as other states.

Despite the consistently high level of CO_2 emissions at the level of 1570–1795 $MtCO_2$ per year, the state and entrepreneurship are taking certain measures to reduce

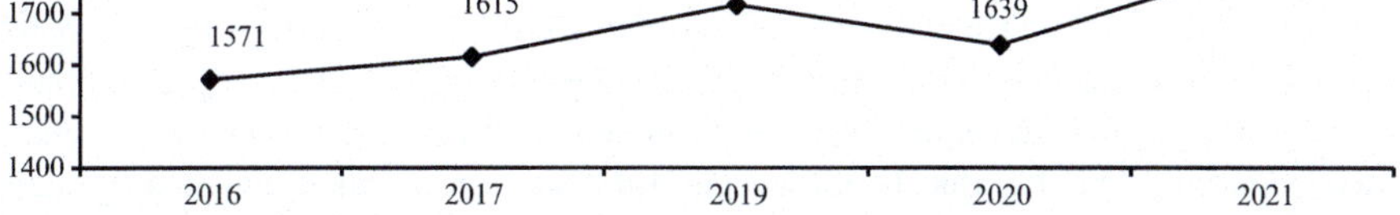

Fig. 1 Dynamics of CO_2 emissions from fuel combustion in Russia, $MtCO_2$. (*Source:* Compiled by the authors based on [2])

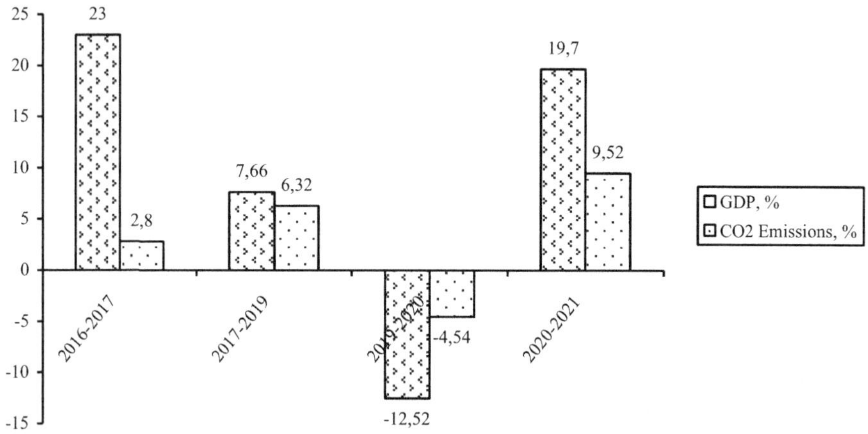

Fig. 2 Level of change in GDP and CO_2 emissions from fuel combustion in Russia, %. (*Source:* Compiled by the authors based on [2, 13])

them through the implementation of the energy transition, the introduction of environmentally friendly waste recycling, the installation of treatment facilities based on the use of digital monitoring tools, etc.

It has been determined that the non-ferrous and ferrous metallurgy of Russia are leaders in the field of CO_2 generation in the state (28%) [10]. Due to the need to solve problems in this area, the growth of CO_2 emissions for these industries has become a situational necessity, adaptation to which has become relevant for enterprises. Ensuring environmentally friendly production has become a prerequisite for stable operation, maintaining the health of personnel and the population living near the enterprises of these industries. In this case, the activities of climate-responsible entrepreneurship are associated with the need to participate in the environmental problems of the state.

An example would be the implementation of a case strategy in the context of managing climate problems at the enterprise CJSC EVRAZ NTMK (the largest producer of ferroalloys, steel, and cast iron) (Table 1).

It was found that this enterprise, demonstrating environmental initiatives, ensures the creation of a clean ecosystem, respectively, it independently solves the problems of environmental pollution in some areas. Namely, in the context of areas related to the reduction of CO_2 emissions associated with production (through the use of new capture technologies created using digitalisation elements); areas of energy efficiency that allow controlling energy costs and reducing the unnecessary negative impact on the climate (introduction of a robotisation tool); directions providing for recycling of waste. It was determined that during 2018, 2019, 2020 and 2021, there has been a gradual introduction of environmental directions that provide climate and economic benefits due to the successful focus on a situational management strategy for results with a commitment to greening and cost optimisation through energy efficiency. It can be stated that despite the costs associated with the

Table 1 Indicators of the implementation of the case strategy of CJSC EVRAZ NTMK

Indicator	Value/characteristic				
	2017	2018	2019	2020	2021
1. Climate-focused event	Designing the implementation of greening goals: reduction of CO_2 emissions from fuel combustion in the production of cast iron: with the help of new capture technologies, it is predicted in 2018–2023 to achieve a reduction in specific CO_2 emissions of 2 tones (corresponding to the equivalent of 1 ton of steel (cast iron))	Installation of CO_2 captures technologies at the blast furnace 7. Improved air quality by 2.5 times compared to the period before the implementation of environmental innovations. Implementation of energy efficiency in process management	Achieving CO_2 emission reductions of 1.97 tonnes (corresponding to the equivalent of 1 ton of steel (iron). Energy efficiency in process control)	Installation of energy-efficient equipment that allows providing own electricity production and its efficient use (lower price is achieved). Capturing CO_2 emissions within the Blast Furnace 7	Operation of energy-efficient equipment. Capture CO_2 emissions (at 2020 levels). Waste recycling (recycling of the metal waste in steel production)
2. Sales income, billion rubles	148	187.5	189	186	297.6
3. Profit from sales, billion rubles	66	108	112.5	64	139.1
4. Effect:					
Economic	Profit through successful management of predictable results	Profit through the successful management of predictable results and the implementation of cost optimisation through energy efficiency	Similar to the previous period	Similar to the previous period	Similar to the previous period

(continued)

Table 1 (continued)

Indicator	Value/characteristic				
	2017	2018	2019	2020	2021
Ecological	Not established, not declared	Reduced CO_2 emissions	Reduced CO_2 emissions	Reduction of CO_2 emissions	Reducing CO_2 emissions, achieving 0% industrial waste (zero environmental impact on industrial waste)
5. Management strategy	Case strategy for results-based management without an actual application of individualised environmental commitments	Case management strategy based on results with a commitment to greening and cost optimisation through energy efficiency	A strategy similar to that used in the previous period	A strategy similar to that used in the previous period	A strategy similar to that used in the previous period

Source: Compiled by the authors based on [3, 4, 10]

Table 2 Indicators of PJSC TMK's case strategy implementation

Names of indicators	Values/characteristics of indicators		
	2019	2020	2021
1. Climate action directions	Installation and maintenance of cleaning filters for capturing CO_2, purification of water resources of territories that border on production	Directions of the previous period	Directions of the previous period. Monitoring, and accounting of industrial waste using robotic technologies, waste processing, and use as raw materials
2. Revenue volume, billion rubles	230	200	321.7
3. Profit from sales, billion rubles	14.8	18	4.6
4. Effect:			
Economic	Profit from the use of a case strategy for managing planned results with a focus on adaptation aspects in the field of greening	Similar to the nature of the economic effect of the previous year	Similar to the nature of the economic effect of the previous year. The reduction in the economic effect is due to the growth of unpredictable costs associated with the impact of the consequences of the crisis after the pandemic
Ecological	Purification of air and water, installation of filters that capture CO_2 emissions that occur during the production process at the main facilities	Similar to the directions of the previous year	Reduction of CO_2 emissions due to the installation of filters: within the structural unit of the Pervouralsk Novotrubny Plant, an 85–99% purification level was achieved, in others – an average purification level; other environmental indicators (reduction of noise level from production, water, and air purification). Continuous reduction of industrial waste through recycling
5. Management strategy	Situational strategy for managing planned results with a focus on adaptive aspects of climate, economic and social growth	Same as the previous year	Same as the previous year

Source: Compiled by the authors based on [1, 6, 10]

installation and maintenance of technologies for capturing CO_2 emissions, the company was able to achieve economic efficiency. The above is related to the success of strategic forecasting of greening management options in the context of the chosen strategy. It is necessary to substantiate the validity of the identification of the strategy indicated by us. That is, since the content of climate reporting [3] indicates clear priorities for the planning periods that were set, and shows the level of their achievement, we concluded that the company is oriented towards this type of strategy.

Let us consider the features of the case experience of PJSC TMK in the implementation of climate initiatives as part of the overall development strategy (Table 2). The company under study is a world leader in the production of pipes for nuclear power and oil and gas enterprises, as well as service in this area.

A study of the analytical and statistical data we have systematised in Table 2 allows us to note that PJSC TMK while declaring a focus on greening, implements it in the context of a situational strategy for managing planned results with a focus on adaptive aspects of climate growth. The conclusion that the strategy used belongs to the indicated type is because the company carries out systematic development planning (for 5 years), highlighting the main boundary parameters of activity. Adaptation parameters are integrated into the planned parameters due to the influence of environmental conditions, which make it impossible to achieve some economic, environmental, social (obligations to maintain employment in the region (territory), income growth, the implementation of measures to create conditions for the realisation of human potential as part of increasing readiness to future changes) goals in time.

As we can see, during the study period, PJSC TMK's enterprise experienced changes in such financial result as profit (14.8 million rubles in 2019, 18 million rubles in 2020, and 4.6 million rubles in 2021). It was established that in 2014, the company's net loss amounted to 24 billion rubles and in 2018 – it was 2.5 billion rubles [6]. Analysing the company's reporting [6], we can state that its activities, like the functioning of all enterprises in this area, are characterised by high material intensity, and, accordingly, dependence on prices for purchased raw materials, materials used in the production of products. The target for the introduction of industrial waste recycling, which began to be implemented in 2021, is aimed at addressing the high dependence on the supply of raw materials, prices for them, which change in the context of a growing global economic crisis. In addition, it is worth noting the positive impact of environmentally friendly waste recycling on the environment.

Let us evaluate the case strategy in the context of the application of some of the areas of greening by CJSC RUSAL (the largest producer of aluminum in the CIS and Russia, including aluminum with a low carbon footprint) (Table 3).

In the course of assessing the strategic implementation of environmental measures (Table 3), it was shown that they allowed CJSC RUSAL to achieve a significant increase in economic efficiency (the level of sales profitability in 2017 was 5% (there were no crisis phenomena), in 2018, 2019, 2020 – 1.8%, 0.48%, 0% respectively (periods of crisis, rising prices for raw materials), in 2021 – 15.9% (period of overcoming the crisis after the pandemic)). Namely, the expansion of production at the expense of aluminum, which does not leave a carbon footprint, made it possible to diversify the range in the direction of reducing the production of traditional types of products that are highly costly due to rising prices for raw materials, materials and energy resources. We can note that the company's strategy formulated for 2021

Table 3 Indicators of the implementation of the case strategy of CJSC RUSAL

Indicator	Value/characteristic				
	2017	2018	2019	2020	2021
1. Greening measures	Declaration of support for environmental initiatives of the state	Declaration of support for environmental initiatives of the state	Declaration of support for environmental initiatives of the state	Declaration of support for environmental initiatives of the state	Introducing low-carbon aluminum into the range, based on innovative inert anode technology, incorporating the use of the latest digital equipment. The production of this type of aluminum leaves no carbon footprint. This type of product is used: for packaging forms, and utensils, as it is safe for the food industry
2. Income from product sales, billion rubles	250	275	312	425	596.2
3. Profit from the sale of products, billion rubles.	12.5	6	1.5	−4	94.9
4. Sales profitability, % (R.3/R.2*100%)	5	1.8	0.48	0	15.9
5. The nature of the effectiveness:					
Economic	Profit generated by management based on predictive results	Profit generated by management based on predictive results. Its decrease is due to the increase in prices for raw materials	Similar to the previous year	Losses due to unpredictable growth in commodity prices	Profit through effective management based on predictive results developed using artificial intelligence tools in forecasting options

(continued)

Table 3 (continued)

Indicator	Value/characteristic				
	2017	2018	2019	2020	2021
Ecological	Not identified	Similar to the previous year	Similar to the previous year	Similar to the previous year	Reducing the carbon footprint in the production of environmentally friendly products, their consumption, and use in subsequent production by other manufacturers
6. Management strategy	Situational strategy for managing planned results	Similar to the previous year	Similar to the previous year	Similar to the previous year	Situational strategy for managing planned results with a focus on expanding the production of environmentally friendly products

Source: Compiled by the authors based on [5, 9, 10]

was realistic, due to the successful implementation of its indicators, including in the field of greening, the effect was obtained according to the planned parameters.

4 Discussion

The presented characteristics and directions for the implementation of case strategies in the field of climate initiatives of Russian enterprises showed the ambiguity of the latter's approach to targets in this area.

One of the important features of the implementation of the situational concept is the primary focus of some enterprises on solving the environmental problems of the territories previously created by their production facilities, and the secondary one is associated with activities based on predictable results achieved through the introduction of innovative digital technologies in the field of energy efficiency and the corresponding cost optimisation. An example of such a practice was the environmentally oriented activity of EVRAZ NTMK JSC, the effectiveness of which is due to the specified case strategy and digital support for the main areas of management. That is, although within the framework of the strategy under study, enterprises adhere to the course of solving national (territorial) environmental problems of a situational nature, they also achieve an economic effect due to constructive management in accordance with the predicted results, which are also related to greening (energy efficiency).

Another feature of the implementation of case experience is related to the introduction by climate-oriented enterprises of a situational management strategy based on results in the context of integrating adaptive aspects of climate, social and economic growth (example: PJSC TMK). The key characteristic of this strategy is the situational adaptation of plans and actual management under the influence of external factors and a general focus on economic performance, the maintenance of which is necessary for further functioning. We can note that the implementation of such a strategy requires high financial, technological, human and organisational potential from enterprises. In the functioning of enterprises that are highly dependent on the resource base, a guideline is needed for the use of integrated digital analytical, predictive solutions that would allow for taking into account the minimum possible changes in the future (tools for analysing large databases, artificial intelligence, integrated into planning programs).

The third feature of the implementation of case experience is related to the development and implementation of a strategy for managing planned results with a focus on expanding the production of environmentally friendly products (example: CJSC RUSAL). We can note that the effectiveness of its implementation is due to an accurate assessment of the possibly high demand within the established price limits for new environmentally friendly products, which is undoubtedly associated with the use of digital technologies (artificial intelligence tools in the prediction of options).

5 Conclusion

Summarising the results of the study, we can state that case strategies related to business climate initiatives can be effective and can lead to a loss of economic stability. The last scene of the development of the event is since modern Russian business operates in the harsh conditions of the internal and external economic crisis, which acquired more complex specifics during the period of overcoming the consequences of the pandemic, in the conditions of the need to conduct business focused on large-scale import substitution. Given the above, the implementation of situational strategies, including those discussed in the study, requires enterprises to focus on a comprehensive analysis of data and the possibility of changing them over time.

As shown in the article, the climate initiatives used by enterprises can contribute to solving serious environmental problems of the state at the level of the territories where production is located. The creation of new economic and environmental solutions also acts as a starting point for the exit of these enterprises from a state that threatens financial stability. We can state that the introduction of a balanced approach to the management of case strategies is a prerequisite for achieving indicators of resistance to threats to functioning.

References

1. Ecostandard (2022) TMK environmental guidelines: education, technology, process. https://journal.ecostandard.ru/eco/keysy/ekologicheskie-orientiry-tmk-prosveshchenie-tekhnologiya-protsess/. Accessed 17 Jan 2023
2. Energystats (2022) CO_2 emissions from fuel combustion. https://energystats.enerdata.net/co2/emissions-co2-data-from-fuel-combustion.html. Accessed 17 Jan 2023
3. Evraz (2022) Our approach to climate change EVRAZ. https://www.evraz.com/upload/iblock/5b4/5b4632992a737d9dc83693b375ce3f16.pdf. Accessed 17 Jan 2023
4. Excheck (2023) Financial statements of CJSC EVRAZ NTMK according to the data of the Federal Tax Service and Rosstat for 2011–2021. https://excheck.pro/company/6623000680-evraz-ntmk. Accessed 17 Jan 2023
5. Excheck (2023) Financial statements of CJSC RUSAL according to the data of the Federal Tax Service and Rosstat for 2011–2021. https://excheck.pro/company/7709329253-rusal. Accessed 17 Jan 2023
6. Excheck (2023) Financial statements of PJSC TMK according to the data of the Federal Tax Service and Rosstat for 2011–2021. https://excheck.pro/company/7710373095-tmk. Accessed 17 Jan 2023
7. Hussain Z, Mehmood B, Khan MK, Tsimisaraka RSM (2022) Green growth, green technology, and environmental health: evidence from high-GDP countries. Front Public Health 9. https://www.frontiersin.org/articles/10.3389/fpubh.2021.816697/full. Accessed 17 Jan 2023
8. Keen MR, Schwarz A-M, Wini-Simeon L (2018) Towards defining the Blue Economy: practical lessons from pacific ocean governance. Mar Policy 88:333–341. https://doi.org/10.1016/j.marpol.2017.03.002
9. Lenta.ru (2021) En+ Group produced aluminum with the lowest carbon footprint. https://lenta.ru/news/2021/04/13/enpl/. Accessed 17 Jan 2023

10. Malceva A (2021) How industries responsible for greenhouse gas emissions are changing. Vedomosti. https://www.vedomosti.ru/partner/articles/2021/06/02/872559-otrasli-parnikovih-gazov. Accessed 17 Jan 2023
11. Pastran A, Colli E, Poclaba C (2021) Sustainable entrepreneurship: a new way of doing business. J Int Council Small Bus 2(2):147–158. https://doi.org/10.1080/26437015.2021.1882915
12. Scarpellini S, Portillo-Tarragona P, Aranda-Uson A, Llena-Macarulla F (2019) Definition and measurement of the circular economy's regional impact. J Environ Plan Manag 62(13):2211–2237
13. Statista (2022) Russia: gross domestic product (GDP) in current prices from 1997 to 2027. https://www.statista.com/statistics/263772/gross-domestic-product-gdp-in-russia/. Accessed 17 Jan 2023
14. Turginbayeva AN, Abildaev ST (2013) Quantitative assessment of regional economy competitiveness (Republic of Kazakhstan case study). Actual Probl Econ 146(8):477–485
15. UNRIC (2022) Green economy: a path towards sustainable development and poverty eradication. https://unric.org/en/green-economy-a-path-towards-sustainable-development-and-poverty-eradication/. Accessed 17 Jan 2023

Customs Regulation in Support of Development of Climate-Responsible Entrepreneurship in Digital Economy Markets

Meerimai Z. Karbekova, Anna A. Grabar, Olga N. Soboleva, Alexander V. Sukhinin, and Asya V. Kotandzhyan

1 Introduction

The customs sphere is a direction used by national, supra-national and inter-government bodies to regulate the movement of resources and commodities across national borders and borders of supra-national unions. This regulation is connected with the protection and priority of support for the national and supra-national interests in the economic, social, technological and climate spheres. Countries that focus on the achievement of climate-related Sustainable Development Goals introduce limitations, norms and tax and customs rates to prevent the growth of climate problems. This concerns also the legislative consolidation of the introduction of customs privileges, which stimulate climate-responsible entrepreneurship that is aimed at the use of the tools of the digital economy allowing for high results in the sphere of ecologisation and economic effectiveness. Depending on the national and international responsibilities and the mission in the climate sphere, rules and standards of commodity movement are adopted at the legislative level and then are regulated at the level of customs bodies. Each country has their specifics in the sphere of customs regulation, including in support of the development of climate-responsible entrepreneurship, which implements green innovations with the use of digital technologies. Gradual negative tendencies of deterioration in the climate at the global level make national governments of many countries reconsider their positions on the

M. Z. Karbekova (✉)
Jalal-Abad College of the Jalal-Abad State University named after B. Osmonov, Jalal-Abad, Kyrgyzstan

A. A. Grabar · O. N. Soboleva · A. V. Kotandzhyan
Vyatka State University, Kirov, Russia
e-mail: grabar@vyatsu.ru; soboleva@vyatsu.ru; usr21823@vyatsu.ru

A. V. Sukhinin
Sebryakovskiy branch of Volgograd State Technical University, Mikhailovka, Russia

E. G. Popkova (ed.), *Smart Green Innovations in Industry 4.0*, Springer Climate,
https://doi.org/10.1007/978-3-031-45830-9_12

introduction of additional privileges and restrictions, including in the customs sphere, which might facilitate the achievement of the environmental SDGs.

In this paper, we formulate the directions for customs regulation that is focused on support of the development of climate-responsible entrepreneurship in digital economy markets. Further in this research, we determine the main aspects of customs regulation in the support of favourable conditions for climate-responsible entrepreneurship activities and find out whether the given regulatory measures influence the improvement of the climate.

2 Materials and Methodology

The problems of the formation of customs policy of countries in the context of the focus on the achievement of environmental goals and support for the development of climate-responsible entrepreneurship in the markets of the digital economy and certain aspects of the emergence of environmental problems, which are solved with the help of the tools of customs policy, were studied in [5–7, 9].

Ruiz [9] presents an analysis of the dynamics of e-waste at the global level and by countries of the world. The author assesses the consequences of the growth of the electronic industry, which is manifested in the stable increase in the volume of e-waste. This raises the risks for the environment and human health at present and in the long term. Lin et al. [5] is devoted to the study of climate, economic and social consequences of China's introducing customs bans on imports of e-waste and of Hong Kong's loyalty to their imports. The authors of [6] dwell on the specifics of managing the imports of e-waste in countries of Africa and describe the legislative and organisational aspects. The provisions of [7] elaborate on the legal basis of the regulation of import and e-waste management in Nigeria.

We can state that despite many works on the given topic, there is still a need for using a complex approach to the research of customs regulation to support the development of climate-responsible entrepreneurship.

In this research, we use a range of methods which allows for the most systemic assessment of all directions and for the achievement of the selected tasks. The index method is used to evaluate the indicators of climate growth of countries. The comparative method is used to analyse the advantages of the government's participation in the customs regulation of the activities of climate-responsible entrepreneurship that aims at digitalisation. The trends method allows determining the main tendencies of national governments' participation in the sphere of customs rules regulation in the context of climate-responsible companies.

3 Results

We shall analyse the selected effective practices of countries in the sphere of customs policy, which is concerned with the stimulation of climate-responsible companies, which activity is connected with the use of digital tools.

An important problem in the sphere of implementation of the UN SDGs is the growth of electronic waste (waste of electronic equipment) (E-Waste). The emphasis on environmental organisations and the UN on the problem of the growth of e-waste is connected with a range of reasons: e-waste contains toxic materials (cadmium, mercury, lead), which are a threat to the environment and human health; and precious materials, which are rarely recycled due to the low level of recycling in this sphere. The waste comes from developed and quickly developing countries to poor regions and countries (e.g. Africa) as used electronic equipment, which very soon becomes e-waste [9]. Table 1 presents the indicators of production and e-waste recycling for the top ten countries in this sphere.

Based on the results of the presented data (Table 1), it is possible to state that China, the USA, India, Japan and Brazil are world leaders in e-waste production. A particular role in this direction belongs to China. With the world's highest level of e-waste production, China ensures only 16% of its recycling (direct recycling by manufacturing companies or recycling based on outsourcing). However, in the 2000s, China was among the countries with the highest imports of e-waste [5]. Individual elements of e-waste were used in the activities of micro-companies and small companies that dealt with the production and repairs of electronics.

Critical phenomena in the sphere of climate management and focus on an increase in the quality of life led to the Chinese government's starting developing laws on the ban of imports of these materials in the 2010s [5]. The customs regulation went into effect in 2019. This allowed ensuring significant improvement of climate in cities that had had bad climate indicators before the introduction of the law. As an example, we can consider changes in the city of Guiyu, which took the path of integration from 2005 to 2017. In 2005, there were around 5000 family

Table 1 Level of production and e-waste recycling in the top ten countries in this sphere

	Country	Recycling e-waste, %	E-waste produced (Kt)
1	Brazil	0	2143
2	China	16	10,126
3	France	56	1362
4	Germany	52	1607
5	India	1	3230
6	Indonesia	No data	1618
7	Japan	22	2569
8	Russian Federation	6	1661
9	UK	57	1598
10	USA	15	6918

Source: Created by the authors based on [9]

companies (workshops) dealing with e-waste recycling in a handicraft way. This led to an increase in the level of toxic fumes (which appear during the recycling of these materials with the use of non-renewable energy sources), burning metal, bad indicators of the population's health and general pollution of the environment.

Under the influence of the public, mass media and foreign environmental organisations and governments, China, together with the implementation of innovations in the customs law on the ban of e-waste imports, started the creation of the National Circular Economy Pilot Industry Park [4]. All micro-companies that had been working in this sphere before were able to conduct their activities on e-waste recycling within this project, with conditions for green recycling (supply of electric energy from renewable sources; subsidies for rent, which allowed purchasing special technologies for CO_2 capture and purification filters for the capture of other polluting substances [8]; paying for educational services for employees which were provided by new types of educational establishments (adapted to dynamic needs of the economy [11]). The main sources of city pollution (micro-companies dealing with e-waste recycling) obtained the status of larger companies or left the market [8].

The Chinese government reached results in the improvement of the law on the strengthening of measures of customs regulation concerning the ban on e-waste imports and adoption of programme documents on the support for green recycling of these materials. This included stimulation of climate-responsible entrepreneurship (integration of competitive micro-companies in medium and large companies in the sphere of green recycling of e-waste of internal origin). However, the conditions of the precariousness of employment were violated [3]. It is peculiar to the activities of uncompetitive micro-companies at the individual level (work of all family members without fixed labour conditions and wages). Though there are studies substantiating the absence of a connection between the growth of digitalisation and an increase in unemployment [2], here we see that the latter grew because of the factor of business's unpreparedness for legislative and environmental innovations.

Countries of Africa have been important the largest volumes of e-waste quite for a long time. Among countries in this sphere, we can mention Nigeria, Tanzania and Ghana, as well as Egypt, Senegal and Kenya [6]. Attention should be paid to the specifics of customs regulation of e-waste imports in Nigeria, which demonstrates certain positive results in this direction, though remaining the world leader in the import of used electric equipment (60,000 thousand tons per year) [1]. The general foundations of customs policy, which involves regulation of e-waste import, are as follows [7]:

- Basel Convention, according to which the exporter must receive from the importing country written consent for importing the materials. Consent is not required for imports of used electric and electronic equipment. Despite such requirements, about 7–20% of used products are actually e-waste, which requires recycling by the country's entrepreneurial sector.
- Bamako Convention, according to which African countries set national norms of the management of trade and customs rules on the imports of e-waste.

- Norms of criminal law on hazardous waste, which depend on the country, prohibit any operations on their imports and placement in the country, but do not contain any provisions on e-waste.
- Law on the assessment of hazardous waste's influence (including e-waste) on the ecology. Though it is a framework regulatory act, it is used during assessment by special bodies that cooperate with customs bodies – assessment of damage dealt to the environment (water resources, land, air, population's health).
- Law on the body performing control over compliance with environmental norms and standards (since 2007). This regulatory act has been applied in recent years concerning the authorities of control over the imports of e-waste.
- Collection of rules on environmental protection from 2009, which is used by national customs bodies to prevent the import of e-waste; a collection of environmental norms).
- Legislative initiatives on the implementation of customs fees on the import of e-waste, which will be imposed on manufacturers of the electric and electronic sectors. In 2023, agreements were reached with Deloitte, Phillips, DELL, HP and Microsoft, which complied with the necessity to pay these fees [12]. The introduction of such customs regulation will allow the country to adopt the measures of stimulation of the activities of climate-responsible entrepreneurship that deals with green recycling of these materials. This includes the purchase of new innovative equipment, technologies with filters and purification installations and the use of renewable energy.

Analysis of the gradual improvement of customs and environmental law in Nigeria and improvement in law enforcement measures showed the formation of preconditions for determining a clear framework for e-waste imports. Though reforms in this direction are continued, the focus on the existing regulatory framework, which is an alternative to hiding operations on e-waste imports, allows the country to implement the UN SDGs in the sphere of climate protection and economic development (support for green companies involve in eco-friendly recycling).

Norway focuses on the minimisation of the import and export of e-waste, to protect the environment in its territory and other countries. The positive practice of Norway is aimed at the prevention of the export of e-waste from electric and electronic products. According to the Norwegian customs law and law on the movement of commodities, the following requirements to the export of electric and electronic products are in effect: these products must be in the packaging; each product of this category is to be tested before export (performed by government and private organisations), with the corresponding protocol on the results of the testing; each product of the studied category is accompanied by a document (e.g. invoice), with information that exported product can be used for its intended purpose; each cargo of this type is confirmed by a declaration stating that it does not contain e-waste [10]. This norm sets an obligation on exporters regarding confirmation of information that electronic and electric products are not e-waste.

The use of some of these norms is directly connected with support for climate-responsible entrepreneurship. For example, the requirement to eco-friendliness of packaging leads to stable demand for packaging products of this category, manufactured by local companies.

4 Discussion

In this research, we were able to formulate a range of features of customs regulation in support of climate-responsible entrepreneurship in digital economy markets.

We should note the targeted approach, which involves the achievement of specific environmental goals (e.g. customs ban on imports of e-waste), which is used by China but does not seek the balance of the regulatory measures' impact on socio-economic justice of regional, territorial and subject development. Customs regulation, which is connected with the achievement of climate goals and is implemented according to Chinese law in the context of the focus on zero imports of e-waste, is based on such an approach. It is rather effective regarding local environmental goals, but there are problems regarding the balance of all goals of sustainable development. Within the application of this approach (e.g. the total ban on imports of environmentally harmful materials), the goals of social and economic spheres for all participants were not taken into account. Accordingly, very soon the government will face problems in the social and economic spheres in regions and at the level of the country. These problems will have to be solved, since they may have negative consequences.

We should mention Nigeria's approach, which includes the legislative adoption of expanded responsibility of manufacturers of electronic and electric sectors on the payment of customs fees for imports of e-waste of their trademarks. This approach is focused on the support for green initiatives of national companies that perform the recycling of these materials. This innovative approach to the system of customs regulation should facilitate the green transfer of Nigeria, even maybe toward zero import of e-waste. Its implementation implies global interaction and public contacts with world manufacturers for substantiating the necessity to adopt responsibility in the sphere of climate protection.

There is also an approach to customs regulation which is aimed to ensure stable demand for the products of national climate-responsible companies that use digital tools (digital modelling of the optimal size of packaging for specific exported products of the electronic and electric sectors). By the example of customs regulation of the export of products of the electric and electronic sectors of Norway, we showed a positive practice with protection from illegal export of e-waste instead of listed products and support for the demand for green companies). Creation of such stimuli motivates national manufacturers that aim at climate-responsible production.

5 Conclusion

Summing up the results of this research, we may note the necessity to search for new means of government's supporting climate initiatives of the entrepreneurial sector, which would better fit the modern conditions and needs for the resolution of climate protection problems and supporting the balance of socio-economic components. Such sphere as customs regulation make a significant contribution to environmental protection, though it is not sufficient in developing countries because of the imperfect legislative and law-enforcement spheres.

In this study, we considered the specific features of customs regulation of the import of e-waste, which is a source of hazardous substances. Some countries, e.g. China, reach prominent results in opposing the import of these hazardous substances. Some countries do not use all capabilities of the government in the sphere of solving the actual environmental problems.

The directions of customs regulation, which is connected with the support of climate-responsible entrepreneurship in the digital economy markets, have large prospects for implementation at the level of countries that strive toward results in this sphere. Since the problem of climate protection is very relevant for developing countries, for they are recipients of many categories of waste that damage ecology (their territories are used as landfills for various types of industrial waste and life activities). There is a need for a global approach to climate protection within these territories, which requires partnerships at the level of international organisations, national governments and business.

References

1. IEEP (2022) Trade in support of circular economy Opportunities between Nigeria and the EU. https://ieep.eu/wp-content/uploads/2022/11/DRAFT_Trade-in-support-of-circular-economy.-Opportunities-between-Nigeria-and-the-EU_IEEP-2022-1.pdf. Accessed 04 Feb 2023
2. Jumambayev S, Turginbayeva A, Moldabekova A (2018) The impact of new technologies on employment in Kazakhstan. Proceedings of the 32nd International Business Information Management Association Conference, IBIMA 2018 – Vision 2020: Sustainable Economic Development and Application of Innovation Management from Regional expansion to Global Growth, 2184–2188. URL: https://ibima.org/accepted-paper/the-impact-of-new-technologies-on-employment-in-kazakhstan/. Accessed 04 Feb 2023
3. Kaliyeva SA, Alzhanova FG, Meldakhanova MK, Sadykov IM, Adilkhanov MA (2018) The Precariousness employment in the Eurasian Economic Space: Measurement problems, factors and main forms of development. J Asian Finance Econ Bus 5(3):157–167. https://doi.org/10.13106/jafeb.2018.vol5.no3.157
4. Li L (2017) Has Guiyu finally shaken off the title of world's e-waste capital? URL: https://www.scmp.com/video/china/2112000/has-guiyu-finally-shaken-title-worlds-e-waste-capital. Accessed 04 Feb 2023

5. Lin S, Man YB, Chow KL, Zheng C, Wong MH (2020) Impacts of the influx of e-waste into Hong Kong after China has tightened up entry regulations. Crit Rev Environ Sci Technol 50(2):105–134. https://doi.org/10.1080/10643389.2019.1619377
6. Maes T, Preston-Whyte F (2022) E-waste it wisely: lessons from Africa. SN Appl Sci 4(72). URL: https://doi.org/10.1007/s42452-022-04962-9. Accessed 04 Feb 2023
7. Okposin AEA (2019) Curbing E-Waste menace in Nigeria: assessing the regulatory framework. Chukwuemeka Odumegwu Ojukwu University Journal of commercial and property law, Cooujcpl 2(1). URL: https://www.nigerianjournalsonline.com/index.php/COOUJCPL/article/download/632/618. Accessed 04 Feb 2023
8. Pinghui Z (2017) China's most notorious e-waste dumping ground now cleaner but poorer. URL: https://www.scmp.com/news/china/society/article/2112226/chinas-most-notorious-e-waste-dumping-ground-now-cleaner-poorer. Accessed 04 Feb 2023
9. Ruiz A (2022) Latest global e-waste statistics and what they tell us. URL: https://theroundup.org/global-e-waste-statistics/. Accessed 04 Feb 2023
10. Toll.no (2023) Electric and electronic products. URL: https://www.toll.no/en/goods/electric-and-electronic-products/. Accessed 04 Feb 2023
11. Turginbayeva A, Smagulova G, Ashirbekova L, Malikova R (2018) Topical issues of management in modern education and ways to solve them. Proceedings of the 32nd International Business Information Management Association Conference, IBIMA 2018 – Vision 2020: Sustainable Economic Development and Application of Innovation Management from Regional expansion to Global Growth, 3643–3647. URL: https://ibima.org/accepted-paper/topical-issues-of-management-in-modern-education-and-ways-to-solve-them/. Accessed 04 Feb 2023
12. UNEP (2023) Nigeria acts to fight growing e-waste epidemic. URL: https://www.unep.org/gef/news-and-stories/press-release/nigeria-acts-fight-growing-e-waste-epidemic. Accessed 04 Feb 2023

Part II
Fighting Climate Change Using Advanced Technology and Green Innovations of Industry 4.0

The Modern Experience and Prospects for the Development of Climate-Responsible Entrepreneurship in the Digital Economy Market in the Sphere of E-Commerce

Oksana V. Shmaliy, Elena B. Ivushkina, Zurakan Sh. Bulanova, and Anna V. Chulkova

1 Introduction

The popularisation of the Internet and the development of e-commerce ensured significant changes in the systems of trade, distribution, logistics, financial operations and marketing. The speed, convenience and innovativeness of digital technologies created additional values of digital trade, which include "purchases without leaving home"; services for searching for information on products; the creation of platforms for trading goods and services in the digital form, etc. Such changes took place due to the synergy of technological innovations, demand from consumers and transformation of trade mechanisms, which were accompanied by the formation of the system of product delivery, the creation of numerous marketplaces and the adaptation of the system of trade, insurance and marketing to new realities.

The sphere of the digital economy ensured substantial advantages of e-commerce, which manifested thanks to saving time, quick access to information, convenience and speed. Given this, a large share of online buyers often do not question the impact of such a form of trade on the environment and perform purchases guided only by economic motives.

O. V. Shmaliy (✉)
Russian Presidential Academy of National Economy and Public Administration (RANEPA), Moscow, Russia

E. B. Ivushkina
Institute of Service and Entrepreneurship (branch) of Don State Technical University, Shakhty, Russia

Z. Sh. Bulanova
Jalal-Abad State University named after B. Osmonov, Jalal-Abad, Kyrgyzstan

A. V. Chulkova
Sebryakovskiy branch of Volgograd State Technical University, Mikhailovka, Russia

E. G. Popkova (ed.), *Smart Green Innovations in Industry 4.0*, Springer Climate,
https://doi.org/10.1007/978-3-031-45830-9_13

Large dynamics of the e-commerce market, which volume exceeded $5.2 trillion in 2021, and the constant increase in the number of Internet users in the world (more than 5 billion people) [12] make the studied sphere an important socio-economic mechanism, which activity influences the environment a lot. Like every sphere of economic activity, e-commerce interacts with the environment and affects the climate situation. This interaction is manifested most vividly in logistics and in the direct sales of products that have a material form. In the first case, the largest carbon footprint is formed by the transport and production of packaging, and in the second case – by companies that physically produce the goods. In both cases, the e-commerce entrepreneurial structures' using climate-responsible allows reducing the pressure on the environment.

Striving toward environmental friendliness of the digital economy and its components is a world trend today. Large entrepreneurs with a substantial influence on the production and distribution of products in the sphere of the digital economy tend to select a long-term strategy and values, contrary to short-term economic effectiveness. An important factor that leads to such actions is climate-responsible consumption. According to the surveys, almost 50% of the British, French and Germans are guided by social and climate values during their purchases. Given this, value-based consumption forms a separate strong segment nowadays [3].

2 Materials and Methods

The methodology of this research consists of approaches and methods of different disciplines. It is based on the interdisciplinary approach, which includes certain elements of the system approach, behavioural economics, marketing and logistics. According to this, the means and methods inherent to IT, economics, logistics, marketing and commercial activities are combined with the values of climate sustainable development and behavioural economics, which allows determining the influence of different factors of e-commerce on climate change. An important role in this research belongs to behavioural economics, which allows for the detailed study of the two main problems of climate-responsible behaviour in the environment of e-commerce: the problem of the "Green Gap", according to which all buyers who share the values of climate-responsible consumption, are not guided by them during online purchases; and the problem of "last-mile delivery", according to which one of the main advantages of e-commerce – the possibility to get quick home delivery – has a significant negative impact on climate change factors.

Given the multidisciplinary character of the problem considered, to disclose it we use the works connected with e-commerce, logistics, behavioural economics, etc. the key works that lie in the basis of this paper are studies devoted to the determination of the essence and architecture of e-commerce [9–11], characteristics of consumer values in the digital economy [3]; critical assessment of the positive and negative effect of e-commerce on the state of the environment [13], understanding of the phenomenon of "last mile delivery" and its influence on climate [4, 7], the problem of "Green Gap" [5, 6], and the ways to solve it through the use of "nudging" and other methods [8]. On the whole, the materials and methods of this research form a certain integrity that allows solving the main task of the research: generalising the specific features and world experience and describing the prospects for future development of climate-responsible entrepreneurship in the sphere of e-commerce.

3 Results

E-commerce is the key element of the digital economy that connects consumers of goods and services with manufacturers or distributors. The OECD treats e-commerce as any form of business relations where the interaction between actors takes place with the help of Internet technologies [9]. This definition is rather wide and it could be applied almost to all components of the digital economy. However, the main difference of the considered sphere is the direction of its processes toward the satisfaction of market needs through the offer of goods and services that have a digital form or are distributed through digital channels of distribution or serving the needs of market participants. Given the character of participants of such relationships and the specifics of the offer, e-commerce is divided into the following components:

- Electronic Data Interchange (EDI)
- Electronic Fund Transfer (EFT)
- e-trade
- e-cash
- e-banking
- e-insurance [11]

The above components of e-commerce prove that its popular treatment – only from the position of trade – is not correct. Moreover, depending on participants that implement online transactions, the considered sphere of the digital economy is also divided by the type of participants of relationships. The following sectors of e-commerce are distinguished:

1. Business-to-Business (B2B)
2. Business-to-Consumer (B2C
3. Mobile Commerce (M-Commerce)
4. Facebook Commerce (F-Commerce)
5. Customer-to-Customer (C2C)
6. Customer-to-Business (C2B)
7. Business-to-Administration (B2A) [10]

Thus, the e-commerce sphere has a rather complex architecture. It is characterised by different types of interactions between market participants and with the external environment. While some sectors of e-commerce ensure an instant positive influence on climate processes, the effect of other sectors requires additional research.

Among the above components of e-commerce, e-trade is the one that most requires physical operations that are connected with supply, production, storing, transportation and delivery. For other sectors, the transition to the digital sphere substantially decreases the need for physical operations and, accordingly, reduced their influence on the environment.

An important direction of studying socially responsible digital entrepreneurship in the e-commerce markets is the identification of mechanisms through which positive changes take place and the concept of climate sustainability is implemented. One of the proposed mechanisms comes from the theory of behavioural economics. The behaviour of consumers in digital markets is rather logical and structured on the whole. At the same time, the focus is made on the so-called Green Gap, when consumers share environmental values and understand the importance of climate-responsible consumption, but, when making decisions on purchases, are guided by motives that imply convenience, price or other advantages that lead to larger damage to the [6].

The main issues that form such behaviour include high prices, difficulties with the determination of the level of products' eco-friendliness, including lack of time for such determination, absence of correct information on the products' influence on the environment, low level of trust in environmental labelling, etc. In general, these issues could be expressed in the form of three main factors:

- Low convenience of climate-responsible purchase compared to traditional purchase
- Lack of knowledge about the influence of the products and the means of transportation on the environment
- Lack of trust in the available information on the climate responsibility of products and environmental labelling [5]

Given the discovered problems, the experience of climate-responsible entrepreneurship suggests the development and implementation of a comprehensive mechanism of the influence on the e-commerce market and consumer behaviour that could be identified as "nudging". According to this mechanism, the information on products' climate parameters is to be provided in the context of the main characteristics of the platforms of online sales. Such information must reflect – clearly and laconically – the parameters of the impact of the product and the means of its delivery and consumption on the environment. The form of such information presentation may contain tools for additional stimulation of climate-responsible behaviour, which includes the possibility of receipt of "eco-friendliness score" and "green points" [5] or other bonuses, which can be converted into image or economic advantages of consumers who are guided by the sustainability principles during online purchases.

The proposed "nudging" toward climate-responsible behaviour can be also built based on the concept of "fast fashion" [8]. In this case, it is important to ensure an increase in motivation for responsible online shopping with the help of PR technologies and the cultivation of the right values.

In the context of assessing the influence of product delivery in the sphere of e-commerce on climate change, it is important to emphasise its substantial advantage compared to a traditional purchase. Here consideration should be given to the fact that the process of purchase consists of three stages – search for products, purchase and return. If the purchase is considered through the prism of these three processes, the carbon footprint of an online buyer is twice as small compared to the traditional buyer. However, around 15% of emissions in the system of online purchases come from the service of fast or urgent delivery [7]. Such a situation is largely connected with the insufficient loading of transport vehicles during fast delivery and the violation of the principles of optimal route building because of the necessity to deliver goods as soon as possible.

These processes are viewed through the lens of the "last-mile delivery" problem, within which the possibility of influence on consumer behaviour to nudge him toward the refusal from fast delivery, given the realisation of harm dealt by such type of delivery, is considered. Studies of this problem show that a large share of online buyers (42%) were not aware of the higher carbon footprint of fast delivery [4]. With this in mind, additional information suggestion to measure the level of influence of different types of delivery on climate allows nudging consumers toward more deliberate purchases from the position of climate responsibility. A similar character of problems is inherent to the sphere of product packaging during online purchases. In this case, much of the packaging material might not be used according to the corresponding desire of the buyer or might be replaced by recyclable stuff.

Thus, the studies of the key problems of climate responsibility of entrepreneurship in the sphere of e-commerce demonstrate the presence of significant potential in the use of different types of mechanisms and stimuli for the decrease in the level of e-commerce elements' burden on the environment. The existing experience of entrepreneurial initiatives of such direction in the sphere of online trade offers a large number of tools, which are shown in Table 1.

As can be seen from the above initiatives, there are measures aimed at the realisation of climate responsibility by consumers, replacement of traditional packaging with renewable materials, the use of alternative sources and optimal planning of routes during delivery of goods, complex use of packaging and related products, stimulation for forests recovery, etc. In a generalised form, the key directions of climate-responsible activities in the sphere of e-commerce are presented in Fig. 1.

4 Discussion

The main directions for reducing carbon footprint through a rational approach to packaging imply the use of renewable materials or partial refusal of packaging. The shipping sphere has reserves of environmental responsibility, which could be reached through the use of fuel with a higher level of eco-friendliness, compensation for emissions through the financing of climate initiatives, optimisation of routes and schemes of shipping and use of innovations in the sphere of transportation and delivery. A separate direction of the influence on climate responsibility, which goes beyond the limits of online trade but has powerful tools for affecting the modelling of product offer through the change in demand, is encouraging consumers to climate responsible

Table 1 Experience of climate-responsible initiatives of entrepreneurial activities in the sphere of e-commerce

Company/brand/online resource	Essence of climate-responsible initiative
Timberland	The company offers to plant a tree if customers choose a 4–8-day delivery instead of regular 4-day delivery
Amazon	Offers bonuses for buyers who choose free delivery ("without haste")
bewustbezorgd.org	Available tool for calculating CO_2 emissions from every package delivery
Packhelp	Offers eco-friendly packaging for online stores, stimulates forests recovery through the corresponding marking of packaging
TOMS	80% of packaging is produced from renewable materials, with text done with soy ink
Sheyn	Uses a simple and elegant design for the packaging of jewellery, with 80% of packaging consisting of renewable materials
H&M	Offers a shopping bag which easily transforms into a clothes hanger
DHL GoGreen	Calculates CO_2 emissions from product delivery and compensates for them by means of its climate projects
GLS ThinkGreen	Offers solutions aimed at the optimal use of resources, decrease in the volume of waste and the need for utilisation

Source: Created by the authors based on [1, 14]

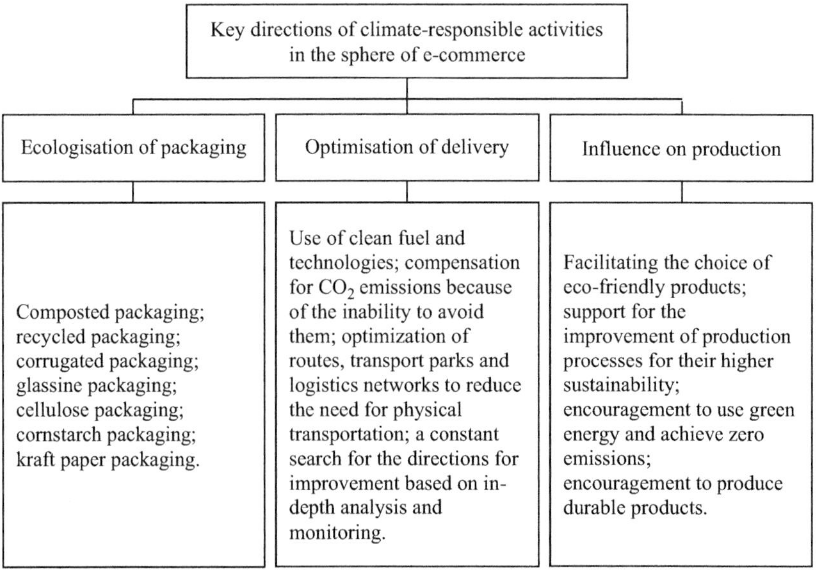

Fig. 1 The key directions of climate-responsible activities in the sphere of e-commerce. (Source: Created by the authors based on [1, 2])

purchases. The main aspect of this group is an increase in the level of awareness of the eco-friendliness of different types of products and means of their delivery, encouraging buyers to choose more eco-friendly options and the offer of compensation for the damage dealt through participation in the initiated climate projects.

The prospects for the improvement of climate-responsible behaviour in the sphere of e-commerce are closely connected with tendencies that are inherent to the entire digital economy. They come from the expectations for further growth of the e-commerce market volume by approximately $500 billion annually [7]. This will increase the impact of the sphere on socio-economic processes and ecology and will require more active actions for the implementation of nature protection measures. Besides, valuable offers on the reduction of e-commerce pressure on the environment could be implemented through the creation of relevant start-ups.

5 Conclusions

The climate responsibility of entrepreneurship in digital markets in the sphere of e-commerce is a rather complex notion, which received a proper scientific and theoretical reflection in the scholarly literature. The main problems elaborated by scholars and practitioners include the determination of tools for modelling of climate responsible behaviour of consumers, search for the ways to optimise product delivery and the minimisation of the negative influence of packaging on climate.

According to this, the modern experience of the development of climate-responsible entrepreneurship is largely oriented toward solving the problems of the "Green Gap" and "last-mile delivery". The generalised study of initiatives, aimed at the resolution of these problems, allowed identifying their main directions, which envisage nudging customers toward more climate-responsible purchases, refusal of fast delivery and excessive packaging. Given the dynamics of the e-commerce market development, it was determined that the influence of climate-responsible initiatives on the state of e-commerce will be growing. The current initiatives will be also developing, with their influence strengthened by successful start-ups, which will be able to offer more innovative and better solutions.

Acknowledgments The research is prepared within the framework of the state assignment "Improvement of legislation in the field of higher education as a factor of safe socio-economic development".

References

1. Byrd (2022) The rise of sustainable e-commerce. https://blog.getbyrd.com/en/rise-of-sustainable-ecommerce. Accessed 20 Jan 2023
2. DHL Express (2022) What the rise in green consumerism means for your e-commerce business. https://www.dhl.com/discover/en-au/e-commerce/e-commerce-advice/what-the-rise-in-green-consumerism-means-for-your-e-commerce-business. Accessed 20 Jan 2023

3. Forrester (2022) B2C marketing in Europe 2022. Forbes. https://www.forbes.com/sites/forrester/2021/10/29/b2c-marketing-in-europe-2022/?sh=7d399b454c98. Accessed 20 Jan 2023
4. Ignat B, Chankov S (2020) Do e-commerce customers change their preferred last-mile delivery based on its sustainability impact? Int J Logist Manag Ponte Vedra Beach 31(3):521–548. https://doi.org/10.1108/IJLM-11-2019-0305
5. Islam MdS, Proma AM, Wohn C, Berger K, Uong S, Kumar V, Smith Korfmacher K, Hoque E (2022) SEER: sustainable e-commerce with environmental-impact rating. https://doi.org/10.48550/arXiv.2209.06156. https://arxiv.org/pdf/2209.06156.pdf. Accessed 20 Jan 2023
6. Joshi Y, Rahman Z (2015) Factors affecting green purchase behaviour and future research directions. Int Strateg Manag Rev 3(1–2):128–143. https://doi.org/10.1016/j.ism.2015.04.001
7. Mangiaracina R, Perego A, Seghezzi A, Tumino A (2019) Innovative solutions to increase last-mile delivery efficiency in B2C e-commerce: a literature review. Int J Physi Distrib Logist Manag. https://doi.org/10.1108/IJPDLM-02-2019-0048
8. Mirbabaie M, Marx J, Germies J (2022) Conscious commerce – digital nudging and sustainable e-commerce purchase decisions. Cornell University. Australasian Conference on Information Systems, arXiv:2202.08696. https://arxiv.org/abs/2202.08696. Accessed 20 Jan 2023
9. OECD (2001) OECD science, technology and industry scoreboard 2001: towards a knowledge-based economy. OECD Publishing, Paris. https://doi.org/10.1787/sti_scoreboard-2001-en
10. Oyedokun GE (2021) Management information system, digital business, and e-commerce/Business policy, Strategy, Governance & Ethics, Aaron & Hur Publishing, Lagos. https://www.researchgate.net/publication/355046537_Management_Information_System_Digital_Business_and_E-Commerce. Accessed 20 Jan 2023
11. Pushko O, Dekhtyar N, Deyneka O, Zhukova T, Syniavska O (2019) Security of e-banking systems: modelling the process of counteracting e-banking fraud. SHS Web of Conferences. https://doi.org/10.1051/shsconf/20196503004
12. Statista (2022) E-commerce worldwide – statistics & facts. E-Commerce. Key Figures of E-Commerce. https://www.statista.com/topics/871/online-shopping/#topicOverview. Accessed 20 Jan 2023
13. Sui D, Rejeski D (2002) Environmental impacts of the emerging digital economy: the E-for-environment e-commerce? Environ Manag 29:155–163. https://doi.org/10.1007/s00267-001-0027-X
14. Tiwari S, Singh P (2011) Environmental impacts of e-commerce. Int Conf Environ Sci Eng IPCBEE 8:202–207. http://www.ipcbee.com/vol8/45-S10040.pdf. Accessed 20 Jan 2023

Features of Climate Risk Management of Entrepreneurship in the Digital Economy Markets in AgroTech

Anastasia A. Sozinova, Natalia M. Fomenko, Elena V. Karanina, and Irina A. Morozova

1 Introduction

Climate risk management is a part of the particular sphere of strategic management. It combines financial, economic and managerial tools, which action is aimed at an increase in the level of manageability of economic processes that are connected with climate phenomena. In the context of risk management, climate change is a factor of uncertainty and destabilisation, which reduce the quality of forecasts, increase the level of volatility and form large-scale risks for the entire mankind.

Agriculture is the economic sector that is most sensitive to climate change. Climate is a factor that determines the development of agriculture. Depending on the climate, agrarian specialisation, competitive advantages, consumer ration, etc. are formed. Given this, anomalous changes in air temperature and precipitations and an increase in the frequency and intensity of droughts and floods [7] are considered the key factors of risks that are connected with climate change. They form an important direction for scientific studies that is researched at the level of global institutions, e.g. the UN, the FAO, the World Bank, etc.

In the digital economy markets, entrepreneurship rather actively reacts to the actual problems on a global scale, proposing solutions based on the combination of the capabilities of innovative technologies and other spheres of activity. In this context, AgroTech is a segment of the market of start-ups and venture capital that ensures the digitalisation of the production and managerial processes in agriculture

A. A. Sozinova (✉) · E. V. Karanina
Vyatka State University, Kirov, Russia
e-mail: aa_sozinova@vyatsu.ru; karanina@vyatsu.ru

N. M. Fomenko
Plekhanov Russian University of Economics, Moscow, Russia

I. A. Morozova
Volgograd State Technical University, Volgograd, Russia

E. G. Popkova (ed.), *Smart Green Innovations in Industry 4.0*, Springer Climate,
https://doi.org/10.1007/978-3-031-45830-9_14

and facilitates the achievement of its stability, including adaptation to climate change. Despite the large progress of AgroTech, agriculture and the sector of food supply remain the least digitalised [11]. On the one hand, this defines a large potential of the sector, and, on the other hand, is a sign of its vulnerability to the current risks. Based on this, a study of the specific features of climate risk management of entrepreneurship in the digital economy markets in the sphere of AgroTech is an important scientific problem, the resolution of which will allow strengthening the impact of digital innovations on the problems of climate change in the least digitalised and climate-resilient spheres of the economy – agriculture and agrifood provision.

2 Materials and Methods

This paper's methodology is based on the system and process approaches. The system approach allows viewing the risks of climate change from the position of their influence on the state of agrifood production and discovering the mutual dependence between the primary and secondary impact of climate change on the considered processes. In its turn, the process approach allows drawing the connection between the influence of climate change, risks for agriculture development and measures aimed at the adaptation to climate factors. According to this, the process of climate change is recognised as a factor that influences a large number of parameters of the agrifood sector's functioning, forms its risks and sets important tasks that could be solved due to targeted actions and approaches. Among the current approaches to the adaptation of agriculture to climate change, we distinguish the sector of the digital economy AgroTech and characterise its essence, directions and specifics from the position of climate risk management.

Given the importance of the considered problems, it is considered based on reports and analytical notes of the global institutions on possible risks for agriculture and the system of supply that are caused by climate change [3, 6–10, 14]. More detailed studies of the influence of climate change on the state of agriculture and food supply from the position of adaptation to its consequences were conducted in the works [2, 13]. Substantiation of the concept of Climate Smart Agriculture and the potential of Hi-Tech Farming from the position of integration of technologies in the systems of managing climate risks in agriculture was done in [4]. A particular direction of researching the specifics of climate risk management of entrepreneurship in AgroTech markets is the works devoted to the analysis of this sector, its structure and venture investment activity [1, 12], as well as the evaluation of the AgroTech sector's impact on climate change [5, 11, 15]. Given the importance and level of disclosure of the considered problem, the main goal of this paper is to identify the specific features and influence of AgroTech on the state of adaptation of the agrifood sector to climate change.

3 Results

The influence of climate change on agriculture is evaluated comprehensively and presented at separate levels or directions. The FAO distinguishes two groups of influence: biophysical (connected with biological, geological or natural & climate processes) and socio-economic (which is a manifestation of the biophysical influences, expressed in the financial dimension or through social implications) [7]. There is also the direct and indirect influence of climate change on agriculture. Direct influence is connected with changes in phenology and calendar, correction of cultivated areas and reduction of soil fertility, changes in water supply and irrigation and determination of the impact of an increased level of CO_2 on crop yield. Indirect influence is a result of previous factors. It can be manifested in the negative dynamics of agricultural production indicators, which are a result of an increase in the number of pests, diseases, invasive species and extreme events, e.g. gales, hail, intense heat and frost [10].

The main risks of agriculture, which are connected with the primary factor of climate change, are manifested in the following:

- Influence of CO_2 on crop yield – covers the results of research that show that in the short-term, carbon compounds may have a positive effect on crop yield.
- Influence of high temperature on specialisation and fertility – in the short-term, this influences the change of agricultural specialisation of regions; in the long-term, this leads to drought and degradation of soil.
- Influence of the change in the amount of precipitation – determines the change in regimes of plant irrigation and is often connected with desertification or water-logging of soils.

The indirect influence of climate change on agriculture has a cascading character [10], which is manifested in the formation of the following risks:

- Growth of threats from pests and diseases – change in temperature regimes and humidity leads to the change in the ecosystem within which the simplest organisms – viruses, insects and bacteria – migrate quickly, causing diseases and death of plants.
- Impact of extreme climate conditions – unpredictable destructive events, which deal damage to crops or conditions of growing crops [3].
- Need to change the specialisation of production and adapt sorts, as well as related risks of the reduction of crop yield and productivity, decrease in production and loss of competitive advantages.
- The necessity for the change in the strategy of management in the sector, correction of the policy of insurance and state regulation.
- Destruction of food supply chains and related risks of famine or violation of nutrition norms, fluctuations of stock indices, growth of the risk of social protests and military conflicts.

Thus, climate change directly influences food systems. The impact of this phenomenon on food production is expected to become more negative, though there might be positive effects in certain regions in the short term. On the whole, the developed trajectories of food system change in the context of climate change imply the loss of biological diversity, degradation of land and water ecosystems, decrease in the level of food security, quality of nutrition, etc. [9].

The generalisation of scientific conclusions and results of the mathematical modelling of climate change impact on the state of agrifood support allows stating that the influence of climate change on agriculture in different countries and regions is not similar. According to this, the key task at the current level of climate change is not just prevention of the change but also adaptation to the new conditions of agriculture, trade and food security support at all levels [13].

In the context of the agrifood sector's adaptation to climate change, several approaches are implemented as of now. Climate Smart Agriculture (CSA), Hi-Tech Farming and AgroTech assign technologies the key role in the implementation of changes. CSA is based on the increase in adaptive capabilities of farmers, support for the stability of agricultural production systems and increase in their effectiveness in the use of resources. Hi-Tech Farming implies the use of new technologies, which are resource-saving and less dependent on the external environment and more favourable for the growth of productivity and effectiveness of production [4]. AgroTech is built on the use of opportunities of the digital economy to solve the current problems of the agrarian sector, including support for a better level of adaptation to the conditions of climate change. The main tools that create the advantages of this approach are formed by digital tools, which combine biotechnologies, sensors, AI solutions, equipment for agriculture, automatisation of production and economic processes, etc. [12].

AgroTech is a market of start-ups and venture capital. It is growing constantly. In 2021, its volumes reached $51.7 billion, of which $18.2 were used for projects in the sphere of Ag Biotech, Farm Management Software & Sensing, Farm Robotics & Equipment, Bioenergy & Biomaterials, Novel Farming Systems, Agribusiness Marketplaces, Midstream Tech and Innovative Food – the so-called upstream sector (from the position of agro-industrial integration, it covers reverse processes aimed at the production of raw materials and its primary processing). Downstream investing ($32.1 billion) is a much larger sector. It covers the spheres of direct agro-industrial integration and includes the following sectors: In-Store Restaurant & Retail Tech, Online Restaurants & Mealkits, eGrocery, Restaurant Marketplaces, Home & Cooking Tech and Cloud Retail Infrastructure [1].

In general, AgroTech start-ups strive toward solving the following tasks: reduction of food waste, CO_2 emissions and wastewater, support for workforce and rational nutrition, elimination of non-transparent supply chains and ineffective distribution of food, tracking the origin of food products, achievement of a high level of effectiveness and farms' profitability, etc. [11].

To better understand the structure of the AgroTech market and its features, let us consider its largest companies and projects that are classified by categories in the form of a table (Table 1).

Table 1 Characteristics of the key sectors of AgriFoodTech start-ups (based on the 2021 data)

Categories	Characteristics	Investments, $ billion	Top companies by the volume of attracted investments
Ag Biotechnology	Internal farm resources for crop research and cattle breeding, including genetics, breeding and animal health	2.6	Pivot Bio, Inari, DNA Script
Agribusiness Marketplaces	Commodity trade platforms, procurement of resources, lease of equipment	1.3	Farmers Business, DeHaat, Agrostar
Bioenergy & Biomaterials	Production and processing of non-food products, processing of raw materials, and pharmaceuticals based on medical cannabis	2.1	Bolt Threads, Spiber, Tripod Preclinical
Farm Management Software, Sensing & IoT	Equipment for data collection, software for decision support, big data analytics	1.2	Gro Intelligence, Samasource, Hologram
Farm Robotics, Mechanisation & Equipment	Agricultural equipment, automatisation, production of drones and other equipment	0.9	BayoTech's, Swift Navigation, EAVision
Midstream Technologies	Food product safety and technologies for quality control, logistics, the technology of processing	3.8	Apeel Sciences, Wiliot, Benchling
Novel Farming Systems	New farming systems	2.3	Bowery Farming Inc., Infarm, Local Bounti
Miscellaneous, e.g. fintech for farmers	Agrofintech start-ups, etc.	0.7	Tarfin, Agro.Club
Innovative Food	Grown meat, new ingredients, vegetable proteins	4.8	Impossible Foods, Nature's Fynd, Perfect Day Foods
In-Store Retail & Restaurant Tech	Assembler robots, 3D printing of food products, POS systems, and monitoring of food products by the IoT	4.2	Trax, Dutchie, SpotOn
Restaurant Marketplaces	Online platforms of technologies that deliver food from various suppliers	3.8	Toast, Olo, ChowNow
eGrocery	Online stores and markets for selling and delivering processes and non-processed agricultural products to consumers	18.5	Furong Xingsheng, goPuff, Gorillas
Home & Cooking Tech	Smart kitchen equipment, food technologies, equipment for food testing	0.4	Tovala, Salted, Good Catch.
Online Restaurants & Mealkits	Start-ups that offer culinary specialities and send ingredients for home cooking	1.2	Liv Up, Liv Up, Not So Dark
Cloud Retail Infrastructure	Ghost kitchens, last-mile delivery services	4.8	CloudKitchens, Nuro, Glovo App

Source: Formed by the authors based on [1]

As can be seen, the directions of the activities of AgroTech companies are very different. Many of them are aimed at reducing the risks of the negative influence of climate change on the state of agrifood provision. This envisages the resolution of the problems of food waste reduction, increase in production volume, reduction of CO_2 emissions, improvement of consumers' access to rational and healthy nutrition, etc. The level of investors' interest in the development of AgTech directions demonstrates an insufficient level of correspondence to the goals of adaptation to climate change. Thus, 35.8% of venture financing was directed in the sphere of eGrocery, which is mostly online stores and food delivery services.

4 Discussion

The influence of the considered sphere on climate change or the increase in the level of adaptation to it is rather low since product delivery to a consumer is often considered very harmful to the environment. In such a situation, the possibility to reduce the volume of food waste, which is offered by online delivery of food products, does not allow for the complete neutralisation of the negative influence of the agrarian sector on climate change [11].

Sectors of AgroTech that have the highest level of climate responsibility include Innovative Food. Replacement of animal meat, production of plant protein and invention of new food ingredients are the directions that have a large potential for the reduction of agriculture's burden on climate and decrease in its negative consequences.

Based on the results obtained, it is possible to note that the level of the influence of each direction or AgroTech project on climate risks has to be studied more thoroughly. Special attention should be paid to the most acute problems and risks to which the AgroTech solutions are least applied as of now. Such research will allow forming concrete offers on the further development of the AgroTech market of the digital economy to ensure higher effectiveness of climate risk management.

5 Conclusions

Climate change is a global problem that has a significant influence on the development of agriculture and food supply in the world. The obtained results on the impact of climate change on the state of the agrifood sector are different for regions and sectors, but all of them characterise changes from the position of risks. Structuring of climate risks from the position of their influence on agrifood production allows identifying their mutual dependence, which implies the cascading influence that substantially increases the primary risks that are peculiar to agricultural production. This is a sign of their possible influence on global socioeconomic processes.

In the context of climate risk management, it was determined that the main direction of managing the studied processes is assuaging the negative consequences of climate change and adapting to the new conditions of agriculture and food supply. Among the approaches to the adaptation of agricultural production to climate change, we distinguished the ones based on the scientific approach and innovations. We elaborated on the sphere of the digital economy – AgroTech – and determined the structure of the AgroTech markets and their specific features. It was found that the correlation between the influence of the current projects and climate risks is low.

References

1. AgFunder (2022) AgriFoodTech: investment report. https://research.agfunder.com/2022-agfunder-agrifoodtech-investment-report.pdf. Accessed 24 Dec 2022
2. Anderson R, Bayer PE, Edwards D (2020) Climate change and the need for agricultural adaptation. Curr Opin Plant Biol 56:197–202. https://doi.org/10.1016/j.pbi.2019.12.006. https://www.sciencedirect.com/science/article/pii/S1369526619301219. Accessed 24 Dec 2022
3. Antyn J, Kimura S, Lankoski J, Cattaneo A (2012) A comparative study of risk management in agriculture under climate change. OECD food, agriculture and fisheries papers, 58. OECD Publishing, Paris. https://doi.org/10.1787/5k94d6fx5bd8-en
4. Bano A, Ali M, Gupta A, Pathak N, Hasan W (2021) Climate smart agriculture and hi-tech farming. Implications for climate smart agriculture. Biotech Books. https://www.researchgate.net/publication/356161853_Climate_Smart_Agriculture_and_Hi-Tech_Farming_Climate_Smart_Agriculture_and_Hi-Tech_Farming. Accessed 24 Dec 2022
5. Dorin A, Penn AS, Siqueiros JM (2022) AgTech that doesn't cost the earth: creating sustainable, ethical and effective agricultural technology that enhances its social and ecological contexts. In: Proceedings of the ALIFE 2022: The 2022 Conference on Artificial Life. ALIFE 2022: the 2022 Conference on Artificial Life. Online. ASME, p 76. https://doi.org/10.1162/isal_a_00544
6. EPA (2022) Global greenhouse gas emissions data. https://www.epa.gov/ghgemissions/global-greenhouse-gas-emissions-data. Accessed 24 Dec 2022
7. FAO (2007) Adaptation to climate change in agriculture, forestry and fisheries: perspective, framework and priorities. Food and Agriculture Organization of the United Nations, 32. https://www.fao.org/documents/card/en/c/1be37c69-0147-44ea-af0b-d1487e223962/. Accessed 24 Dec 2022
8. FAO (2018) Upscaling climate smart agriculture. Lessons for extension and advisory services. Occasional papers on innovation in family farming. Food and Agriculture Organization of the United Nations, Rome
9. IPCC (2022) Climate change 2022: impacts, adaptation, and vulnerability. In: Pörtner H-O, Roberts DC, Tignor M, Poloczanska ES, Mintenbeck K, Alegría A, Craig M, Langsdorf S, Löschke S, Möller V, Okem A, Rama B (eds) Contribution of working group II to the sixth assessment report of the intergovernmental panel on climate change. Cambridge University Press, Cambridge, UK and New York, p 3056. https://doi.org/10.1017/9781009325844
10. Jacobs C, Berglund M, Kurnik B, Dworak T, Marras S, Mereu V, Michetti M (2019) Climate change adaptation in the agriculture sector in Europe (No. 4/2019). European Environment Agency (EEA). https://www.eea.europa.eu/publications/cc-adaptation-agriculture. Accessed 24 Dec 2022
11. Lieb Th (2022) Food and agtech investments don't stack up with climate needs. GreenBiz. https://www.greenbiz.com/article/food-and-agtech-investments-dont-stack-climate-needs. Accessed 24 Dec 2022

12. Malinova P (2022) Agtech and the future of sustainable food production. PitchBook blog. https://pitchbook.com/blog/agtech-and-the-future-of-sustainable-food-production. Accessed 24 Dec 2022
13. Nelson GC, Valin H, Sands RD, Willenbockel D (2013) Climate change effects on agriculture: Economic responses to biophysical shocks. PNAS 111(9):3274–3279. https://doi.org/10.1073/pnas.1222465110
14. WBG (2021) Climate Change Action plan 2021–2025. Supporting Green, resilient, and Inclusive Development. World Bank Group. https://openknowledge.worldbank.org/handle/10986/13192. Accessed 24 Dec 2022
15. Weber H (2022) 7 investors discuss how agtech can solve agriculture's biggest problems. TechCrunch+. Accessed 24 Dec 2022

The Role of Robotisation in the Development of Climate-Responsible Entrepreneurship in Developing Countries' Digital Economy Markets

Victor P. Kuznetsov, Tamara A. Ergeshova, Andrey A. Lezhebokov, and Lyudmila A. Shvachkina

1 Introduction

The development of the digital economy and its components ensures the transformation of a large number of current global and regional processes. Among them, climate change is, perhaps, the most important as of now, since it objectively influences the future of the entire planet. Problems that are linked to such change are considered from the position of neutralisation of negative influence and increase in the level of adaptation to changing conditions. The use of the potential of the digital economy to solve the designated tasks is an important direction for the development of climate-responsible entrepreneurship.

According to this, within the search for the directions for the most rational use of the potential of the above tools, the focus is often made on developed countries, which impact on climate change is the largest. However, from the position of the consequences of such impact, these countries have wide capabilities for its neutralisation and avoidance of the most acute climate issues. Despite the existence of mechanisms and tools for preventing the most critical consequences of climate change, which are connected with drought, famine, natural cataclysms, etc., the

V. P. Kuznetsov (✉)
Minin Nizhny Novgorod State Pedagogical University, Nizhny Novgorod, Russia

T. A. Ergeshova
Jalal-Abad State University named after B. Osmonov, Jalal-Abad, Kyrgyzstan

A. A. Lezhebokov
North-Caucasus Federal University, Stavropol, Russia
e-mail: alezhebokov@ncfu.ru

L. A. Shvachkina
Institute of Service and Entrepreneurship (Branch) of Don State Technical University, Shakhty, Russia

E. G. Popkova (ed.), *Smart Green Innovations in Industry 4.0*, Springer Climate,
https://doi.org/10.1007/978-3-031-45830-9_15

least protected regions are the ones with a lower level of incomes – developing countries [13].

Thus, the search for the mechanisms and tools for assuaging the negative consequences of climate change for this group of countries is an important scientific issue, which has significant potential in the sphere of the digital economy and consists in the search for the methods of effective use of tools of robotics in developing countries in the context of the development of climate-responsible entrepreneurship and formation of the digital economy.

2 Materials and Method

This paper's methodological framework consists of several levels of research on the selected problem. It involves the identification of risks connected with climate change and their specification from the position of developing countries; the study of the most current directions for the possible influence of climate change or its consequences; the study of the potential of solving the given problems with the help of the digital economy tools, including robotics; study of the experience of using robotised systems in climate-responsible entrepreneurship and substantiation of the directions of their dissemination. To ensure the integrity of such a research structure, we use the systems approach, which allows for the coordination of climate issues and the potential of robotics given the conditions and specifics of developing countries. The methods that supplement the methodological framework and allow solving other specific tasks include analysis, generalisation and observation. The proposed methodology allows determining the role and potential of robotics for ensuring the development of climate-responsible entrepreneurship in developing countries.

Given the large importance of the considered problem, the related provisions are presented in the documents of global institutions that work with the problems of climate change – UNCTAD [12–14] and the World Bank [15–16]. Besides, given the multi-level character of the research, its provisions are disclosed in scientific works that dwell on the specifics of the influence of climate change in developing countries [5], consider the essence and specifics of robotics as an element of the digital economy [7], propose the concepts of the use of robots to achieve the Sustainable Development Goals and ecology [1, 6, 8], assess the role of robotics and other tools of the digital economy in the achievement of the SDGs [2, 10, 11], study the practical experience of using robots for managing climate risks [3, 4] and assess the impact of their use on the reduction of the consequences of climate change [9]. Thus, the main goal of this research is to search for opportunities for the effective use of tools of robotics in developing countries in the context of the development of climate-responsible entrepreneurship and the formation of the digital economy.

3 Results

Under the conditions of the development of the digital economy, climate responsibility is one of the key concepts aimed at the achievement of the Sustainable Development Goals. Despite the high level of its institutionalisation, it has different features in different spheres of economy and regions of the world. These features define the character of dissemination and implementation of actions and measures aimed to solve the problems of the negative influence of climate on social and economic processes.

In the architecture of the global economy, developing countries form a separate group, which is peculiar for the low level of income per capita and the related problems of lower living standards, lower level of technical equipment and innovativeness and the presence of significant socioeconomic and climate risks.

The activity aimed to fight climate change is characterised by a range of features from the position of developing countries. First of all, this is because developing countries were not the strongest drivers of the climate crisis. Thus, 46 least-developed countries, in which 14% of the world's population lives, account for only 1.1% of world CO_2 emissions [13]. At that, the carbon footprint of these countries' population is at least eight times smaller compared to countries with a high level of economic development. Despite the small contribution to the global climate crisis, developing countries feel the negative influence of climate change to the greatest extent. Thus, over the last fifty years, 69% of all deaths from climate disasters took place in countries with a low level of development [13].

The reasons for such a situation have a complex character. They are connected with the geographic location of many countries and with their historical past, as well as the financial and institutional ability to oppose climate risks. According to this, such countries require international support for the financing of measures for adaptation to climate change [5] and the use of available technological tools to solve the given problems and achieve the SDGs.

According to the global mechanisms, which were implemented based on the Paris Agreement, one of the tools of support for actions and measures on the management of climate risks is technological support through the Global Environment Facility or other UN programmes. Two main types of technologies for which support is provided within the above tools are production and energy use (reduction of the consequences of climate change) and agriculture (adaptation to climate change) [14].

Within the mechanism of the support for nationally determined contributions, the needs for the implementation of climate-smart capabilities are financed; they are implemented in the form of low-carbon transport, alternative energy, GIS mapping of climate risks and green and sustainable development of cities and communities [15]. Within the adaptation to climate change, the focus is made on consideration of risks linked to climate changes and extreme weather conditions through the protection of agriculture, management of the impact of sea level rise and an increase in infrastructure sustainability [5].

In the digital economy, heavy emphasis is made on the comprehensive use of technological capabilities of modern science and practice to ensure the climate responsibility of entrepreneurship in developing countries. From the position of digital technologies, such actions involve the use of the potential of Blockchain, Big Data, Artificial Intelligence (AI), Cloud Computing, 3D printing, the Internet of Things (IoT), automation and robotics. From the above list, the use of robotics to solve problems of climate change in developing countries is the least studied. It is rarely studied as a separate category of research in the modern literature [7].

The closest – by the character of problems and aspects of their resolution – ones are the use of robotics for environmental purposes [6] or the study of the environment [2]. Therefore, the main advantages of robotics that can be used to decrease the consequences or adapt to climate change are as follows:

- Use of autonomous robots, in particular drones, in previously unreachable places, to collect data on climate and related processes.
- Remote sensing of the atmosphere, land or water environment and high-performance computation that uses complicated statistics and climate models.
- Possibility to use robots within networks (robot swarms), which are coordinated with the help of the Global Command and Control System (GCCS).

Besides, to solve climate issues, the use of robots can be useful for processing certain types of resources (e.g. plastic); reduction of the volume of waste or optimisation of the food chain; service and maintenance of mechanisms or products that are not exploited due to defects; comprehensive monitoring of conditions of agricultural production; management of the capacities of production of alternative types of energy; optimisation of transport flows and logistics; monitoring of the state of the environment and climate parameters; increase in the quality of education, etc. The main advantages that are offered by robotics within the set tasks include endurance, strength, prevision, succession and speed in scientific research, sensor capabilities and computation capacities [2, 6].

On the whole, a robot is treated as a multi-component device that requires multidisciplinary knowledge, e.g. sensing, an action mechanism and intelligent control. The concepts of robotisation could be implemented in pure robotics systems and in hybrid systems [2]: robot-human, robot-AI, robot-big data, etc. According to the capabilities of robotics in the formation of digital entrepreneurship, its solutions are used for a wide range of social problems, including traffic safety, demographic processes, labour efficiency and energy efficiency and climate change. From the organisational position, such influence is ensured with the help of start-ups, traditional entrepreneurial initiatives or public-private partnerships, which also includes the interaction of scientific and government establishments. The most well-known projects and companies that use robotics to deal with climate issues are given in Table 1.

As we see, robotised systems that are connected with climate change are used to solve different types of tasks, which are directly or indirectly linked to the monitoring of the state of the environment, adaptation to climate change or reduction of its negative consequences. The projects that deal with monitoring include ARGO, The

Table 1 Characteristics of the most well-known robotised systems connected with climate change

Project/ company	Essence and potential from the position of climate-responsible entrepreneurship
ARGO	Global system consisting of 3000 profiling floats that measure temperature and salinity of upper 2000 m of the ocean to measure temperature, electric conductivity and pressure.
Climate Robotics	The project aimed at sequestering CO_2 emissions deep underground for a long period.
Drought resilient crops	The project uses 3D technology to study agricultural plants and provide a better understanding of the reasons due to which certain plants can grow during a drought. The project aims to obtain data necessary for the development of new plants that are resilient to droughts.
GrowBot	A robot that can plant trees ten times quicker than a specially trained person.
The Plantoid	The project uses robots to find polluting substances in the air and water. Imitating the behaviour and functioning of plants, they integrate "plants-robots" with sensors that can find polluting substances and understand their influence on the environment.
ClimateRobo	Autonomous intelligent mobile robots for climate purposes, which notifies people about weather conditions based on the data on the environment.
Aerobotics	The project deals with analytics of aerial images, using machine learning for multi-spectral images, to ensure understanding and competitive advantage for farmers.
Apollo Agriculture	Develops robots for the effective production of biochar, to absorb CO2 and improve soil, starting from urban land.

Source: Formed based on [3, 4, 6, 9]

Plantoid and ClimateRobo. The projects that aim at adaptation to climate conditions include Drought resilient crops and Aerobotics. The projects aimed at the reduction of negative consequences of climate change are the following: Climate Robotics, GrowBot and Apollo Agriculture.

Most of the described projects and companies were created and work on the territory of developed countries. Given this, an important task of the development of climate-responsible entrepreneurship in the digital economy markets is ensuring the quick transfer of new robotised technologies for their effective use in developing countries to reduce the critical risks of the damage dealt by climate change to the economy and population of developing countries.

4 Discussion

The results of the research contain the generalisation of issues connected with the influence of climate change on socioeconomic processes in the world with a focus on developing countries. A large share of the problems that are peculiar to this group of counties can be solved or reduced due to the use of the potential of robotics. A more detailed assessment of the impact of robotics on the management of risks caused by climate change requires additional studies that are based on the monitoring of the use of relevant robots, formation of a system of indicators that would allow determining their influence on climate and climate-related processes and development of models that can forecast and model the directions for the most effective use of robots for the designated range of problems.

Apart from this, additional attention should be paid to the aspects of the research connected with the transfer of technologies of robotics, since most of them were developed in developed countries, and the current system of transferring the leading solutions to developing countries is often accompanied by a time lag, limitations in the application, etc. Examples of the dissemination of digital technologies to countries with a lower level of development show that the speed of obtaining access to technologies and their implementation often is of key importance for the resolution of current problems, which influence is not limited by the borders of individual countries.

The currently available tools that can ensure the most optimal access to the mentioned technologies include the concept of open innovations, which can be implemented within the UN Framework Convention on Climate Change, the Climate Technology Centre and Network or other mechanisms based on an open network, open demand for technical support, open transfer of technologies and financing and open knowledge for the public.

5 Conclusion

Even though the influence of developing countries on the global climate crisis is insignificant, many of them feel the consequences of climate change to a larger extent than the rest of the world. Given the lower level of financial and institutional ability of this group of countries in the resolution of problems caused by climate change, an important task for global institutions and private entrepreneurial structures is the search for mechanisms and tools that can reduce the negative consequences or risks caused by such circumstances. Within the implementation of climate-responsible entrepreneurship and the digital economy, robotics has a large potential in solving the given problems. The existing project and companies that use robotics to deal with climate issues focus primarily on the monitoring of the environment, adaptation of economic systems to climate change or reduction of negative consequences. An increase in the influence of robotics on processes that are connected with climate risks in developing countries implies the ways for the quick transfer of relevant technologies and their use in view of the specifics of the technological and institutional environment.

References

1. Alves Filho SE, Filgueira Burlamaqui AM, Aroca RV, Gonçalves LMG, de Lima Sa ST (2018) Green Robotics: concepts, challenges, and strategiesю. IEEE Lat Am Trans 16(4):1042–1050. https://doi.org/10.1109/TLA.2018.8362135
2. Bugmann G, Siegel M, Burcin R (2011) A role for robotics in sustainable development? 1–4. https://doi.org/10.1109/AFRCON.2011.6072154. https://www.researchgate.net/publication/261247263_A_role_for_robotics_in_sustainable_development. Accessed 25 Jan 2023
3. Climate Robotics (2022) Climate Robotics: our technologies. https://www.climaterobotics.com/our-technologies. Accessed 25 Jan 2023
4. Duckett T, Pearson S, Blackmore S, Grieve B (2018) Agricultural Robotics: the future of robotic agriculture. UK-RAS White papers. https://arxiv.org/ftp/arxiv/papers/1806/1806.06762.pdf. Accessed 25 Jan 2023
5. Georgieva K, Gaspar V, Pazarbasioglu C (2022) Poor and vulnerable countries need support to adapt to climate change. International Monetary Fund.: https://www.imf.org/en/Blogs/Articles/2022/03/23/blog032322-poor-and-vulnerable-countris-need-support-to-adapt-to-climate-change. Accessed 25 Jan 2023
6. Grémillet D, Puech W, Garcon V, Boulinier T, Le Maho Y (2012) Robots in Ecology: welcome to the machine. Open J Ecol 2:49–57. https://doi.org/10.4236/oje.2012.22006
7. Michalec O, O'Donovan C, Sobhani M (2021) What is robotics made of? The interdisciplinary politics of robotics research. Humanit Soc Sci Commun 8:65. https://doi.org/10.1057/s41599-021-00737-6
8. Pearson S, Camacho-Villa T, Valluru R, Gaju O, Rai M, Gould I, Brewer S, Sklar E (2022) Robotics and autonomous systems for net zero agriculture. Curr Robotics Rep 3:1–8. https://doi.org/10.1007/s43154-022-00077-6. https://www.researchgate.net/publication/360253928_Robotics_and_Autonomous_Systems_for_Net_Zero_Agriculture. Accessed 25 Jan 2023
9. Pell S (2021) Are Robot farmers the perfect "comrade" in the war against climate change? Robotical. URL: https://robotical.io/blog/are-robot-farmers-the-perfect-comrade-in-the-war-against-climate-change/?currency=USD. Accessed 25 Jan 2023
10. Rutenberg I, Gwagwa A, Omino M (2021) Use and impact of artificial intelligence on climate change adaptation in Africa. In: Oguge N, Ayal D, Adeleke L, da Silva I (eds) African handbook of climate change adaptation. Springer, Cham. https://doi.org/10.1007/978-3-030-45106-6_80
11. Samadi M (2019) Implementation of an autonomous intelligent mobile robot for climate purposes. Int J Ad Hoc and Ubiquitous Comput 31:200–218. https://doi.org/10.1504/IJAHUC.2019.10022345
12. UNCTAD (2021) Frontier technology adoption in developing countries A measurement framework and proposed questionnaire. UNCTAD/DTL/STICT/2021/5 United Nations Conference on Trade and Development. https://unctad.org/system/files/official-document/dtlstict2021d5_en.pdf. Accessed 25 Jan 2023
13. UNCTAD (2022a) The least developed countries report 2022. The low-carbon transition and its daunting implications for structural transformation. United Nations Conference on Trade and Development. URL: https://unctad.org/webflyer/least-developed-countries-report-2022. Accessed 25 Jan 2023
14. UNCTAD (2022b) Technology assessment in developing countries – a proposed methodology. UNCTAD/TCS/DTL/INF/2022/4. United Nations Conference on Trade and Development. https://unctad.org/system/files/official-document/tcsdtlinf2022d4_en.pdf. Accessed 25 Jan 2023
15. World Bank Group (2020) 3 things you need to know about ambitious climate action in developing countries. World Bank, Washington, DC. World Bank. https://www.worldbank.org/en/topic/climatechange/brief/3-things-you-need-to-know-about-ambitious-climate-action-in-developing-countries. Accessed 25 Jan 2023
16. World Bank Group (2021) World Bank Group climate change action plan 2021–2025: supporting green, resilient, and inclusive development. World Bank, Washington, DC. World Bank. https://openknowledge.worldbank.org/handle/10986/35799. Accessed 25 Jan 2023

The Concept of Smart Risk Management of Climate-Responsible Entrepreneurship in Digital Economy Markets with Reliance on AI

Altynai D. Duishenalieva, Jypar S. Orozmamatova, Aleksandr A. Solovev, Galina A. Markeeva, and Artem V. Lukomets

1 Introduction

Entrepreneur's activity is always connected with risk, which is based on uncertainty. Any decision of entrepreneur is connected with uncertainty to a certain extent. There are very few events in the socioeconomic environment which course can be predicted with a 100% certainty. Therefore, all spheres of human activity are characterised by higher (human behaviour, interaction with natural environment) or lower (use of industrial equipment, material resources management) level of uncertainty.

In its turn, uncertainty is connected with a risk, which, generally, can be characterised as a possibility of an uncertain situation – damage or losses to a subject of entrepreneurial activity. To assess the scale of losses, different indicators are used: financial, quantitative, managerial, etc. To assess the level of risk, caused by climate change, all the above indicators can be used depending on the character of events; they can cover financial indicators (destructions or losses that are dealt due to non-standard climate events), human losses (casualties of natural cataclysms), biological

A. D. Duishenalieva (✉)
Kyrgyz State Technical University named after I. Razzakov, Bishkek, Kyrgyzstan

J. S. Orozmamatova
Osh Technological University named after M.M. Adyshev, Osh, Kyrgyzstan

A. A. Solovev
Federal Accreditation Service, Moscow, Russia

G. A. Markeeva
MGIMO University, Moscow, Russia

A. V. Lukomets
Federal scientific center "V.S. Pustovoit All-Russian Research Institute of Oil Crops", Krasnodar, Russia

E. G. Popkova (ed.), *Smart Green Innovations in Industry 4.0*, Springer Climate,
https://doi.org/10.1007/978-3-031-45830-9_16

losses (disappearance of plant species and animals, change of ecological environment), etc.

Thus, risks connected with climate change are objective factors of entrepreneurial activity, which reached a high level of institutionalisation through their recognition and creation of global programmes and initiatives that are aimed at their removal or neutralisation of negative consequences. The main initiatives that define the programme goals of tackling climate change are determined by the UN Sustainable Development Goals [14] and the Paris Agreement [16]. According to the Paris Agreement, global goals of fighting climate change until 2050 are aimed at limiting the level of global temperature rise by 1.5–2.0 °C compared to the pre-industrial age temperature. Such initiatives form additional demands and, accordingly, risks for entrepreneurs, which are connected with the search for the possibilities to achieve entrepreneurial goals with the simultaneous compliance with the Sustainable Development Goals, achievement of decarbonisation indicators and compliance with sustainability principles. Many of such tools lie in the sphere of the digital economy.

2 Materials and Method

The methodology of this article is connected with the provisions of risk management, digital economy, social responsibility, entrepreneurship and institutionalism. Therefore, the key parameters of risk management are expressed through its concept of smart management; the capabilities of the digital economy are considered from the position of AI; the criteria of social responsibility are formed within the concepts of Corporate Climate Responsibility (CCR) and Responsible Business Conduct (RBC); entrepreneurship is shown from the position of the combination of economic and socially-responsible goals; institutionalism is demonstrated though determination of general global values, expressed in the form of the Sustainable Development Goals and the Paris Agreement. The methods utilised are based on the systems approach and include observation, analysis, synthesis, generalisation, etc.

The scientific problem, which is studied in this article, was considered in the world scientific and expert literature in a fragmentary manner, in the following aspects only: substantiation of the provisions of smart risk management [4, 7, 11, 12]; identification of the tasks of climate responsibility and determination of the level of their execution [2, 10]; description of the interconnections between digital technologies and climate change [6]; determination of the essence and components of artificial intelligence [13]; study of the influence of artificial intelligence on climate change [1, 5, 8]; substantiation of the opportunities and problems in the use of artificial intelligence to tackle climate change [3, 9].

Given the gaps in the existing literature on the research topic, the main goal of this study is to elaborate on the possibilities of the use of AI tools in the activity of entrepreneurial structures in digital markets in the context of the smart risk management context.

3 Results

As of 2022, the schedule of decarbonisation of the world economy was in jeopardy. According to the UN, the world is by 1.2 °C warmer compared to the pre-industrial age [15]. Responsibilities accepted by national governments are not observed in full, so the excess of the level of climate warming by 2 °C is very realistic in the mid-term. Thus, when assessing risks, entrepreneurs should expect more intensive droughts, disastrous floods, forest fires and hurricanes.

The expected negative consequences of climate change require from entrepreneurs around the world the use of rational tools of risk management. Such tools should not be based on passive evaluation of possible alternatives and their probability – they should be used in advance and involve all possible measures to avoid not only risks but also the formation of their preconditions. Such an approach to risk management is implemented within smart risk management.

The concept of smart risk is derived from clear statements that risks are an objective phenomenon that is inherent to entrepreneurial structures. An entrepreneur cannot entirely avoid risks, but he can determine the list and volume of risks he can accept and create conditions for protecting his business from these risks [11]. Climate change in this context belongs to a particular category of risks. It has large-scale (very often, global) character, and, from the position of an individual entrepreneur, cannot be localised or removed. Thus, most of the tools aimed at the management of climate risks involve the compensation of losses and damage or partial alleviation of consequences. Management that is aimed at risk prevention requires the coordinated activity of many entrepreneurs, which are united by common goals and principles.

The basis for the coordination of the activity of entrepreneurs for tackling climate change is the concept of Corporate Climate Responsibility (CCR) and Responsible Business Conduct (RBC). The impact of the first concept is focused on "tracking and informing of emissions, determination of the targeted indicators of waste reduction, reduction of own emissions..." [2], while the impact of the second concept involves the fight against environmental threats and achievement of global environmental goals, including supply chains and business relations [10].

As for the achievement of tasks that cover complex interactions and processing of a large array of information, the digital economy market has advantages that allow ensuring better communication between market subjects, as well as collection and processing of information about the environment. In this context, AI technologies – as the components of the digital economy – have a large potential, which allows tracking relevant signals connected with climate change under the conditions of information noise, recognising regularities between human activity and climate phenomena and creating forecasts and models of climate change and its consequences [1].

Under the conditions of quick expansion of AI in different spheres of human activity, its main task is supplementing or replacing the elements of human thinking with machine tools or algorithms. There is no unified definition of AI in the scientific and expert literature. Most often, it is understood in the following way:

- Sphere of informatics that develops intellectual computer systems, which have abilities inherent to human mind: understanding language, learning, thinking, etc. [13].
- Algorithm that allows a computer to solve complex tasks with the help of the methods of machine learning or automatic removal of templates from data [5].
- Machine system, which can make forecasts, recommendations or solutions for the set of goals designated by a human [9].

We think that the first definition is the most complete, since the notion of "intellectual computer systems" includes such categories as algorithm and machine system, and the focus on the capabilities of human mind allows for the most precise characterisation of the definition of intellect for artificial computer systems.

The use of AI for solving the problems of sustainable development has advantages that are inherent to digital technologies; their essence lies in the expansion of powerful tools of receipt, analysis and manipulation of information, as well as the search for templates that allows modelling and forecasting scenarios in different contexts. According to this, scholars agree that AI may bring "substantial" or larger benefit to studies of sustainable development, pave the way for further efforts in this sphere and facilitate the creation of more sustainable and fair world due to the achievement of the Sustainable Development Goals [6].

Thus, the tasks of smart risk management include the study of the situation, forecasting of its further course, improvement of operational effectiveness, forecasting of the state of technical servicing of business and the possibility to use scientific experiments and approximated simulations with coordinated with AI tools and allow achieving significant results in the context of the influence on the environment, including climate change (Table 1).

As we see, the tools of AI are coordinated with the tasks of smart risk management and ensure an important effect on the processes connected with climate change. The results obtained allow for a more precise evaluation of the dynamics of natural and economic processes, determining the reserves of their optimisation, increasing the effectiveness of their use and accelerating the innovative development of economic systems based on a better understanding of natural and economic processes, imitation, modelling or approximation of their dynamics based on the multiple choice opportunities.

4 Discussion

Study of the issues of using AI to tackle climate problems in the context of smart risk management is an important scientific direction, which covers such topics as the possibility of interpretation (explanation why forecasts are correct), quantitative determination of uncertainty and learning (understanding when the available data are not enough for decision making). However, despite the quick development, AI is not an independent technology and fully depends on public decisions. It is society

Table 1 Coordination of tasks and technologies of AI from the position of their influence on processes connected with climate change

Tasks	AI technologies	Influence on processes connected with climate change
Gathering relevant information in the environment of raw data	Analysis of images; recognition of visual forms; understanding and analysis of text in natural language; collection and/or generation of information	Receipt of information on the state of the environment by analysing large arrays of raw data (geo-spatial image, text documents, data from sensors), which allows tracking sources of CO2 emissions, collecting information on the characteristics of building's effectiveness and tracking defosteration
Forecasting	Intellectual systems of information security; business analytics	Development of forecasts on wind energy production, transport demand and emergencies based on historical data models, which allows for the optimisation of the energy system, planning of infrastructure and fight against natural disasters
Increasing operational effectiveness	Robotics, business analytics	Increasing the effectiveness of the interaction of elements and systems of the real world through improvement and optimisation of their coordination, which forms the preconditions for industrial heating and cooling systems management, consolidation of cargo transportation and reduction of waste in food industry
Technical services based on forecasts	Robotics; business analytics	Improvement of infrastructure security, reduction of costs and increase in energy efficiency of systems
Accelerating scientific experiments	Collection and / or generation of information; robotics, expert systems	Accelerating the process of scientific inventions, e.g., based on past experiments – to offer future experiment, which might be more successful, e.g., development of clean technologies such as batteries or next-generation solar elements
Approximation of simulations	Collection and/or generation of information; robotics, expert systems	Accelerating simulations that require calculations and are used for the modelling of climate physics or I engineering systems. AI could help coordinate the parts of climate models and models of energy system optimisation and improve the use of urban planning tools to facilitate real-time decisions

Source: Created by the authors based on [3, 5, 13]

that defines tasks to be solved by AI and sets limitations and principles of its use, as well as assesses the expedience and effectiveness of its use in resolution of the designated tasks.

There are a lot of debatable issues in this sphere. They involve the determination of top-priority goals for the use of AI, development of criteria and principles of its use, as well as the system of assessment of effectiveness of its application. The key provisions that are discussed nowadays are the issue of ethical character of the use of AI and determination of the limit of its independence in the interaction with human, given the ability to learn and develop; the issue of energy efficiency, in view of large energy consumption in operations of information processing, modelling, etc.; possibility to use AI for solving strategic tasks in which mankind has not yet reached significant results, including climate change.

Despite the large number of unsolved or partially solved tasks, the use of AI technologies to prevent climate change or reduce consequences within the concept of smart risk management is treated positively by almost all scholars. In this context, in the case of correct determination of principles and criteria of the use of AI, the damage it might deal is lower than benefits from its potential use.

5 Conclusion

From the position of the concept of smart risk management, climate-responsible entrepreneurship, which functions in digital economy markets, has a large potential to solve climate-related tasks. Its advantages are based on technological features and include the possibility for better identification, collection and processing of information due to the means of filtering of information noise and the use of templates. They involve the creation of more precise forecasts and their interpretation, based on constant analysis of previous results and elimination of previously discovered inaccuracies. An improvement of operational effectiveness and technical maintenance of economic systems due to the use of AI tools facilitate the better coordination of economic & financial and production processes, as well as tracking of technical features of the use of infrastructure and equipment. The maximum progress of scientific thought in this case is achieved due to acceleration and increase in scientific results and modelling of processes that are connected with climate change. The described capabilities conform to the criteria of smart risk management and allow not only for better adaptation to potential consequences of climate change but also for the improvement of the capabilities of global and local economic systems to prevent such changes due to the achievement of a higher level of decarbonisation, increase in the energy efficiency level and eco-friendliness of technologies.

References

1. Barnes EA, Hurrell JW, Ebert-Uphoff I, Anderson C, Anderson D (2019) Viewing forced climate patterns through an AI Lens. Geophys Res Lett 46:13389–13398. https://doi.org/10.1029/2019GL084944
2. CCRM (2022) Corporate Climate Responsibility Monitor. https://newclimate.org/resources/publications/corporate-climate-responsibility-monitor-2022. Accessed 25 Jan 2023
3. Coeckelbergh M (2021) AI for climate: freedom, justice, and other ethical and political challenges. AI Ethics 1:67–72. https://doi.org/10.1007/s43681-020-00007-2

4. Holmes A (2017) Smart risk management: a guide to identifying and calibrating business risks. Wiley, 288 P
5. Kaack LH, Donti PL, Strubell E, Rolnick D (2020) Artificial intelligence and climate change: opportunities, considerations, and policy levers to align AI with climate change goals. https://eu.boell.org/en/2020/12/03/artificial-intelligence-and-climate-change. Accessed 25 Jan 2023
6. Leal Filho W, Yang P, Eustachio JHPP, Azul AM, Gellers JC, Dinis MAP, Kozlova V (2022) Deploying digitalisation and artificial intelligence in sustainable development research. Environ Dev Sustain. https://doi.org/10.1007/s10668-022-02252-3
7. Lin J, Jia S, Deng J (2017) Smart risk management with financial big data, IEEE/SICE International Symposium on System Integration (SII), Taipei, Taiwa, 60–65. https://doi.org/10.1109/SII.2017.8279189
8. Maher H, Meinecke H, Gromier D, Garcia-Novelli M, Fortmann R (2022) AI is essential for solving the climate crisis. https://www.bcg.com/publications/2022/how-ai-can-help-climate-change. Accessed 25 Jan 2023
9. OECD (2022) Artificial intelligence & responsible business conduct. https://mneguidelines.oecd.org/RBC-and-artificial-intelligence.pdf
10. OECD (2023) Responsible business conduct. OECD Guidelines for Multinational Enterprises. https://mneguidelines.oecd.org/environment/. Accessed 25 Jan 2023
11. Rael R (2008) Smart risk management: a guide to identifying and reducing everyday business risks. American Institute of Certified Public Accountants. Inc, New York, 163 P
12. Rael R (2017) Smart risk management: a guide to identifying and calibrating business risks. Wiley
13. TADVISER (2023) Iskusstvennyj intellekt (II) [Artificial intelligence (AI)]. TADVISER: Gosudarstvo. Biznes. Tehnologii [TADVISER: State. Business. Technologies]. https://www.tadviser.ru/index.php/Продукт:Искусственный_интеллект_(ИИ,_Artificial_intelligence,_AI). Accessed 25 Jan 2023
14. UN (2022) THE 17 GOALS. UN. Department of Economic and Social AffairsSustainable Development. https://sdgs.un.org/goals. Accessed 25 Jan 2023
15. UNEP (2022) 10 ways you can help fight the climate crisis. United Nations Environment Programme. https://www.unep.org/ru/novosti-i-istorii/istoriya/10-sposobov-borby-s-klimaticheskim-krizisom. Accessed 25 Jan 2023
16. UNFCCC (2015) The Paris agreement. What is the Paris agreement? https://unfccc.int/process-and-meetings/the-paris-agreement/the-paris-agreement. Accessed 25 Jan 2023

Improvement of Green Entrepreneurship Planning in Digital Economy Markets with the Help of Big Data to Increase Climate Resilience

Ilham M. Saipidinov, Aida T. Ajibekova, Farida T. Artykbaeva, and Victoria N. Ostrovskaya

1 Introduction

Entrepreneurship that deals with green projects can be considered innovative. It also adopts responsibility for the future of society, connected with tackling climate change. Not all initiatives of green entrepreneurship ensure an increase in climate resilience. The final result depends on the potential of companies and their desire to protect the climate. Small and medium enterprises, the share of which is more than 90% in the entire private sector of the economy in certain countries (UK), declare the focus on sustainable development only formally. They need it to build a status of a reliable business partner in the national and international markets [2]. However, SMEs may realise very important projects, which allow for the measures to fight climate change. An important place in green entrepreneurship projects belongs to the digital economy tools since they can allow assessing the forecast results of economic activities, including the achievement of certain parameters of climate resilience. A special interest here is posed by Big Data, which is used in the analysis and further planning of the indicators of activities.

This research is aimed at determining the possible directions of improving the planning of green entrepreneurship in digital economy markets with the help of Big Data, which also involves an increase in the level of climate resilience. For this, the following tasks are solved: determining the dynamics of the level of environmental efficiency of the leading countries in this sphere, which describes the results of climate policy implementation at the national level (including with the participation of entrepreneurship), determining the reasons for transformations; assessing the

I. M. Saipidinov · A. T. Ajibekova · F. T. Artykbaeva
Osh State University, Osh, Kyrgyzstan

V. N. Ostrovskaya (✉)
Center of Marketing Initiatives, Stavropol, Russia

E. G. Popkova (ed.), *Smart Green Innovations in Industry 4.0*, Springer Climate,
https://doi.org/10.1007/978-3-031-45830-9_17

directions of using Big Data to improve the planning of green entrepreneurship, which is aimed at the increase in climate resilience level.

2 Materials and Method

There are scientific works in which the directions of environmental activity of the entrepreneurial sector of certain countries, including with the use of digital markets' capabilities, are studied. These works include [2, 6–8].

In [2], the authors dwell on the results of the evaluation of British small and medium enterprises' participation in the projects of sustainable social and environmental development and show the reasons for their focus on green growth. The authors of [6] assess the role of digitalisation, including Big Data, in the achievement of the goals of environmental sustainability of territories. Nevavuori et al. [8] considered the issues of using digital tools in the agrarian sector, and studied the main features of the focus on digital economy markets to raise crop yield. Müller and Jensen [7] presented an analysis of the use of Big Data in companies of various sectors of the Danish industry and proves the impact of the focus on this tool in the context of an increase in value-added and green growth.

The above works, like many other studies on this topic, dwell on the problems at the level of individual countries, which predetermines the necessity to perform comprehensive research.

In this work, the authors use the methods aimed at determining and analysing the parametric data on green growth, an indicator of implementation of Big Data by the entrepreneurship of countries (including green entrepreneurship that is aimed at an increase in climate resilience). The methods used include the statistical method (use of official statistical data on the studied indicators), the index method (assessment of a range of indicators with the help of indices) and the comparative method (comparison of the indicators' values in dynamics). The method of complex analysis was used to determine the directions of the application of Big Data in the improvement of the planning of green entrepreneurship in the considered countries.

3 Results

Determination of a list of countries that demonstrate a high level of the Environment Performance Index (Table 1) will allow assessing the state of the national involvement in the resolution of the problem of climate resilience (also within the activities of green entrepreneurship).

High results in 2022 are observed with Austria, Denmark, Finland, Iceland, Luxembourg, Malta, Slovenia, Sweden, Switzerland, and the United Kingdom. All these countries – with a high level of environmental efficiency – belong to the category of developed countries and are located in Europe.

Table 1 Dynamics of the Environment Performance Index for 2018, 2020 and 2022

	Environment Performance Index			Change	
Country	2018	2020	2022	2018–2020	2020–2022
Austria	8	6	8	−2	+2
Denmark	3	1	1	−2	0
Finland	10	7	3	−3	−4
France	2	5	–	+3	–
Germany	–	10	–	+10	–
Iceland	–	–	10	–	–
Ireland	9	–	–	–	–
Luxembourg	7	2	6	−5	+4
Malta	4	–	4	–	–
Norway	–	9	–	–	–
Slovenia	–	–	7	–	–
Sweden	5	8	5	+3	−3
Switzerland	1	3	9	+2	+6
United Kingdom	6	4	2	−2	−2

Source: Prepared by the authors on the basis of [1]

Table 2 The level of application of the use of Big Data and analytics by the business environment of selected countries

	Use of Big Data and analytics index			Change	
Country	2018	2020	2022	2018–2020	2020–2022
Austria	22	36	44	+14	+8
Denmark	7	12	6	+5	−6
Finland	30	15	15	−15	0
France	42	47	43	+5	−4
Germany	41	46	52	+5	+6
Iceland	25	19	17	−6	−2
Ireland	28	18	18	−10	0
Luxembourg	16	38	46	+22	+8
Malta	No data	No data	No data	–	–
Norway	14	6	7	−8	+1
Slovenia	33	28	35	+5	+7
Sweden	9	7	14	−2	+7
Switzerland	24	25	25	+1	0
United Kingdom	11	23	19	+12	−4

Source: Prepared by the authors on the basis of [3–5]

The possible influence of the use of the tool of the digital economy – use of Big Data and analytics – to improve the planning of green entrepreneurship and raise climate resilience is shown in Table 2. The importance of this indicator is that it shows coverage in the market of digitalisation in the context of entrepreneurship.

Further on, an analysis of the main directions of the use of Big Data and analytics by the business environment in the selected countries in the context of improvement of the planning of green entrepreneurship and increase in climate resilience is presented.

Agricultural areas of Sweden accounted for 7.4% of the total area of the country in 2018 and 2022 [12]. Since 1961 (10.4%), this indicator was gradually reducing. Sweden is dependent on the import of the agrarian sector products from the USA and countries of Europe. The COVID-19 pandemic led to the temporary closure of borders, restrictions on movement and requirements of social distancing, which, in its turn, led to problems with the use of the workforce in importing countries. This predetermined the necessity to ensure food security. At present, agricultural companies in Sweden are focused on the prevention of climate change, namely CO2 emissions, production waste that pollutes water and the atmosphere and an increase in the use of natural resources. Agrarian companies in Sweden that work in the sphere of grain and barley production use technologies of digital neural networks combined with robotisation (drones) to perform the monitoring of crop yield, climate conditions and level of consumption of resources (water) [6]. These digital neural networks perform distant sensing of the conditions of agricultural production [8]. These systems of monitoring function based on the integration of materials from the system that is based on Big Data analysis (information on the state of certain agricultural plants, norms of irrigation and protection of plants); robotised equipment that transfers data to the common analytical centre, in which, with the help of previously conducted machine learning, a decision on the correctness of plant care is made, or a conclusion (recommendation) is formed on the necessity to change the processes. It is peculiar for the maximum coverage of information on the factual state of the processes of agrarian production management, which minimises errors. Thus, remote sensing and analytics of Big Data received from the studied objects and the data on standards in this sphere ensure the prevention of the excessive use of water resources and the burden on soil and water from the use of fertilisers and pesticides.

Denmark holds leading positions in the sphere of improvement of climate resilience, which is largely facilitated by green entrepreneurship. An increase in the level of companies' involvement in the achievement of the environmental SDGs is ensured with the help of Big Data. This tool of the digital economy is used by small, medium and small companies. In this case, systems for analysis, planning and monitoring of processes are used. In 2022, Denmark was ranked second in the world by the level of achievement of the UN SDGs [10] and first in the world by environmental performance (Table 1). The economic activity of Danish entrepreneurial subjects is conducted according to the main SDGs, which stimulates green growth. The systems of analysis, planning and monitoring of processes, which are based on the use of Big Data and analytics, are built in the main management systems (software products), which are adapted to each sector of the economy [7]. Development and maintenance of these management systems are performed by IT companies, which also support the digitalisation of business in other countries of Northern Europe. On the whole, Denmark's approach to the planning of green entrepreneurship through the use of Big Data and analytics is similar to the Swedish one [9].

The United Kingdom demonstrates high results in the sphere of improvement of green entrepreneurship based on the use of Big Data and analytics. Sectors of the economy which demonstrate the most sustainable green entrepreneurship in the UK include the sphere of financial technologies, healthcare, production based on circular technologies and renewable energy.

Attention should be also paid to companies in the sector of financial technologies in the UK, which attract consumers who support green initiatives to the implementation of the UN SDGs in the sphere of increase of climate resilience. An example is the activity of the company Tred, which was founded in 2019, and achieved by the end of 2021 the growth of revenues by 122%. Its financial product (green debit card) gained large popularity in the market. This product allows customers to assess their carbon footprint (based on analysis of the data on transactions), receive recommendations on the reduction of the level of negative impact on climate, receiving recommendations on investments in the green economy, managed by Tred [11]. Analysis of this company's activity showed that implementation of systems based on the use of Big Data and analytics allows for the planning of the following: a segment of consumers to be covered by the company's offer (green debit card); contents of commercial offers that are presented within the product; effectiveness of investing in certain green projects with the help of modelling of the level of incomes; own profitability.

The market of green financial technologies in the UK also has the following companies, which are worth mentioning: Treecard (product: sustainable, wooden debit card), which is aimed at sustainable development of forests; Neutreeno (product: software for retailers that work in the fashion sector), which is aimed at involving consumers' to compensate for the negative impact on the environment from the use of a certain product [11]. Based on the approaches to managing the policy of product promotion, it is possible to assume that these companies use the scheme of application of systems based on the use of Big Data and analytics, which is similar to the one used by Tred.

The mentioned British financial companies are focused on such a direction of the use of Big Data and analytics as modelling of internal (for consumers involved with climate initiatives) and external sustainability (for their own goals of sustainable development).

4 Discussion

Analysis of the effectiveness of the improvement of planning of green entrepreneurship with the help of Big Data for the purpose of climate resilience allowed distinguishing the following directions in this sphere.

The direction of remote sensing and analysis of Big Data on the processes of production management is especially important for use in the short term. Application of this tool is important for the control and further planning of energy efficiency, which reduces energy consumption and CO2 emissions (in the case of non-renewable energy sources), as well as excessive spending of natural resources (water). For long-term planning, they can be used for strategies with forecast estimated indicators of ecologisation.

It should be noted that such direction as modelling of internal and external sustainability in the sphere of tackling climate change and supporting economic growth is peculiar to companies that work in the financial sector. As a rule, the use of Big Data and analytics allows for the systematisation and analysis of information and presentation of a certain decision within imitation models created based on the use of machine learning function. The choice of these models is aimed at the creation of variants of the behaviour of consumers and company under the conditions of uncertainty and risks to the environment as well as threats to economic stability. This direction of using the digital tool may lead to the effect of an increase in climate resilience and economic transformations, which is a sign of its market attractiveness at the current stage.

5 Conclusion

Based on the analysis of the results obtained, it is possible to conclude that the tool of digital economy markets – use of Big Data and analytics – has great prospects under the conditions of the development of various spheres of green entrepreneurship. A high level of this tool implementation is realised mainly in developed countries, which is due to their financial and technological potential and business's readiness for the introduction of innovations and growth of its needs in the aspect of improving the planning of green projects that are aimed at sustainable climate growth. To use such a tool at the level of developing countries or mass implementation at the level of SMEs, it is necessary to accelerate public-private partnerships, which would ensure support in the creation of software and apps that would use Big Data for the planning of indicators of activities, including in the sphere of climate growth.

The discovered directions of the implementation of Big Data may facilitate the use of other digital tools, which ensure the achievement of climate goals, e.g. robotisation (drones, automatic lines, chat bots, etc).

References

1. BYJUS (2023) Environment Performance Index (EPI) 2022. https://byjus.com/free-ias-prep/environment-perfomance-index-epi/. Accessed 22 Jan 2023
2. Crossley RM, Elmagrhi MH, Ntim CG (2021) Sustainability and legitimacy theory: The case of sustainable social and environmental practices of small and medium-sized enterprises. Bus Strateg Environ 30(8):3740–3762. https://doi.org/10.1002/bse.2837

3. IMD (2018) IMD world digital competitiveness ranking 2018. https://www.imd.org/globalassets/wcc/docs/imd_world_digital_competitiveness_ranking_2018.pdf. Accessed 22 Jan 2023
4. IMD (2020) IMD world digital competitiveness ranking 2020. https://www.imd.org/globalassets/wcc/docs/release-2020/digital/digital_2020.pdf Accessed 22 Jan 2023
5. IMD (2022) IMD world digital competitiveness ranking 2022. https://www.imd.org/centers/world-competitiveness-center/rankings/world-digital-competitiveness/. Accessed 22 Jan 2023
6. Mondejar ME, Avtar R, Diaz HLB, Dubey RK, Esteban J, Gómez-Morales A, Hallam B, Mbungu NT, Okolo CC, Prasad KA, She Q, Garcia-Segura S (2021) Digitalization to achieve sustainable development goals: steps towards a Smart Green Planet. Sci Total Environ 794:148539. https://www.sciencedirect.com/science/article/pii/S0048969721036111#bb0570 Accessed 22 Jan 2023
7. Müller SD, Jensen P (2017) Big data in the Danish industry: application and value creation. Bus Process Manag J 23(3):645–670. https://doi.org/10.1108/BPMJ-01-2016-0017
8. Nevavuori P, Narra N, Lipping T (2019) Crop yield prediction with deep convolutional neural networks. Comput Electron Agric 163(104859). https://doi.org/10.1016/j.compag.2019.104859. https://www.sciencedirect.com/science/article/abs/pii/S0168169919306842. Accessed 22 Jan 2023
9. OECD (2022) Denmark. Key facts on SME financing. https://www.oecd-ilibrary.org/sites/866dae42-en/index.html?itemId=/content/component/866dae42-en. Accessed 22 Jan 2023
10. Sdgindex (2022) Sustainable development report 2022. From crisis to sustainable development, the SDGs as roadmap to 2030 and beyond. https://www.sdgindex.org/. Accessed 22 Jan 2023
11. Walde L (2022) 17 of the UK's top sustainable companies. https://www.beauhurst.com/blog/top-sustainable-companies/. Accessed 22 Jan 2023
12. Worldbank (2022) Agricultural land (% of land area) – Sweden. https://data.worldbank.org/indicator/AG.LND.AGRI.ZS?locations=SE. Accessed 22 Jan 2023

The System Approach to the Management of Climate Responsibility of Entrepreneurship in Digital Economy Markets Based on the Internet of Things

Olga A. Boris, Rashid O. Tazhiyev, Galina A. Markeeva, and Nikolay I. Litvinov

1 Introduction

Climate change is one of the key challenges for mankind in the twenty-first century. The intense development of technologies over the recent centuries substantially changed our civilisation, ensuring progress in all spheres of human life activities, including the improvement of living standards, system of transport and communications, level of healthcare, culture, etc. However, extensive nature use, which was the basis of this development, led to an increase in risks, a large part of which are connected with climate change and concern the destruction of sustainable ecosystems, frequent natural disasters, impoverishment of flora and fauna, etc. Fast heating of the planet creates risks for all living beings and is to be strictly controlled. This heating is connected with greenhouse gases, which are the products of the burning of carbon compounds in the processes of production and consumption of energy, agriculture, etc.

Static models [13] show that even in the case of stabilisation of the volume of greenhouse gas emissions at the current level, the process of increase of air temperature by 2 °C (determined by the Paris Agreement) [10] will not stop. That is why the problem of reduction of greenhouse gas emissions is very acute. In this context, the task of searching for effective tools that can help solve this problem is very important.

O. A. Boris (✉)
North-Caucasus Federal University, Stavropol, Russia

R. O. Tazhiyev
Al-Farabi Kazakh National University, Almaty, Kazakhstan

G. A. Markeeva
MGIMO University, Moscow, Russia

N. I. Litvinov
Volgograd State University, Volgograd, Russia

E. G. Popkova (ed.), *Smart Green Innovations in Industry 4.0*, Springer Climate,
https://doi.org/10.1007/978-3-031-45830-9_18

According to research by Ericsson, the digital sphere has a unique potential to ensure the transition to a low-carbon economy. Digital solutions will be able to ensure the reduction of CO_2 emissions by 15% by 2030 [1]. Here an important task is not only the implementation of low-carbon innovations but also the formation of an effective system of monitoring that would allow for clear tracking of volumes and sources of emissions, determination of the influence of CO_2 emissions reduction on the environment and measuring the level of decarbonisation in different spheres, countries or regions. At present, thanks to digitalisation, such capabilities are ensured by IoT (Internet of Things) technologies, which are an important component of entrepreneurship in digital economy markets.

2 Materials and Method

With the reliance on the multidisciplinary approach, we use the multidisciplinary view of the problem of climate change from the position of its resolution due to the use of the tools of entrepreneurship, the digital economy and technologies, as well as social and climate responsibility. The conceptual unity within the system approach to climate-responsible entrepreneurship in the digital economy with a special focus on the Internet of Things is ensured by mutual complementarity and description of the mentioned components. According to this, the methodological basis of the research consists of the system approach, the methods of analysis, synthesis and structuring, as well as the graphical method.

Given the interdisciplinary character of the research, we analysed provisions of scientific works. Large attention in the literature was paid to the issues of development, evolution and prospects for digital entrepreneurship [5, 6], digital responsibility from the position of climate change [2], influence of digital technologies on climate change [7], activities on tackling climate change [12], possibility to use digital technologies to fight climate change [1, 3, 4], the essence, structure and specifics of the IoT [9], practical aspects of the use of the IoT to solve applied tasks [11] and the influence of the IoT on environmental resilience [8] and climate change [13].

The goal of this research is to study the theoretical and practical aspects of managing climate-responsible entrepreneurship in digital economy markets in digital economy markets based on the Internet of Things. To reach this goal, we shall elaborate on the essence and features of climate-responsible entrepreneurship, the digital economy and the IoT from the position of their interaction with each other and with the elements of the environment.

3 Results

The system approach requires the study of provisions inherent to each research sphere separately and envisages the coordinated reflection of their interconnections in the context of the considered problem. Thus, the structure of this work is characterised by the gradual deepening of scientific research on the features and problems of entrepreneurship and digital entrepreneurship, climate change and climate responsibility, markets of the digital economy and digital technologies based on the IoT, as well as their comprehensive presentation within one system, which takes into account the specifics of each component.

Entrepreneurship in digital economy markets has its specific features. On the whole, it is not considered separately from its traditional treatment and is aimed at the "recognition and use of market opportunities through their transformation into goods and services, which is connected with risks and involves profit-making" [4]. From this point of view, the search for new opportunities is one of the key functions of entrepreneurship. When digital technologies offer more opportunities for the creation of goods and services, as well as an increase in their value, digitalisation of entrepreneurship becomes a common tendency, which determined the direction of development of different sectors of the economy.

The natural environment for entrepreneurship is inextricably connected with uncertainty. It is in uncertainty that entrepreneurs find ideas for business and realise them. Innovations, which are an inseparable part of entrepreneurship, are also realised in conditions of uncertainty. However, the implementation of digital innovations in the entrepreneurial community changed the nature of this uncertainty and attitude toward it. The main changes connected with the digitalisation of entrepreneurship are an expansion of limits from the position of spatial and time parameters, and a separation of the product's function from the form and contents of the carrier. Value of the product can develop even after its presentation in the market, introduction or purchase (upgrade of operation systems, download or access to new products, etc.). Besides, the digitalisation of entrepreneurship raise the level of uncertainty and non-linearity of processes, increased the capabilities of business in different sectors and strengthened their influence on markets.

From the position of the system approach, the above processes are internal and external factors that influence the digital market's development. The basis of the digital market is determined by three main elements of digital technologies: digital artefacts, platforms and infrastructure. Artefacts offer defined value or functionality for the consumer and have a form of digital component, media content or programme. Artefacts from the functionality of such components of the digital market as the Internet of Things. The digital platform is used as a joint foundation for placing digital products. Digital infrastructure is a factor of support for the digital economy. Its functioning takes place with the help of digital infrastructure, which includes the following tools and systems of digital technologies:

- Cloud computing
- Data analytics

- Online communities
- Social media
- 3D print, etc. [6]

A combination of the elements of the digital economy, which is based on entrepreneurship, creates conditions for a constant search for new opportunities and values in almost all spheres of human life. Values and effects constantly expand and are overlapped with previous ones, which allows developing the components of the digital economy in different directions. Thus, the use of the tools of analytics together with Big Data, obtained with the help of the digital platform, IoT network or the system Earth Observation (EO), allows for monitoring and processing of previously unprocessed information in the spheres of healthcare, transportation, marketing and interaction with the environment.

Due to the fact that climate change becomes more visible and based on the realisation of the critical impact of climate on mankind, fighting these changes becomes more relevant. The scale and nature of the problem of tackling climate change are unlocked at several levels, which cover the alleviation of consequences (reduction of emissions) or adaptation (preparation for inevitable consequences). The reduction of greenhouse gas emissions is directly connected with changes in the systems of energy supply, transport economy, construction, and industrial and agricultural production. Adaptation is largely oriented toward the planning of resilience and is largely oriented toward the planning of resilience and natural disaster management, as well as the development of abilities in understanding nature and the mechanism of natural disasters [7].

A combination of the ideas of entrepreneurship and tackling climate change forms the approach of climate-responsible activities, which develops according to the principles of socially responsible entrepreneurship or the concept of sustainable development. Under the conditions of the digital economy, such a vision is aimed at the use of the capabilities of digital technologies to alleviate the consequences of climate change or adapt to them. In such case, entrepreneurs not only limit greenhouse gas emissions or modernise supply chains – they also create new values, which might have a limited or decisive influence on climate change management.

Today, despite the wide integration of digital technologies with the processes of tackling climate change, there is a range of problems that require priority decisions and solutions. These problems include operations connected with the collection, reflection and use of data that concern the following:

- Absence of standardisation of criteria and forms of data that are a necessity to assess the impact of climate change on natural and socioeconomic processes.
- Low resolution (time and spatial) of most of the technologies that are used for recording processes and phenomena.
- Insufficient completeness or uncertainty of data obtained from natural environment or production systems.
- Absence of transparency in reports on climate responsibility of entrepreneurs or government institutions.
- Other technical barriers [7].

According to the above problems, the emergence of such technology as the IoT and EO might ensure new achievements in the collection of data for monitoring of reduction of climate change. This particularly concerns situations in which traditional means of monitoring and analysis of data are expensive and labour-intensive [3]. Apart from the direct use, it is expected that the IoT will allow inventing new materials with unlimited information [12], which will significantly expand the opportunities and raise the effectiveness of climate-responsible activities of companies in digital economy markets.

The Internet of Things is a system of interconnected physical devices, which have in-built software and sensors to optimise information with the help of the Internet. Throughout its application, the IoT demonstrated exceptional results from the position of collection of information, analysis of data, reporting and forecasting for the integration of the results obtained in the systems of future planning. The elements of the IoT are technological elements, programmes and components of networks, which include smart devices, networks and gates, which unify all elements of the system into one software, which performs the functions of structuring and storing of information, as well as software of final user [9].

Development and capabilities of the IoT in the environment of climate change are implemented at six main levels, which cover the physical level (devices), network level, level of data (databases), level of analytics (visualisation), level of programme (integration) and level of security and management. At the lowest level, sensors and software allow generating or gathering large volumes of information, which, due to systematisation and processes at the next levels, acquire the form of reports, conclusions, elements of visualisation, etc. This is the basis for important decisions on climate change and influences on this process.

Within climate-responsible entrepreneurship, their actions are aimed at the monitoring of processes connected with climate change, which cover the volume and intensity of precipitations, direction and strength of wind, air temperature, level of CO_2, melting of glaciers, intensity and area of fires, drought, etc. The IoT is integrated into different spheres of entrepreneurial activities, which include smart homes and cities, unmanned cars, online commerce and production, supply chain management, etc. Together with IT systems and tools, an important role in the formation and functioning of the IoT belongs to the tools of data processing, namely SAS; Apache Hadoop, Tableau, TensorFlow, BigML, Knime, Apache Flink, PowerBI, DataRobot, Python, Trifacta, Minitab, R, MicroStrategy, Google Analytics, SPSS and other. All these elements interact with the structure of the IoT and form a system of climate-responsible entrepreneurship based on the IoT (Fig. 1).

Each element of the IoT is an element of an open system, according to which, information with different characteristics circulates at the inputs and outputs of this system, and necessary operations are performed with it. In certain cases, the outgoing stage at the lower level of the IoT structure is an incoming stage for a higher level. The links that are shown in Fig. 1 characterise the targeted action and functionality of the system of climate-responsible entrepreneurship, its components and tools and the related spheres and processes. Such structuring allows for a clear identification of processes within which the generation of data takes place, tools for data

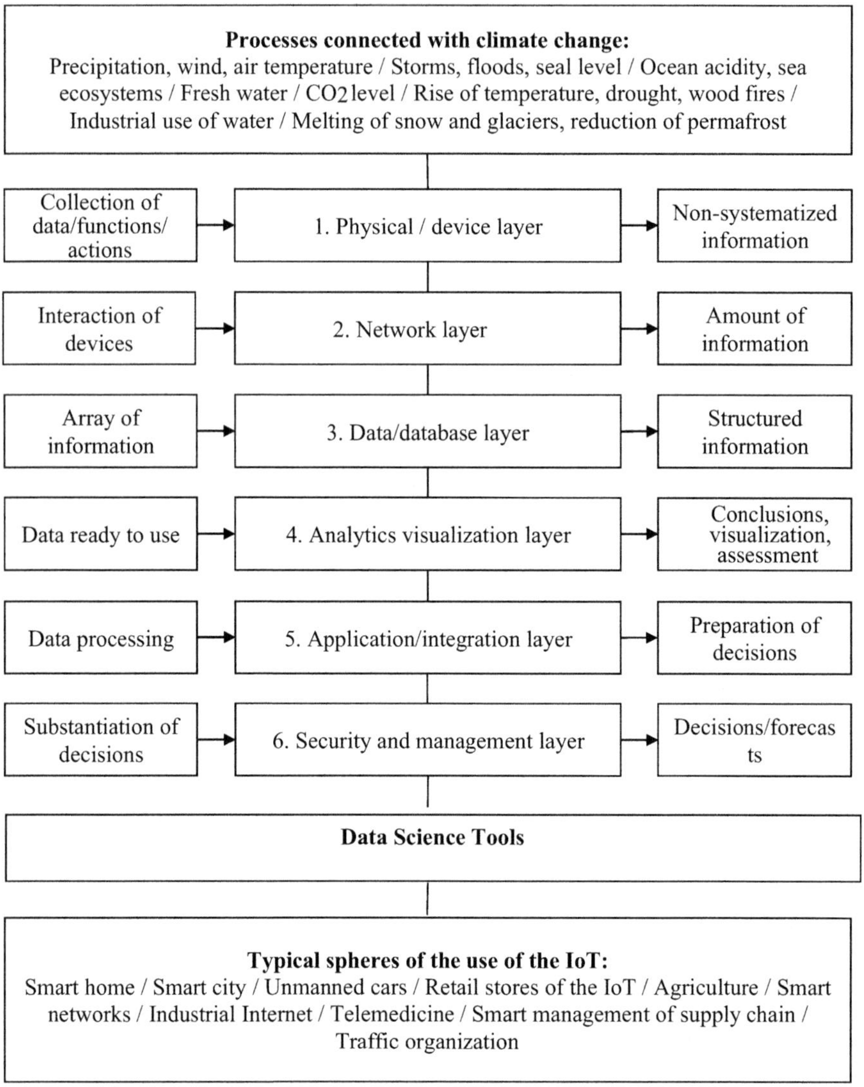

Fig. 1 System of climate-responsible entrepreneurship based on the IoT. (Source: Compiled by the authors based on [7, 9, 11, 13])

processing and spheres of the use of the IoT in which the information is used to achieve a positive effect in the context of the fight against climate change.

4 Discussion

The system approach to managing the climate responsibility of entrepreneurship in digital markets based on the Internet of Things is a rather complex scientific concept, the interpretation of which depends on the specifics of the studied problem or scholars' view. The presented approach, which relies on the consideration of the system of climate responsibility based on the IoT, from the position of its interaction with processes that are connected with climate change, tools of Data Science and spheres of the use of the IoT, may acquire other forms and models. In this context, from the position of climate change management, a prospective direction is the combination of IoT tools with the tools of Earth Observation (EO).

In this case, it is possible to create a digital ecosystem which would ensure the exponential increase in the number of connected devices and sensors, as well as the integration of the received and processes data into the systems of effective management of climate change with the help of a wide range of scientific tools. Such a view may be developed through the expansion of the limits of the IoT to the level of the Internet of the future, cognitive IoT and social IoT. Though these offers involve different connections between data and analysis, they include the necessity to connect the IoT with analytics to transform data into knowledge [3].

Besides, it is worth mentioning that technologies are not a replacement for social interactions, they only supplement them, increasing or adding up to the individual or social capabilities of a certain group of people. Therefore, technologies cannot solve climate problems without the active involvement of humans. This concerns primarily the necessity to take the required climate-responsible measures and the formation of the relevant policy. This is proven by the retrospective of the considered problem, which shows that most of IoT tools have been available for quite some time, but risks of climate change still grow. Thus, further research of the system approach to the management of climate responsibility of digital entrepreneurship with the use of the IoT requires active consideration of the mechanisms of the interaction between technologies and social groups.

5 Conclusion

Digital entrepreneurship – similarly to traditional entrepreneurship – deals with the search and use of opportunities under conditions of uncertainty. Due to digital technologies, uncertainty acquires other forms and substantially expands entrepreneurs' capabilities from the position of influence on different processes. From the position of the system approach, the digital market is based on artefacts, platforms and infrastructure. Digital artefacts form the functionality of the IoT, in particular its lower level, which includes devices and sensors. The capabilities of digital entrepreneurship and the IoT are widely used in the context of tackling climate change. Their action is mostly aimed at the issues of collection, accumulation and processing of information, which can be used to alleviate the consequences of climate change or adapt to them. This activity takes place within the concept of climate responsibility. The system of climate-responsible entrepreneurship is formed based on six levels of the IoT, which are aimed at detailed monitoring of processes connected with climate change, processing of information, including with the use of Data Science tools and development of solutions that are used in the spheres of active application of the IoT, which include smart homes and cities, supply chain management, etc.

References

1. Ericsson (2023) The journey to Net Zero starts here. Technology for climate action. URL: https://www.ericsson.com/en/about-us/sustainability-and-corporate-responsibility/environment/climate-action. Accessed 29 Jan 2023
2. George, G. & Schillebeeckx, S.J.D. (2021). Digital sustainability and its implications for finance and climate change. Macroecon Rev, XX(1), 103–109. Research Collection Lee Kong Chian School Of Business. https://ink.library.smu.edu.sg/lkcsb_research/6721. Accessed 29 Jan 2023
3. Hsu A, Khoo W, Goyal N, Wainstein M (2020) Next-generation digital ecosystem for climate data mining and knowledge discovery: a review of digital data collection technologies. Front Big Data 3(29). https://doi.org/10.3389/fdata.2020.00029
4. Hull C, Hung Y-T, Hair N, Perotti V, Demartino R (2007) Taking advantage of digital opportunities: a typology of digital entrepreneurship. IJNVO 4:290–303. https://doi.org/10.1504/IJNVO.2007.015166
5. Kraus S, Palmer C, Kailer N, Kallinger FL, Spitzer J (2019) Digital entrepreneurship: a research agenda on new business models for the twenty-first century. Int J Entrep Behav Res 25(2):353–375. https://doi.org/10.1108/IJEBR-06-2018-0425
6. Nambisan S (2017) Digital entrepreneurship: toward a digital technology perspective of entrepreneurship. Entrep Theory Pract 41(6):1029–1055. https://doi.org/10.1111/etap.12254
7. Rolnick D, Priya L, Kaack LH, Kochanski K, Lacoste A, Sankaras K, Ross AS, Milojevic-Dupont N, Jaques N, Waldman-Brown A et al (2022) Tackling climate change with machine learning. Association for Computing Machinery 55(2). https://doi.org/10.1145/3485128
8. Salam A (2020) Internet of things for environmental sustainability and climate change. https://doi.org/10.1007/978-3-030-35291-2_2
9. UNext Jigsaw (2022) Layers of IoT that you should know in 2022. UNext Jigsaw. https://www.jigsawacademy.com/blog/cloud-computing/layers-of-iot-that-you-should-know-in-2022/. Accessed 29 Jan 2023
10. UNFCCC (2015) The Paris agreement. What is the Paris agreement? https://unfccc.int/process-and-meetings/the-paris-agreement/the-paris-agreement. Accessed 29 Jan 2023
11. Vadakkanmarveettil J (2022) 12 Major applications of IoT you should know. UNext Jigsaw. https://www.jigsawacademy.com/top-uses-of-iot/ Accessed: 29.01.2023
12. Venkataraman S (2020) Fighting climate change. https://doi.org/10.2139/ssrn.3837541
13. Yamini (2022) How can IoT combat with climate change? Analitics steps. https://www.analyticssteps.com/blogs/how-can-iot-combat-climate-change Accessed 29 Jan 2023

Development of the E-government System to Support Climate-Responsible Entrepreneurship in the Markets of the Digital Economy

Anastasia A. Sozinova, Alexander V. Bespyatykh, Victoria V. Kotlyarova, and Gulnura B. Dzhumabaeva

1 Introduction

Climate-responsible entrepreneurship in the markets of the digital economy becomes effective in the case of the existence of attractive conditions of the market environment, with the government's participation in the support of environmental projects and society's (consumers') readiness for responsible consumption. The attraction of the population to green programmes is also possible under the conditions of the government's activities in the sphere of promoting the country, territory and society's course toward the prevention of climate change. Accordingly, the government – represented by the bodies of central and local authorities of different levels – becomes the key subject of modern climate initiatives. Government bodies, according to their competences, ensure the adoption, regulation and support for environmental parameters, connected with the activity of various sectors of the economy, which are represented by large, medium small and micro companies.

Therefore, national governments should ensure constant and stable work on the interaction with companies, including the ones involved with environmentally important projects that influence climate and preservation of territorial ecosystems. Governments' involvement with the globalisation processes leads to large-scale challenges, which may grow due to national problems and specifics and become a cause for difficulties in the traditional work of governments. Global pandemics, protests and natural cataclysms often make direct interaction at the level of

A. A. Sozinova (✉) · A. V. Bespyatykh
Vyatka State University, Kirov, Russia
e-mail: aa_sozinova@vyatsu.ru; bespyatyh@vyatsu.ru

V. V. Kotlyarova
Institute of Service and Business (branch) Don State Technical University, Shakhty, Russia

G. B. Dzhumabaeva
State Tax Service under the Ministry of Finance of Kyrgyzstan, Bishkek, Kyrgyzstan

E. G. Popkova (ed.), *Smart Green Innovations in Industry 4.0*, Springer Climate,
https://doi.org/10.1007/978-3-031-45830-9_19

government bodies – entrepreneurship impossible. Countries that are aimed at stable communication between the government and companies and provision of certain administrative services create favourable conditions for the achievement of these goals. In this direction the function of e-government is formed; its quality depends on the level of the country's digitalisation and the corresponding regulatory framework. The level of digitalisation implementation in the sphere of e-government is connected with such indicators and simplicity of functional support, Internet access among population and business, quality software for management systems and use of robotisation for solving problems without human participation. A high level of e-government is important for climate-responsible business since at the current stage there are problems in the sphere of climate protection, which neglect may lead to environmental catastrophes.

Thus, the purpose of this work is to identify the directions for the development of the e-government system that would ensure stable and effective support for climate-responsible entrepreneurship. Further in this work we consider countries with a high level of development of the e-government system and identify effective directions for the formation of the e-government system that support the environmental initiatives of business subjects.

2 Materials and Method

Directions of the formation of e-government, including in the context of support for climate-responsible entrepreneurship in markets of the digital economy, were studied in [7, 8, 11] and other works. In [8], the authors dwell on the history of the development and effectiveness of e-government in Estonia and its impact on the creation of attractive conditions for business and participation in the country's sustainable development. In [11], the authors determine the government's support for the projects of the circular economy, which includes processing production waste by production companies that are oriented toward environmentally responsible recycling. Kim et al. [7] dwelled on the specifics of the government's role in the regulation of the problems of energy efficiency of commercial offices in Australia, using the functions of the e-government system. Despite a large number of works on the topic of this research, there is still a need for a systemic approach to the assessment of the directions of the influence of e-government on the effectiveness of climate-responsible entrepreneurship activities in markets of the digital economy.

To achieve the goal of this research, we use the method of systematisation – to generalise the materials and determine the directions of the influence of e-government on the activities of the entrepreneurial sector; statistical method – to identify the statistical indicators of implementation of the function of e-government at the level of countries; the trend method – to determine the main spheres of formation of climate-responsible entrepreneurship in markets of the digital economy; the complex method – to work with the data obtained with the help of the above methods. Thus, we formulate provisions on the main directions of the influence of e-government on the projects of climate-responsible entrepreneurship, which is oriented toward digitalisation.

3 Results

Digital economy tools are used in the sphere of entrepreneurship and at the level of state management. The level of their implementation in digitalisation of government services and communications between government and business and population shows the government's striving toward openness in governance. Let us consider the list of countries that have high indicators of the e-government system implementation (Table 1).

As shown in Table 1, the highest results in the sphere of creation and development of the e-government system are observed in Denmark, South Korea, Estonia, Finland, Australia, Sweden, the United Kingdom, New Zealand, the USA and the Netherlands. Attention should be also paid to Kazakhstan, which was able to make it in the top 30 countries with a high level of e-government system.

Let us dwell on the main top-priority directions for the development of the e-government system, which involve support for the projects of climate-responsible business in the digital economy.

Table 1 E-government index for selected countries

	E-government, rank					Dynamics of change, rank			
Country	2018	2019	2020	2021	2022	2018–2019	2019–2020	2020–2021	2021–2022
Australia	2	2	5	5	5	0	3	0	0
Austria	16	20	15	15	15	4	−5	0	0
Canada	14	23	26	26	26	9	3	0	0
Denmark	9	1	1	1	1	−8	0	0	0
Estonia	13	16	3	3	3	3	−13	0	0
Finland	5	6	4	4	4	1	−2	0	0
Iceland	26	19	12	12	12	−7	−7	0	0
Japan	11	10	14	14	14	−1	4	0	0
Kazakhstan	29	33	27	27	27	4	−6	0	0
Netherlands	7	13	10	10	10	6	−3	0	0
New Zealand	8	8	8	8	8	0	0	0	0
Norway	18	14	13	13	13	−4	−1	0	0
Singapore	4	7	11	11	11	3	4	0	0
South Korea	3	3	2	2	2	0	−1	0	0
Sweden	6	5	6	6	6	−1	1	0	0
United Kingdom	1	4	7	7	7	3	3	0	0
USA	12	11	9	9	9	−1	−2	0	0

Source: Compiled by the authors based on [1–5]

In 2019, Denmark adopted responsibilities on participation in the international climate process. At the legislative level, the following goals were set: reduction of CO_2 emissions by 70% by 2030 and the achievement of full carbon neutrality until 2050 [6, 10].

This reform is being implemented at the level of the central government in the context of the determination and regulation of general indicators and measures on the reduction of CO_2 in the activities of entrepreneurship. The main measures are implemented in a decentralised manner. As an example of managing this process, let us consider Aalborg Municipality, with the following stages and indicators:

1. Development of financial cooperation with the European international fund, which provides subsidies to projects on the reduction of CO_2 emissions. It is implemented at the level of Aalborg Municipality and the fund, with the agreement's adopting the level of subsidies for various models of climate-responsible companies' participation in this project. Thus, companies that passed the test on the assessment of the striving toward the reformation of business according to the needs of the course toward climate protection may apply for reimbursement of 50% of costs on the planning of a green business model of development, which includes such directions as forecasting of the indicators of eco-friendly recycling of waste (creation of business plan); development of forecasts for the achievement of energy efficiency of business processes; creation of plans for the digitalisation of processes that will allow reducing energy costs, etc. Another aspect of support involves grants to companies with a high potential for improvement of the environment (50% of costs for the implementation of measures to reduce CO_2 emissions are returned).
2. Announcement of the project in various media, including informing the subjects of small and medium entrepreneurship on the website of Aalborg Municipality, acceptance of applications from companies, making decisions on the participation of climate-responsible companies in this project. These procedures were implemented at the level of functioning of online government services. Despite the COVID-19 restrictions, local authorities of this municipality received applications from 200 SMEs, 60 of which received financing for the development of green business models. Twenty-five companies received grants to cover 50% of the costs of the measures on the reduction of CO_2 emissions (installation of equipment for CO_2 capture, energy transition to renewable energy, etc.) [6].

Over 2020–2021, this project allowed for the reduction of CO_2 emissions in the region by 900 tons, which is a good result [6]. Such an indicator is a sign of the successful functioning of the e-government system and support for climate-responsible entrepreneurship in the achievement of environmental goals.

Estonia is one of the leaders in the development and improvement of e-government. According to [8], the e-government system in Estonia involves simple and convenient conditions for digital communication with the population and entrepreneurship; the maximum possible number of services for business in the digital sphere: support for various business initiatives that is implemented at the level of robotises communication. Digitalisation allows doing business from any location in the country. In [11], the authors analyse the mass implementation of eco-friendly recycling of waste by Estonian industrial companies with the help of special digital equipment. Implementation of economic activities of climate-responsible entrepreneurship, connected with recycling, demonstrates ecological effectiveness. In 2015, the volume of waste processing equalled 9390 thousand tons, and the volume of waste landfills – 12,221 thousand tons; in 2021, these indicators were 13,404 thousand tons and 6517 thousand tons, accordingly [9]. As we see, the volume of waste landfills reduced by more than two times due to the initiatives of climate-responsible entrepreneurship. This was also supported thanks to the participation of the government (at the local levels) in the development of projects, assessment of results, recommendations on the options for business organisation and online consulting.

In Australia, the law on disclosure of data on the state of energy efficiency of buildings was passed in 2010. According to it, the annual mandatory inspection applies to commercial buildings larger than 1000 sq. metres [7]. This inspection is conducted to eliminate excessive spending on energy resources, which pollutes the environment due to the use of electric energy from non-renewable sources; determine the level of the energy transition to autonomous production of electric energy with the use of alternative sources, which does not produce CO_2 emissions; identify achievements in the sphere of waste management in the exploitation of buildings and activity of offices. According to the procedure of such inspection, companies that manage commercial buildings systematise information on the measures in the energy sphere using automatized systems of accounting; the data from these systems are filed in the form of a report to local authorities at the end of the year; incoming data are processed automatically, with further formation of assessment, recommendations and requirements on the removal of violations. This system of control is implemented with the help of the function of the e-government system, facilitating the country's course toward energy efficiency, reduction of CO_2 emissions from the activity of commercial real estate and reduction of waste that are to be processed at waste landfills. This approach must be realised not only to ensure the achievement of climate goals but also to support entrepreneurship, for the interaction with the government in the context of compliance with recommendations enables business to reduce its costs and, accordingly, to raise the profitability of activities [7]. Commercial buildings that are aimed at energy efficiency are attractive for tenants, for their rent is lower compared to commercial buildings smaller than 1000 sq. metres that do not cooperate with the government in this direction. During the operation of this law, the number of commercial buildings in the programme of assessment of energy efficiency grew from 1700 in 2011 to 10,299 in 2020 [7]. This is a sign of the positive effect of the e-government system in the achievement of climate goals (energy efficiency, decrease in CO_2 emissions, reduction of waste) and in economic goals (increase in the number of offices of various companies in energy-efficient buildings).

4 Discussion

The research results allowed us to distinguish the leading directions for the development of the e-government system to support climate-responsible entrepreneurship.

One of the important directions is programmes of ecologisation, which involve financial guarantees for companies for participation. This can be realised at the level of a partnership: e-government – climate-responsible entrepreneurship – funds that finance projects. This may stimulate regions' transition to the established climate indicators within adopted terms (example of the resolution of the problem of CO_2 emissions reduction in the Danish municipality). The advantages of the use of the e-government system include the possibility to provide public services during the period of pandemic restrictions.

Another important aspect is the e-government system's adopting a range of organisational functions and tasks of climate-responsible entrepreneurship. This is connected with the SMEs' inability to perform certain functions and tasks. For example, the design of circular production must be conducted with the application of the digital tools of Big Data and robotisation, which is rather complex – technically and economically – for small and medium enterprises. In this case, design with the help of the e-government system solves this problem, which also makes a positive effect on the reduction of ecological footprint.

Also, the participation of the government in the remote regulation of buildings' energy efficiency through e-government is very important. This measure helps implement the national policy of decarbonisation and reduction of production waste, as well as supports for economic effectiveness of climate-responsible subjects of entrepreneurship, due to the decrease of energy expenditures. This measure is most effective in the case of adoption of the corresponding laws and organisational support for electronic interaction between the participations of the programmes on buildings' energy efficiency.

5 Conclusion

In this work, we analysed the ways to apply the e-government system to support climate-responsible entrepreneurship in the markets of the digital economy. We demonstrated the unquestionable advantages of implementing the digitalisation of public services in the context of the resolution of complex climate problems and issues of direct interaction between public officials and business.

The systematised directions for the positive development of the e-government system in the context of support for climate-responsible companies should be a model for countries with an open economy, which seek to achieve the environmental SDGs, protect the climate and increase their GDP due to the "soft" policy of environmental development. To achieve climate and economic effectiveness, there is a need for systemic communication at the government–business environment level. Countries that do not have financial capacities for the implementation of large-scale projects of e-government in different spheres, including in support of climate-responsible entrepreneurship, may seek external sources and start working on pilot projects, which should become test models of digitalisation in this sphere.

References

1. IMD (2018) IMD world digital competitiveness ranking 2018. https://www.imd.org/globalassets/wcc/docs/imd_world_digital_competitiveness_ranking_2018.pdf. Accessed 27 Jan 2023
2. IMD (2019) IMD world digital competitiveness ranking 2019. https://www.imd.org/globalassets/wcc/docs/release-2019/digital/imd-world-digital-competitiveness-rankings-2019.pdf. Accessed 27 Jan 2023
3. IMD (2020) IMD world digital competitiveness ranking 2020. https://www.imd.org/globalassets/wcc/docs/release-2020/digital/digital_2020.pdf. Accessed 27 Jan 2023
4. IMD (2021) IMD world digital competitiveness ranking 2021. https://www.imd.org/globalassets/wcc/docs/release-2021/digital_2021.pdf. Accessed 27 Jan 2023
5. IMD (2022) IMD world digital competitiveness ranking 2022. https://www.imd.org/centers/world-competitiveness-center/rankings/world-digital-competitiveness/. Accessed 27 Jan 2023
6. Interregeurope (2021) Green transition in Danish SMEs through low carbon business models and financial implementation aid. https://www.interregeurope.eu/good-practices/green-transition-in-danish-smes-through-low-carbon-business-models-and-financial-implementation-aid. Accessed 27 Jan 2023
7. Kim S, Lim BTH, Oo BL (2022) Energy consumption and carbon emissions of mandatory green certified offices in Australia: evidence and lessons learnt across 2011–2020. Sustainability 14:13773. https://doi.org/10.3390/su142113773
8. Mechitov A, Moshkovich H (2021) Estonia – a small giant of e-Government. https://www.researchgate.net/publication/355003366_ESTONIA_-_A_SMALL_GIANT_OF_e--GOVERNMENT. Accessed 27 Jan 2023
9. Stat.ee (2022) Waste and circular economy. https://www.stat.ee/en/find-statistics/statistics-theme/environment/waste-and-circular-economy. Accessed 27 Jan 2023
10. Timperley J (2019) Denmark adopts climate law to cut emissions 70% by 2030. https://www.climatechangenews.com/2019/12/06/denmark-adopts-climate-law-cut-emissions-70-2030/. Accessed 27 Jan 2023
11. Vihma M, Moora H (2022) Potential of circular design in Estonian SMEs and their capacity to push it. Agri 3:94–103

Consequences of the Development of International Digital Platforms for the Environment

Vladimir V. Shapovalov

1 Introduction

The popularity of the Internet and mobile devices changed the way customers and service providers interconnect with each other. Goods and services are being promoted internationally via digital channels.

Innovations improved people's quality of life. Search engines provide required information to customers and enable businesses to promote their products and services online. Social networks, cloud technologies and machine learning increased productivity and widened customer value propositions. Technologies accelerate customer demand and offer convenient access to information, goods and services.

The COVID-19 pandemic boosted digital transformation. Services and goods offered by electronic platforms help to save time while allowing customers to compare prices and get goods and services online with better quality. Big data and new technologies push businesses to discover new opportunities.

Automatic language translation and the Internet removed geographical and language barriers. Digitalisation changed the landscape of global competition. The largest technology companies have lots of client data and achieve growth using the economy of scale effect. Investors are ready to buy shares of technology companies in the absence of positive financial results.

Ecosystems globally promote local products and deliver them physically by post or in electronic form in the case of remote services, music, films or software. Consumers get significant advantages from electronic platforms [11].

Simultaneously, it is important to study how the boost of global digital consumption affects the ecological agenda, particularly in the countries where necessary electronic equipment is being produced and disposed of.

V. V. Shapovalov (✉)
MGIMO University, Moscow, Russia
e-mail: v.shapovalov@inno.mgimo.ru

E. G. Popkova (ed.), *Smart Green Innovations in Industry 4.0*, Springer Climate,
https://doi.org/10.1007/978-3-031-45830-9_20

2 Methodology

The theoretical and methodological basis of the research is the key issues of the works of Russian and foreign authors on the problems of digitalisation changes and its impact on today's society. During the research, the author applied such research methods as analysis, synthesis, induction, deduction, modeling, as well as the index method of studying socio-economic phenomena.

The author also uses data from the UN [12] and Skolkovo [10] and reports from technological companies and consulting agencies.

3 Results

3.1 Digitalisation and Energy Consumption

The technological revolution and the emergence of digital platforms created a need for large amounts of energy for communications purposes and the functioning of electronic equipment. It also significantly increased the demand for this equipment. High energy consumption and the growth of electronic equipment production considerably affect the environment.

Energy consumption is the main contributor to greenhouse gas emissions. Industry, agriculture and individuals are the key energy consumers, with transport accounting for only one-fifth of total emissions [14]. Equipment required by digital service providers is a notable energy consumer.

The largest energy-consuming countries are listed below:

1. China – 146 billion kWh in 2020
2. The USA – 88 billion kWh in 2020
3. India – 32 billion kWh in 2020
4. Russia – 28 billion kWh in 2020
5. Japan – 17 billion kWh in 2020
6. Canada – 14 billion kWh in 2020
7. Germany – 12 billion kWh in 2020
8. Iran – 12 billion kWh in 2020
9. Brazil – 12 billion kWh in 2020
10. South Korea – billion kWh in 2020 [18]

However, the environment is affected not only by energy overconsumption but also by electronic equipment production and waste.

3.2 *Electronic Industry*

Digital platforms would not exist without the Internet, mobile devices and electronic equipment.

The revenues of the electronic industry are estimated at $1 trillion as of 2022 [3], with semiconductors production accounting for more than 50% of it. Semiconductors are mainly produced in the USA and Taiwan.

The demand for electronics and semiconductors is increasing with the emergence of digital technologies and platforms. Electronics is a global industry: raw materials and sub-parts come from different countries.

The COVID-19 pandemic affected the global supply chains. In particular, the global automotive industry was the one that suffered the most [17].

Electronic equipment and smartphones required for the existence of digital platforms consume energy at all stages of their lifecycle: production, usage and utilisation. In 2021, 1.35 billion smartphones were produced and shipped worldwide [9]. Its manufacturing has the strongest impact on the environment. For example, semiconductors production factories consume 5% of Taiwan's electricity and significant amounts of water [2].

The semiconductor manufacturing and electronics industry require various chemical elements coming from locations all around the world and affecting the environment of local countries.

3.3 *E-waste*

Electronic products create tons of waste at the end of their life. E-waste consists of computers, cell phones, TVs and other electrical and electronic products.

As per UN calculations, the world each year annually produces about 50 million tons of electronic and electrical waste (e-waste). It is estimated to reach 120 million tons per year by 2050.

Only one-fifth of e-waste is legally recycled. The rest is being informally recycled – much of it using human labour in developing countries. E-waist contains lots of chemical components. Hand recyclers expose themselves to hazardous and toxic substances such as mercury, lead and cadmium [15]. Batteries represent the biggest issue for the environment [19]. E-waste also affects soil and groundwater.

In 2021, 57.4 million tons of e-waste were created worldwide [8], of which only 17% were properly documented. Thus, the rest was not properly and legally recycled.

In 2020, European Commission identified 30 critical raw materials used for EU strategic technologies [4].

China is the largest producer of e-waste. According to the UN, Asia is leading in the generation of e-waste, with 24.9 million tons of e-waste, compared to 13.1 million tons of e-waste in the Americas and 12 million tons of e-waste in Europe [5].

Table 1 lists the environmental impact and health hazards of e-waste [7].

Table 1 Health hazards and environmental impact generated by e-waste

E-waste component	Process followed by informal sectors	Potential health hazard	Potential environmental impact
Cathode ray tubes (CRT)	Breaking the copper yoke, removal of the copper yoke, and openly discarding it in landfills	Silicosis, commonly known as Miner's phthisis Lacerations from CRT glass Inhalation of phosphor-containing cadmium or other metals; skin contact with it	Leaching of heavy metals such as lead, barium, and others into groundwater and the release of toxic phosphorus
Printed circuit boards (PCBs)	De-soldering and separating computer chips	Inhalation of lead and tin Inhaling potential brominated dioxin, cadmium, mercury, and beryllium	Probable emission of the exact substances into the atmosphere
Processing of dismantled PCB	Open incineration of PC boards	Toxicity of tin, lead, brominated dioxin, beryllium, cadmium, and mercury inhalation in employees and nearby residents	Brominated dioxins, beryllium, cadmium, and mercury inhalation, as well as tin and lead pollution of the nearby environment, including surface and groundwater
Gold-plated substances and other chips	Chemical stripping of contaminants along riverbanks with nitric and hydrochloric acid	The eyes or skin coming in contact with acid can cause lifelong damage. Inhalation of acid mists and fumes, chlorine, and sulfur dioxide gases can cause respiratory irritation and lead to serious consequences	Direct release of hydrocarbons, heavy metals, brominated chemicals, and other pollutants into the river and its banks. Acidification of the river, which kills fish and plants.
Plastics from the computer and peripherals	Lower temperature melting and shredding	Workers living in the vicinity of the burning operation may be exposed to hydrocarbons, brominated dioxin, and PAH.	Brominated dioxins, heavy metals, and hydrocarbons emissions into the atmosphere
Smelting of secondary steel, copper, and precious metals	Steel or copper is recovered from waste in the furnace	Dioxins and heavy metals may be exposed to workers	Dioxins and heavy metals are released into the environment, contaminating the air
Wires	Open incineration for the recovery of copper	Workers living in the vicinity of the burning operation may have been exposed to brominated and chlorinated dioxin and PAH	Discharge of hydrocarbons and ashes, including PAHs, into the air, water, and soil

Source: Compiled by the author based on [19]

3.4 Role of Transnational Corporations

Transnational corporations are the key players in globalisation. They are also responsible for the environmental impact. Technological companies often have strong lobbying capacities and solid financial influence.

Transnational corporations like Alphabet, Apple, Microsoft and Amazon participate in the global competition and provide technological infrastructures (telecom services and satellites), equipment (computers and smartphones) and means of communication (Internet access and social media).

Transnational corporations in the electronic industry can influence ecology and customers' behaviour. However, these companies can be affected by environmental issues themselves. Resources depletion, natural disasters or pandemics can cause supply chain interruption. Noncompliance with the environmental agenda would have a social and public opinion impact.

Therefore, transnational corporations have to adapt their governance to the environmental agenda.

3.5 Regulation Issues

Respective policies should be implemented to reduce the environmental impact of electronic platforms.

Energy-saving and water recycling technologies implementation should be stimulated in the semiconductor industry. The collection of e-waste should be ordered with specialised licensed companies. Controls and penalty measures will help fight the creation of illegal dumps.

Consumers must be regularly informed about the consequences of excessive energy consumption while using electronic platforms.

The United Nations defined the Sustainable Development Goals (SDGs), which are to be achieved by 2030 [16]. Most goals are directly linked to environmental issues and cover water, energy, industry, innovation, infrastructure, consumption and production issues.

UN also introduced the StEP [6] (Solving the E-waste Problem) Initiative. UN Environment Management Group (EMG) issued a report [13] stressing the necessity to collaborate on e-waste issues with international organisations and non-state actors. E-waste monitors have provided respective statistics on the matter since 2017 [1].

ISO and the Institute of Electrical and Electronic Engineers Society provide norms and standards for electronic products (ISO14000, ISO/TC297). Interpol follows up on crime in e-waste management.

4 Conclusions

The development of electronic platforms and related consumer habits is a great challenge to the environment. Digital platforms stimulate the development of communication networks, which increases the pace of globalisation and the need for energy. International transportation and consumption promoted by platforms affect the environment.

In this chapter, the author discussed the impact of human activity on the environment and the type of solutions in terms of adaptation, mitigation and transformation to be defined and implemented for the green economy. The implementation of these measures is essential for sustainable operation. The author illustrated the importance and impact of the electronics industry, which is an integral part of the digital economy.

The increase in energy consumption and the growth of electronic waste significantly affect the environment. The digital transformation that changes business processes and consumers' habits and lifestyles should not threaten sustainable development.

References

1. Baldé CP, D'Angelo E, Luda V, Deubzer O, Kuehr R (2022) Assessment of E-waste flows. In: Global transboundary E-waste flows monitor 2022. FOEN, Bern. Retrieved from https://api.globalewaste.org/publications/file/286/Global-Transboundary-E-waste-Flows-Monitor-2022.pdf. Accessed 22 Dec 2022
2. Belton P (2021) The computer chip industry has a dirty climate secret. The Guardian. Retrieved from https://www.theguardian.com/environment/2021/sep/18/semiconductor-silicon-chips-carbon-footprint-climate. Accessed 21 Nov 2022
3. Deloitte (2022) Digital transformation efforts accelerate. In: Semiconductor industry outlook 2022. London, UK. Retrieved from https://www2.deloitte.com/us/en/pages/technology-media-and-telecommunications/articles/semiconductor-industry-outlook.html. Accessed 15 Nov 2022
4. European Commission (2020) Critical raw materials for strategic technologies and sectors in the EU: a foresight study. Retrieved from https://rmis.jrc.ec.europa.eu/uploads/CRMs_for_Strategic_Technologies_and_Sectors_in_the_EU_2020.pdf. Accessed 7 Dec 2022
5. Forti V, Baldé CP, Kuehr R, Bel G (2020) The Global E-Waste Monitor 2020: Quantities, flows, and the circular economy potential. UNU; ITU, and ISWA, Bonn, Geneva, Rotterdam. Retrieved from https://www.itu.int/en/ITU-D/Environment/Documents/Toolbox/GEM_2020_def.pdf. Accessed 26 Dec 2022
6. ITU, StEP Initiative, & WEEE Forum (2022) Complementary actions. In: Global and complementary actions for electronics extended producer responsibility: A thought paper for International E-waste Day 2022. ITU, Geneva, pp 7–16. Retrieved from https://www.step-initiative.org/files/_documents/publications/Global%20and%20complementary%20actions%20for%20electronics%20extended%20producer%20responsibility_final.pdf. Accessed 15 Dec 2022
7. Mihai F, Gnoni M, Meidiana C, Ezeah C, Elia V (2019) Waste electrical and electronic equipment (WEEE): flows, quantities, and management – a global scenario. In: Prasad MNV,

Vithanage M (eds) Electronic waste management and treatment technology. Butterworth-Heinemann, Oxford, UK, pp 1–34. https://doi.org/10.1016/B978-0-12-816190-6.00001-7
8. Rosanne O (2021) This year's E-waste to outweigh Great Wall of China, paper presented at the World Economic Forum. Retrieved from https://www.weforum.org/agenda/2021/10/2021-years-e-waste-outweigh-great-wall-of-china/#:~:text=This%20year%2C%2057.4%20million%20tonnes,the%20Great%20Wall%20of%20China. Accessed 3 Dec 2022
9. Shirer M (2022) Smartphone shipments declined in the fourth quarter but 2021 was still a growth year with a 5.7% increase in shipments. IDC. Retrieved from https://www.idc.com/getdoc.jsp?containerId=prUS48830822. Accessed 18 Nov 2022
10. Skolkovo Foundation (2022) Distant & Digital will discuss the transformation of law and intellectual property. Retrieved from https://sk.ru/news/distant-and-digital-will-discuss-the-transformation-of-law-and-intellectual-property/. Accessed 11 Nov 2022
11. SKOLKOVO Moscow School of Management (2021) Number of ecosystems in Russia to grow. Retrieved from https://www.skolkovo.ru/en/news/number-of-ecosystems-in-russia-to-grow/. Accessed 8 Nov 2022
12. UN DESA (2022) Online connectivity improves, but digital inclusivity remains a challenge, new UN survey shows. Retrieved from https://www.un.org/en/desa/online-connectivity-improves-digital-inclusivity-remains-challenge. Accessed 9 Nov 2022
13. UN Environment Management Group (2017) E-waste across the United Nations System. In: United Nations system-wide response to tackling E-waste. UNEP, Geneva, pp 18–41. Retrieved from https://unemg.org/images/emgdocs/ewaste/E-Waste-EMG-FINAL.pdf. Accessed 19 Dec 2022
14. UNEP (2022) Global progress on adaptation planning. In: Adaptation Gap Report 2022: Too Little, Too Slow – Climate adaptation failure puts world at risk. UNEP, Nairobi, Kenya, pp 9–16. Retrieved from https://www.unep.org/resources/adaptation-gap-report-2022. Accessed 13 Nov 2022
15. UNEP (January 24) UN report: Time to seize opportunity, tackle challenge of e-waste. Retrieved from https://www.unep.org/news-and-stories/press-release/un-report-time-seize-opportunity-tackle-challenge-e-waste. Accessed 25 Nov 2022
16. UNSD (2022) Climate action. In: The sustainable development goals report 2022, pp 52–53. Retrieved from https://unstats.un.org/sdgs/report/2022/The-Sustainable-Development-Goals-Report-2022.pdf. Accessed 9 Dec 2022
17. Wayland M (2021) Chip shortage expected to cost auto industry $20 billion in revenue in 2021. CNBC News. Retrieved from https://www.cnbc.com/2021/09/23/chip-shortage-expected-to-cost-auto-industry-210-billion-in-2021.html. Accessed 17 Nov 2022
18. World Population Review (2023) Energy consumption by country 2023. Retrieved from https://worldpopulationreview.com/country-rankings/energy-consumption-by-country. Accessed 14 Nov 2022
19. Zhao Y, Pohl O, Bhatt AI, Collis GE, Mahon PJ, Ruther T et al (2021) A review on battery market trends, second-life reuse and recycling. Sustain Chem 2(1):167–205. https://doi.org/10.3390/suschem2010011

Economic and Legal Aspects of Environmental Quality Management in Industry 4.0

Galina M. Golobokova

1 Introduction

Studies devoted to managing the quality of the natural environment and the search for mechanisms of environmental protection and nature restoration activities to reduce the damage from anthropogenic activities are currently becoming relevant. Such studies pay special attention to the economic justification for taking measures to ensure the safety of the operation of industrial facilities. Imperfect environmental management mechanisms lead to a violation of the ecological situation and climate change in the territory where industrial complexes are located.

In 1989–1990, Soviet Russia carried out a large-scale economic experiment to improve the mechanism of nature management, which consisted in charging for environmental pollution [2]. However, during this experiment, it was found that, despite these principles, the area of land polluted by emissions from industrial enterprises is still large and amounts to more than 62 million hectares, more than 1.0 million hectares are disturbed, and more than 3.0 million hectares are occupied by landfills [3]. The use of extensive methods for solving environmental problems of water supply (e.g., in the Republic of Crimea, the activity on the "development" of 50 billion rubles from the federal budget intensified in 2009) also causes a deterioration in the environmental situation [4].

Particularly acute environmental problems manifested themselves in the northern and eastern regions of Russia, where the natural environment has a low level of recovery after environmental damage from industrial activity.

According to the analysis, the oil and gas producing regions of the Far East experience environmental risks that are human-made or caused by natural disasters (e.g., earthquakes, floods, landslides, hurricanes, tornadoes, and volcanic eruptions).

G. M. Golobokova (✉)
Republican Science Research Institute of Intellectual Property, Moscow, Russia
e-mail: golobokova@rniiis.ru

E. G. Popkova (ed.), *Smart Green Innovations in Industry 4.0*, Springer Climate,
https://doi.org/10.1007/978-3-031-45830-9_21

When conducting a retrospective analysis of the development of the mining industry in the North-East of Russia, it was found that the use of only economic (limitation of environmental costs) or only environmental (cessation of production activities) approaches does not solve the problem of reducing the level of environmental pollution. For this purpose, the author proposes applying an integrated approach using economic, environmental, and legal mechanisms. An urgent task is also the development and application of innovative technologies to solve global environmental problems and climate change problems. Another important task is the formation of a promising scientific view on the consequences of the transition to Industry 4.0, which considers the negative consequences of digitalization and changes in the level of environmental pollution in Industry 4.0.

The view from the opposite side also deserves attention, which allows reflecting the contribution of Industry 4.0 to the fight against climate change and determining the prospects for harmonizing these priority areas for developing the world economy [10]. These issues have not been ignored by the corporate sector. The largest foreign multinational companies in the oil and gas and other resource-producing industries, which are powerful pollutants, have long been transforming their own business models to combat or adapt to climate change [8].

2 Materials and Method

The works of many researchers are devoted to topical issues of environmental protection, including A. N. Averchenkov [2]; N. V. Komov [3]; and K. N. Trubetskoy, Yu. P. Galchenko, and L. I. Burtsev [9]. The issues of environmental and economic problems and the assessment of environmental safety are considered in the work of I. G. Polyanskaya and A. V. Malysheva [6]. Environmental problems of the oil industry are discussed in the work of A. K. Ambartsumyan [1]. Environmental problems of the mining industry are touched upon by K. N. Trubetskoy, Yu. P. Galchenko, and L. I. Burtsev [9] and Zh. R. Sakenova, D. S. Dosymbek, and A. K. Kalmandabayeva [7]. The impact of Industry 4.0 on climate change is considered in the work of E. B. Zavyalova and E. G. Popkova [10]. An overview of the practice of countering climate change in oil and gas companies is provided in the work of E. A. Starkova and E. A. Shamanina [8]. The works of V. N. Lopatin [4] and I. G. Polyanskaya and A. V. Malyshev [6] are devoted to the issues of innovative development, intellectual property, and digitalization for solving environmental problems.

The data of the Ministry of Natural Resources and Ecology of the Russian Federation were used as an information base for this research [5]. When conducting a scientific study, the authors considered the practical experience of scientists from the Baikal State University (BSU), Irkutsk (in the oil industry); the Irkutsk State Technical University (IrSTU), Irkutsk (in the mining industry); and the Republican Science Research Institute of Intellectual Property (RNIIIS), Moscow (in innovation, intellectual property, and digitalization).

3 Results

As a result of the research, it was found that the environmental damage caused to the natural environment as a result of anthropogenic activities affects climate change, and, accordingly, the population's health. High pollution levels of fresh surface waters in the territory of the Russian Federation were noted in 2020 at 130 water bodies. There were 2783 cases of high pollution levels for 35 pollutants and six water quality indicators. The analysis performed by the author based on data from the Ministry of Natural Resources and Ecology of the Russian Federation [5] showed that the total volume of pollutant emissions from stationary sources in 2021 amounted to 6968.1 thousand tons, which is 213.3 thousand tons more than in 2020. The relative contribution of this industry to environmental pollution was 40.5% (Table 1). In 2021, the absolute contribution of enterprises for the production and distribution of electrical and other types of energy in the same period amounted to 3061.3 thousand tons, which is 170.4 thousand tons more than in 2020. The relative contribution of these industries amounted to 17.8%.

According to the author, the growth of pollution indicators is caused by inefficient management of the quality of the natural environment. Therefore, the author carried out a search for economic mechanisms to prevent and eliminate environmental damage through environmental legislation and the protection of intellectual property.

As a result of the research, it was found that socio-political and organizational-economic mechanisms can be attributed to economic mechanisms. The author proposes using an approach based on the use of the assimilation potential and an integral indicator characterizing the anthropogenic impact.

The assimilation potential (AP) is the limiting level of the ability to restore the quality of the natural environment disturbed by anthropological activity, determined by the amount of pollution that can be allowed to be released into the atmosphere and water bodies without causing environmental damage to the environment.

The integral (complex) indicator (IP) is a complex of three types of component indicators that characterize the impact of anthropogenic impact on environmental pollution (environmental) on the cost of maintaining and restoring the normal state of natural resources (economic) and public health (social):

Table 1 Emissions of pollutants

The period for which the analysis was carried out	Emissions from mining, thousand tons	Change to the previous year, %	Emissions in the production and distribution of electrical and other types of energy, thousand tons	Change to the previous year, %
2019	4956.4	1.02	3004.2	1.08
2020	6754.8	1.36	2890.9	–0.96
2021	6968.1	1.03	3061.3	1.06

Source Compiled by the author

$$IP = Environmental + Economic + Social.$$

Of the economic methods, this chapter analyzes the method of levying fines when using natural resources, which does not play the role of a motivator for the protection of the natural environment. Based on the analysis of foreign experience, the work found that the most effective economic mechanisms include the mechanisms of natural rent and payments for violating the limits of permissible pollution of the natural environment, which the author personally met during business trips to the University of Alaska, Anchorage, USA, and at the Training Center, Tokyo, Japan. It is proposed to use this approach to sell limits on pollutant emissions for Russian regions. As shown by the analysis and study of the specifics of the mining industry of the Magadan Region, the economic and organizational mechanism can also be industry clusters as elements of the future innovation infrastructure that allow combining the efforts of several mining enterprises to solve common environmental problems.

As legal mechanisms in the work, along with measures of environmental legislation, it is proposed to use the protection of intellectual property, which is created and used in the environmental complex, and the commercialization of which allows enterprises to accumulate funds for environmental protection measures through licensing and lending against the security of intellectual property.

According to the analysis, the environmental problems of the oil industry are caused by the fact that the production activity for oil and gas production is a constant source of technogenic danger and the occurrence of accidents, the number of which is growing due to the increase in the degree of oil extraction from the subsoil. This is despite the introduction of new technologies in oil production in fields where 50–70% of the oil still remains in the depths, as well as in the development of hard-to-recover reserves [6].

This chapter found that climate change is occurring in oil production zones. This change is caused by environmental damage to territories and water bodies due to an increase in the accident rate at obsolete technological facilities and pipelines. However, as is found, Russian entrepreneurs do not show interest in introducing innovative developments to protect the environment. When installing oil-producing equipment, innovative products account for only about 1.5%, which is ten times lower than the world average [1]. The development of oil fields is hampered by the lack of necessary investments for geological exploration and the development of new industrial complexes with the appropriate infrastructure.

For example, it was found that the search for and production of oil on the shelf of the Sea of Okhotsk in the Magadan Region is vital for the region because it reduces its dependence on the supply of petroleum products and gives impetus to the development of the economy. Although the oil production project at the Magadan-1 and Magadan-2 sites is considered highly profitable, the damage to the natural environment from the development of fields is estimated at $4.3 billion.

The environmental problems of the mining industry are justified by the environmental risks of the implementation of mining projects. Pollution of the atmosphere and water bodies in the areas of mining enterprises is due to emissions of dust, sulfur dioxide, carbon monoxide, and nitrogen oxides.

An analysis of the activities of mining industries in the economy of the northern territories of Russia and the USA showed that an important production factor is the quality of the natural environment, which must be preserved with the full use of the natural resource potential of the territory, as well as environmental protection costs (Table 2).

As follows from Table 2, the volume of pollutant emissions into the atmosphere in the Magadan Region in 2021 decreased by 20.42 thousand tons compared to 2020. This was achieved by increasing the cost of environmental protection by 79,733 thousand rubles compared to 2020, and by 342,995 thousand rubles, compared to 2019.

Simultaneously, the Republic of Sakha (Yakutia) saw a sharp increase in the volume of pollutant emissions into the atmosphere in 2021 by 104.2 thousand tons compared to 2020. In our opinion, the reason for this increase is a significant reduction in environmental protection costs by 9,555,071 thousand rubles.

In 2021, the Chukotka Autonomous Area managed to keep environmental pollution indicators at a fairly low level because expenses for environmental protection were increased by 527,089 thousand rubles compared to 2020.

The mechanisms of natural rent, which forms more than 80% of the revenue part of the territory's budget, have been studied as a result of the author's work at the Institute of Social and Economic Research of the University of Alaska, USA. Simultaneously, in the Magadan region, the activities of mining enterprises

Table 2 Volumes of pollution and environmental protection costs of the northern regions of the Russian Federation

Years	The volume of emissions of pollutants into the atmosphere, thousand tons	Discharge volume of polluted waters into reservoirs, million cubic meters	Environmental protection costs, thousand rubles
	Magadan Region		
2019	89.42	4.9	1,003,438
2020	69	4.6	1,083,171
2021	67.7	4.36	1,346,433
	Republic of Sakha (Yakutia)		
2019	299.6	83.3	11,586,607
2020	297	74.3	19,574,974
2021	401.2	75.4	10,019,903
	Chukotka Autonomous Area		
2019	18.9	2.92	266,509
2020	18.3	2.93	258,371
2021	18.8	2.94	785,460

Source Compiled by the author based on the Ministry of Natural Resources and Ecology of the Russian Federation [5]

lead to the creation of environmental risk zones because imperfect environmental measures and the low restorative capacity of the territory do not protect against environmental damage.

According to statistics, each inhabitant of the region has an average of 284 kg of harmful substances, which annually grows by 4% and almost doubles the average Russian level. The short period of development of alluvial deposits in the mining industry of the region (no more than 20–25 years) has led to an unstable environmental and economic situation. Therefore, more than 200 rivers in the basin of the upper reaches of the Kolyma River have lost their fishery importance, and it takes from 50 to 70 years for their natural restoration. The study of the territory of the Magadan Region showed a violation in the process of gold mining of 89% of the lands. There are "lunar landscapes" everywhere in the places of development of deposits and the violation of local flora and fauna due to transport movements and overburden operations. According to biologists, the natural restoration of these "lunar landscapes" will require at least 25 years.

In recent years, the field of subsoil use of the Magadan Region saw a significant increase in funding from 71 million rubles to 412.2 million rubles per year from the federal budget for geological exploration, the introduction of new processing equipment and innovative technologies, and the solution of environmental problems and rational management of nature.

However, the largest share of innovation spending is on technological innovation (57%) related to developing new processes. The acquisition of new production and environmental technologies (including patent and non-patent licenses) accounts for only 2% of the costs. The priority mechanisms for the mining industry in Russian regions include the economic mechanism of natural rent and payment for pollution and the environmental mechanism for installing efficient equipment, making it possible to reduce emissions of harmful substances and clean up pollutants [7].

Therefore, the priority environmental measures for the mining industry include installing effective equipment for cleaning from pollutants and developing measures to reduce emissions of harmful substances to the maximum permissible or temporarily agreed standards [7]. In oil production, it is proposed to ensure environmental protection through measures aimed at the rational use of rocks and waste from their processing [9].

According to the author, the elimination of environmental problems will also be facilitated by the innovative development of extractive industries, when the intellectual property can be created and commercialized in various ways during the development of environmental protection technologies. To use legal mechanisms in the work, it was proposed to provide legal protection for each innovative development for the purpose of subsequent commercialization.

To activate these processes, the author proposes improving the methodological and regulatory framework, giving it a motivational character for managing the quality of the natural environment in Industry 4.0.

4 Conclusions

Thus, as it was established in the work, environmental pollution as a result of industrial activity has reached its limit value, and the confrontation between the economy and ecology has become a dangerous omen in today's world. Industry 4.0 causes the intensification of production and the market desire to increase the profitability of business activities in any way, regardless of the damage caused to the natural environment.

This chapter considers the ecological situation and the experience of solving ecological and economic problems in Russia and abroad. This chapter considers the environmental situation and the experience of solving environmental and economic problems in Russia (in the Magadan Region) and abroad (Alaska, USA). As the analysis of the functioning of industrial facilities in the northern regions of Russia and the state of Alaska (which have the same natural and climatic conditions) has shown, the mechanisms used to protect and restore the natural environment in extractive industries differ sharply. This chapter found that mining companies in Russia, unlike in the USA, do not pay due attention to environmental protection measures, which worsens the state of land quality.

To change the situation, the author searched for mechanisms for managing the quality of the natural environment from the standpoint of economic, environmental, and legal aspects. The author proposes measures to ensure the environmental safety of the operation of subsoil use facilities.

The use of the research results will contribute to the achievement of the effectiveness of implementing the economic potential of Industry 4.0, including the commercialization of the results of intellectual activity.

References

1. Ambartsumyan AK (2010) Modern factors of competitiveness of oil and gas companies – technologies and staff. Russ J Entrep 7-2:100–105
2. Averchenkov AN (1995) Ecological policy in the transition period: problems and solutions. Vopr Ekon [Questions of Economics] 2:150–159
3. Komov NV, Loiko PF, Zhirov AA (1994) On measures to prevent soil degradation in Russia. Soil Sci 10:5–9
4. Lopatin VN (2020) Digitalization, standardization and intellectual property market as a condition of competitiveness within the framework of the European integration for the period until 2030. Intellect Prop Law 4(62):10–24
5. Ministry of Natural Resources of Russia and Lomonosov Moscow State University (2022) On the state and protection of the environment of the Russian Federation in 2021 [State report]. Ministry of Natural Resources of Russia, Moscow. Retrieved from https://www.mnr.gov.ru/docs/gosudarstvennye_doklady/gosudarstvennyy_doklad_o_sostoyanii_i_ob_okhrane_okruzhayushchey_sredy_rossiyskoy_federatsii_v_2021_/. Accessed 27 Jan 2023
6. Polyanskaya IG, Malyshev AV (2022) Methodological approaches to assessing the level of digital transformation of the oil and gas industry. In: Ecological and economic safety of

mining regions: collection of scientific articles. Institute of Economics of the Ural Branch of the Russian Academy of Sciences, Yekaterinburg, pp 294–313
7. Sakenova ZR, Dosymbek DS, Kalmandabaeva AK (2016) The main aspects of ensuring environmental safety at mining enterprises with open pit mining. Eurasian Union Sci 1–2(22):106–108
8. Starikova EA, Shamanina EA (2021) Corporate practice of implementing measures to combat climate change in the Russian oil and gas companies. In: Zavyalova EB, Popkova EG (eds) Industry 4.0: exploring the consequences of climate change. Palgrave Macmillan, Cham, pp 221–233. https://doi.org/10.1007/978-3-030-75405-1_20
9. Trubetskoy KN, Galchenko YP, Burtsev LI (1998) Environmental protection during subsoil development. Bull Russ Acad Sci 68(7):629–637
10. Zavyalova EB, Popkova EG (eds) (2021) Industry 4.0: exploring the consequences of climate change. Palgrave Macmillan, Cham. https://doi.org/10.1007/978-3-030-75405-1

Part III
Decarbonization and Sustainable Economic Development Based on ESG Principles

Applied Solutions for ESG Management of Entrepreneurship in Digital Economy Markets with the Help of Blockchain

Ainura M. Khamzaeva, Vera I. Menshchikova, Meerimai Z. Karbekova, and Zhanar N. Tazhiyeva

1 Introduction

Amid the challenges of the twenty-first century, entrepreneurship and management go beyond the limits of their initial perception. As of now, the key goals of business structures are rarely considered separately from the Sustainable Development Goals, which cover the environmental, social and governance spheres. In the theory and practice of management, these theories create a stable imperative of the ESG concept, which is considered at the level of global and regional institutions and transnational and local companies. It is a tool for the formation of value orientations for entrepreneurship and management.

Implementation of the given provisions in the system of management based on stable business models is the form of companies' managers' reaction to numerous external challenges and events, which influence the sensitivity of employees, customers and society on the whole to the problems of climate change and the growth of inequality between social groups. According to this, profit-making in business is more often combined with the achievement of environmental and social benefits from the position of stakeholders [3].

A. M. Khamzaeva (✉)
Osh Technological University named after M.M. Adyshev, Osh, Kyrgyzstan

V. I. Menshchikova
Tambov State Technical University, Tambov, Russia

M. Z. Karbekova
Jalal-Abad College of the Jalal-Abad State University named after B. Osmonov, Jalal-Abad, Kyrgyzstan

Z. N. Tazhiyeva
Al-Farabi Kazakh National University, Almaty, Kazakhstan

E. G. Popkova (ed.), *Smart Green Innovations in Industry 4.0*, Springer Climate,
https://doi.org/10.1007/978-3-031-45830-9_22

ESG concept becomes popular in the business environment, though it is treated differently by economic agents: some consider it a legal business strategy, others – an independent initiative or a tool of marketing promotion [12]. A popular treatment of this concept is the one from the position of the value of sustainable and coordinated development, which takes into account economic, environmental, social and managerial benefits and is an investment philosophy, aimed at the long-term growth of the value of assets and property [10].

In any case, the use of the ESG concept in the management of entrepreneurship urges managers toward the implementation of the Sustainable Development Goals in the system of management, which ensures the socially important changes in the environment that is aimed only at profit-making. In such conditions, satisfying the non-commercial types of activities in the sphere of ecology, social development or governance through business resources requires the use of additional tools, which might increase the potential of this sphere.

Technological innovations are among such tools. According to the theory of ecological modernisation, technologies help distinguish the aggravation of the state of the environment from economic growth. In this case, solutions are sought that can bring benefit from both positions – environmental and economic sustainability. In this context, technological progress suggests a wide spectre of information, social and production innovations, which include technologies in the spheres of additive production, micro-factories, nanotechnologies, the Internet of Things, unmanned vehicles, the sharing economy and blockchain. Each of these technologies influences the stability of organisations and their supply chains [8].

In the above list of technologies, blockchain is considered a tool for the resolution of important global problems, in particular, the fight against climate change. Due to its specifics, the technology of blockchain can ensure better conditions of production, distribution and use of energy, including the monitoring of energy load, generation or distribution [15], achieve a higher level of data verification on the activities of companies in all sectors of the economy and ensure the traceability of supply chains.

2 Materials and Method

The methodology of this research includes approaches from different scientific spheres, which cover management, the basics of entrepreneurship, economic theory and technical sciences. In their totality, they form a basis for a comprehensive study of the issues of research and generalisation of the applied examples of using solutions in the sphere of ESG with the use of blockchain. The methods of observation and generalisations allow identifying the examples of using blockchain to solve the problems of the environmental, social and governance character; the systems approach allows coordinating the results obtained and systematising them in the form of an integrated map, created with the help of the graphical method.

Studies of the examples of using blockchain technologies for supporting ESG management of entrepreneurship in digital economy markets are at the initial stages of scientific cognition. During the work on this chapter, we analysed provisions based on the use of the relevant scientific and expert publications and reports devoted to the following issues: theoretical and practical substantiation of the specifics of the ESG concept and the state and prospects for its implementation [5, 10], issues of formation and use of ESG reporting as a tool of management [2, 9], substantiation of the opportunities and practice of using blockchain technology in the context of sustainable development of entrepreneurship and realisation of the ESG concept [3, 12, 13], generalisation of the examples of the use of blockchain in the environmental [7, 11, 14], social [1] and governance [4] spheres and supply chain management [6, 8].

The key purpose of this research is to identify and analyse information on the use of applied blockchain-based solutions in the sphere of ESG management of entrepreneurship in digital economy markets. This involves the generalisation of theoretical and practical provisions of the ESG concept and the determination of the directions for its interaction with blockchain within the environmental, social and governances sectors of sustainable development.

3 Results

ESG principles – as the tools of sustainable development – began to be used officially in 2004 [10]. At present, their coordinated use in the practice of management and entrepreneurship is fully implemented as a concept that offers new horizons of responsibility of business in the environmental, social and governance spheres. In this case, the environmental sector covers the issues connected with CO2 emissions and other forms of environmental pollution, energy efficiency, rational use of water resources, waste management, biodiversity protection, determination of mutual dependence of business and natural ecosystems and implementation of environmentally friendly innovations.

The social direction envisages the resolution of the problems of social support for employment of different groups of the population, including children, forced or compulsory labour, healthcare and production and consumption safety, fight against discrimination, provision of equal opportunities for education, fight against poverty, supply chain management and development of communities. The management (governance) direction is aimed at the creation and compliance with the code of behaviour and responsibility of business, ensuring the transparency of information disclosure, fight against corruption, involvement of interested parties in the resolution of different problems, support for the rights of owners and shareholders, etc. [10].

The technology of blockchain has been widely used since 2008. After the implementation in the financial market, it gradually entered other sectors of the economy, including supply chains, governance, real property, the energy sector and healthcare. All advantages of the technology are based on the use of a unique structure of data, which involves the creation of a chain of blocks. Parameters of the blocks are recorded in other connected blocks, after which they cannot be changed. Such features form the advantages of blockchain, which include decentralisation and database security, transparency and safety of information, as well as the possibility to create smart contacts [6, 8].

Given the obvious advantages of blockchain, the creation of applied solutions for ESG management of entrepreneurship in world markets is implemented in different directions, which are aimed at solving the following problems:

- Tackling climate change
- Smart management of energy
- Formation of a stable supply chain
- Rational waste management
- Organisation of production based on the principles of sustainability
- Attraction of investments to solve sustainability problems
- Increase in the level of social involvement of economic agents and innovations in the resolution of environmental, social and governance problems

The available applied blockchain solutions allow implementing almost all Sustainable Development Goals within the ESG concept: SDG 2, SDG 4, SDG 6, SDG 7–13 and SDG 15.

The general interconnection of the applied blockchain solutions that are to solve the tasks of ESG management with the Sustainable Development Goals is shown in Fig. 1. The figure has a form of a map, where directions and applied blockchain solutions are grouped and placed in the corresponding proximity to the Sustainable Development Goals.

To better understand the mechanism of the effect of blockchain technologies on ESG management of entrepreneurship in digital economy markets, we shall dwell on the specific features of certain solutions. Sun Exchange and GreenX are kind of ecosystems of blockchain, which action is aimed at the coordination of interests and capabilities of investors with the construction of solar power plants around the world, especially in regions with a high level of solar radiation. Due to the proper verification of data through the connection with crypto platforms, the process of investing is agreed with tax or credit subsidies and stimulates private and institutional investors to invest in green energy.

Blockchain solutions for tackling climate change involve the support for payment for climate-responsible projects by using a part of funds received due to a high level of CO2 emissions. Most often, such measures are implemented in the form of tokenised carbon credits, which function with the help of the blockchain platform. This direction is implemented within the projects 1PLANET Marketplace, KlimaDAO and Taucan.

From the position of sustainable investments, the Swinca project allows anyone to invest in real property around the world. Due to blockchain, such investments are performed without financial establishments, which simplifies the procedure of concluding smart agreements and the process of investing. Within social interaction, the

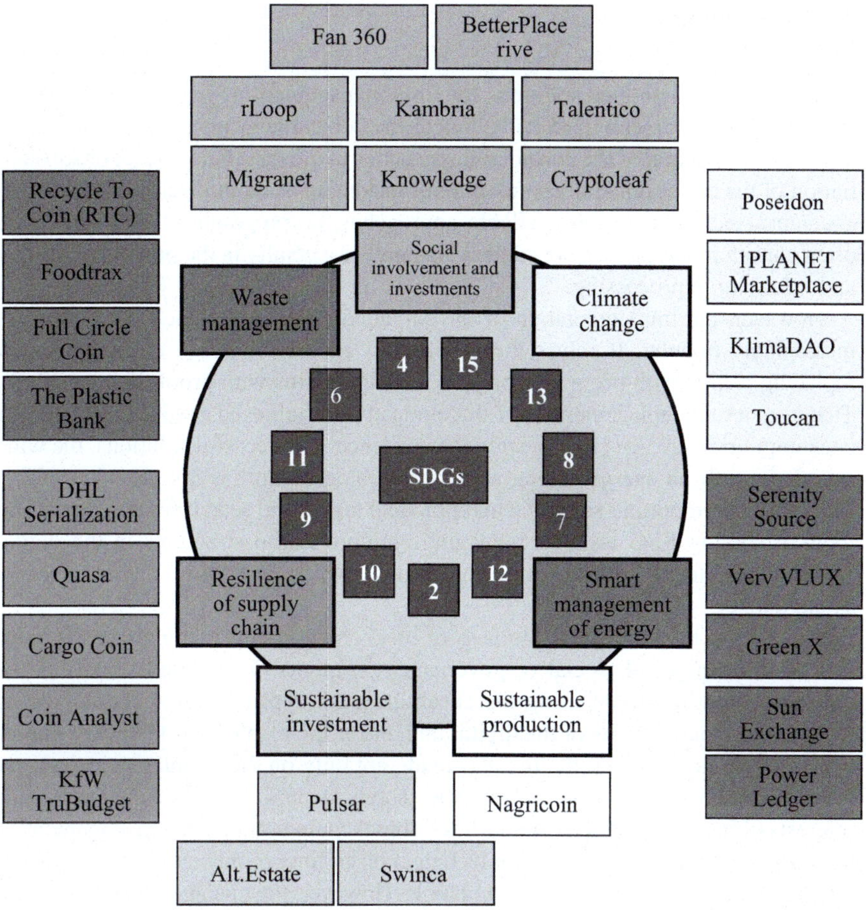

Fig. 1 Map of applied blockchain solutions to the problems of ESG management, grouped by the directions and the Sustainable Development Goals. (*Source* Created by the authors based on [1, 3, 7, 13])

cryptographic project Knowledge allows creating an integral ecosystem, within which some users of the system offer their knowledge, experience and information, while other users use them on a paid basis or exchange knowledge through the system of tokens. The project The Plastic Bank strives to attract residents of third world countries to the gathering of ocean plastic. In exchange for collected plastic waste, the project offer material reward in the form of mobile communication plans, money, cryptographic tokens, etc.

Similar mechanisms of using blockchain technology to solve the managerial problems of the environmental, social or governance character are used within other applied solutions. The capabilities of such solutions are significantly raised if blockchain is combined with other digital technologies, including cloud storages, sensor meters, IoT devices and artificial intelligence [14].

4 Discussion

Blockchain-based applied solutions for ESG management in entrepreneurial structures in digital economy markets belong to the debatable group of scientific issues. The main discussion of the considered issues lies in the insufficient level of substantiation of the environmental usefulness of blockchain as an innovative tool of management. According to the available approaches, cryptographic technologies are often considered as intense consumers of energy, especially in the process of mining and information processing.

However, the implementation of the so-called sustainable blockchain becomes increasingly popular. It solves the problem of excessive energy consumption by replacing chains with proof of work (PoW) with chains with proof of stake (PoS). Thanks to a such replacement, the Ethereum network achieved a reduction of energy consumption by 99.9% [13]. Despite the existence of successful solutions, the wide introduction of an energy-saving approach to blockchain has not yet taken place. Therefore, an important scientific and practical task is the search for the means and tools of solving ESG tasks of managing entrepreneurial structures in the digital economy by using blockchain technologies, which are based on an energy-saving model.

Apart from this, the successfulness of blockchain in contributing to the resolution of environmental, social or governance problems of mankind actualises the search for new means of using the capabilities of cryptographic technologies, to involve more participants of the digital market for the resolution of socially important tasks. In this context, the focus is made not only on the popularisation of successful applied solutions but also on the search for new means of integrating the interests of business and the capabilities of blockchain around ESG management. A large share of such tasks lie within technical disciplines, connected with the development of new technical solutions and tools. However, the managerial and entrepreneurial aspect of the problem, aimed at determining the directions for the use of available technologies and ensuring the support for their development and implementation, as well as expansion of their capabilities, also possesses a significant scientific and practical value and requires further research.

5 Conclusion

The ESG concept is an important element of the system of entrepreneurial structures management, which covers traditional and digital markets. This concept envisages entrepreneurial structures' resolving the tasks of the social, environmental or governance character, most of which do not belong to the category of profit-making but contain demands and interests of stakeholders and may be considered by entrepreneurs as an element of a business strategy aimed at sustainability – as a voluntary response to social problems or a tool of marketing promotion.

Active development of digital technologies creates additional possibilities for management systems, including the ESG sphere. The key tool of the digital economy that has a very large potential in solving the tasks of the environmental, social and governance character is blockchain. Due to the high level of reliability, transparency, decentralisation and possibility to service smart contracts, blockchain solutions are widely implemented in the spheres of tackling climate change, development of alternative energy, creation of sustainable supply chain, waste management, sustainable production, sustainable investing, social involvement and innovations. The existing applied solutions in these spheres allow achieving the Sustainable Development Goals and create integral blocks, united by common goals and technical solutions.

References

1. Accenture (2022) Blockchain for social impact. URL: https://www.accenture.com/us-en/services/blockchain/blockchainforgood-index. Accessed 24 Jan 2023
2. Berg F, Kölbel JF, Rigobon R (2022) Aggregate confusion: the divergence of ESG ratings. Rev Financ 26(6):1315–1344. https://doi.org/10.1093/rof/rfac033
3. Calandra D, Secinaro S, Massaro M, Dal Mas F, Bagnoli C (2022) The link between sustainable business models and Blockchain: a multiple case study approach. Bus Strateg Environ:1–15. https://doi.org/10.1002/bse.3195
4. Clavin J, Duan S, Zhang H, Janeja VP, Joshi KP, Yesha Y, Erickson LC, Li JD (2020) Blockchains for government: use cases and challenges. Digital Gov: Res Pract 1(3). Article 22, 21 P. https://doi.org/10.1145/3427097
5. Cornell B (2021) ESG preferences, risk and return. Eur Financ Manag 27:12–19. https://doi.org/10.1111/eufm.12295
6. Friedman N, Ormiston J (2022) Blockchain as a sustainability-oriented innovation? Opportunities for and resistance to Blockchain technology as a driver of sustainability in global food supply chains. Technol Forecast Soc Chang 175:121403. https://doi.org/10.1016/j.techfore.2021.121403
7. Gauri P (2022) How can blockchain help in the environmental crisis? How can blockchain help in the environmental crisis? The Economic Times. URL: https://economictimes.indiatimes.com/markets/cryptocurrency/crypto-weekly-authored-article/articleshow/91999466.cms?utm_source=contentofinterest&utm_medium=text&utm_campaign=cppst. Accessed 24 Jan 2023
8. Kouhizadeh M, Sarkis J (2018) Blockchain practices, potentials, and perspectives in greening supply chains. Sustainability 10(10):3652. MDPI AG. URL: https://doi.org/10.3390/su10103652
9. Lanxin J, Yu G, Wenjun Y, Jun D (2022) Blockchain-based life cycle assessment system for ESG Reporting (May 28, 2022). URL: https://ssrn.com/abstract=4121907 or https://doi.org/10.2139/ssrn.4121907. Accessed 24 Jan 2023
10. Li T-T, Wang K, Sueyoshi T, Wang DD (2021) ESG: research progress and future prospects. Sustainability 13(21):11663. https://doi.org/10.3390/su132111663
11. Novikova K (2019) Top 5 blockchain projects for ecology. Examples of latest blockchain projects aimed at maintaining the ecology and environmental protection. DigitalForest Blog. URL: https://digiforest.io/en/blog/5-blockchain-eco-cases. Accessed 24 Jan 2023

12. Pérez JM (2023) How blockchain can maximise the potential of ESG initiatives. FinBoot Tech. URL: https://www.finboot.com/post/how-blockchain-can-maximise-the-potential-of-esg-initiatives. Accessed 24 Jan 2023
13. Renton S (2022) 2023: the year blockchain becomes a sustainability solution. Opinion. Consensus Magazine. Coin Desk. URL: https://www.coindesk.com/consensus-magazine/2022/12/13/sustainable-blockchains/. Accessed 24 Jan 2023
14. SAF (2022) Blockchain for sustainable energy and climate in the Global South. Use cases and opportunities. Social Alpha Foundation. United Nations Environment Programme (UNEP). URL: http://www.socialalphafoundation.org/wp-content/uploads/2022/01/saf-blockchain-report-final-2022.pdf. Accessed 24 Jan 2023
15. UNEP-CCC (2022) The Emissions Gap Report (EGR) 2021. UNEP, UNEP Copenhagen Climate Centre. URL: https://www.unep.org/resources/emissions-gap-report-2021. Accessed 24 Jan 2023

Peculiarities of Financing ESG Investments in the Russian Market

Anna A. Gamilovskaya, Yana A. Kalugina, Mariia N. Koniagina, Arkadii G. Mikhailov, and Dmitry V. Sidorkin

1 Introduction

The increasing importance of the ESG approach to investing has occurred gradually over the past 15 years. The foundation was laid in 1970–1980, when the world community, led by the UN, recognized several environmental and social problems as significant and requiring action by economic actors. The concept of sustainable development was introduced, within the framework of which the principles of responsible investment (UN PRI) were formulated in 2006. These principles boiled down to the fact that when carrying out pre-investment analysis, as well as when making investment decisions, investors should consider environmental, social, and governance aspects; the objects of investment (public companies) should disclose information about their activities in the field of sustainable development and ESG approach. However, according to the World Economic Forum report "The Global Risks Report 2020" [23], from 2007 to 2011, purely financial risks were most important at the global level: the development of crises, widening income gaps, unemployment, imbalances in financial systems, price risks, and others. That is, the active implementation of the ESG approach was not a priority in the struggle for global stability.

A. A. Gamilovskaya (✉)
Financial University under the Government of the Russian Federation, Moscow, Russia

Y. A. Kalugina · M. N. Koniagina
St. Petersburg State Marine Technical University, St. Petersburg, Russia

A. G. Mikhailov
Financial and Investment Holding "Leader Consult", St. Petersburg, Russia
e-mail: a.mihailov@lc-rus.com

D. V. Sidorkin
Saint Petersburg University of the Ministry of Internal Affairs of the Russian Federation, St. Petersburg, Russia

E. G. Popkova (ed.), *Smart Green Innovations in Industry 4.0*, Springer Climate,
https://doi.org/10.1007/978-3-031-45830-9_23

The transition period began in 2012 and lasted until 2017; the role of social risks increased significantly. The increase was largely due to the geopolitical instability of certain regions, which led to forced migration and its consequences. The environmental group appeared in the structure of the most significant risks [11]. In response to new challenges, in 2015, the UN formulated 17 Sustainable Development Goals 2030 (SDGs), which were endorsed by all member countries and formed the basis of the ESG approach to investment.

The absolute primacy of environmental and social risks began in 2018. These groups of risks lead in terms of the level of potential damage and the probability of their realization. Of the 13 risks in the critical zone in 2021 [24], six are environmental, and four are social, which substantiates the importance and relevance of work in the field of sustainable development. In this regard, the issue of financing the ESG transition becomes relevant.

The experience of other countries on the issue of financing ESG investments is quite extensive and is actively developing. The USA and the EU are the undisputed leaders. As of 2020, the USA accounted for 48% of sustainability assets, the EU accounted for 34, and Japan accounted for 8% [20]. The Russian Federation engaged in this process with a significant delay; active work began only in 2018. By the time the first financial ESG asset was issued in Russia, such assets were circulating in the world for 585 million dollars.

Recently, the consequence of high geopolitical tensions has been the reorientation of Russian sales markets from European to Asian platforms. Thus, the integration of production chains and B2B interactions with one of the leaders of the ESG transition has significantly weakened. Simultaneously, cooperation with EU countries was one of Russia's most significant incentives for the ESG transition. The above factors raise the question of determining the current level of the development of ESG financing in Russia, as well as its sufficiency, diversity, and prospects.

2 Materials and Method

The financing of ESG investments in Russia has been considered in the works of the following authors: O. S. Belokrylova [8], N. I. Ivanova [12], L. S. Kabir [13, 14], O. I. Klyutchnikov [15], I. N Makarov [16], and B. N. Porfiriev [18].

The methodological basis of the research is presented by the following methods: analysis of normative-legal base [1–7] and statistical data, schematic modeling of institutes of organization of financial support of ESG investments, classification of financial instruments of ESG investments realization, and graphic method of information presentation.

3 Results

The theoretical essence of ESG investments is derived from three words: environmental, social, and governance. In other words, ESG investments are investments of money in assets that implement their business on the principles of environmental friendliness, social responsibility, and quality management.

In the Russian Federation, there is an alternative approach to this abbreviation, developed with the participation of the Financial University under the Government of the Russian Federation. According to this approach, ESG is interpreted as EHG:

- E – the environmental component, characterized by the degree of impact on the environment
- H – the human resources policy (level of wages, availability of social package, and participation in charitable projects)
- G – the government (economic sustainability, tax, history and business reputation, and availability of social investments) [9]

However, for the most part, the market follows a global approach.

3.1 Institutional Approach to Studying the Financing of ESG Investment in the Russian Market

The functioning of any industry is built on three basic components:

- The regulatory framework that governs the relationship in the industry
- The institutions that have the appropriate powers
- The set of tools with which these powers are implemented

Regulatory and legal support for ESG financing in the Russian Federation is represented by a number of acts, the main of which are presented in Table 1.

The basic regulatory document is the "Taxonomy of Green and Adaptation Projects and Requirements for the Verification System" (approved by the Decree of the Government of the Russian Federation on September 21, 2021 No. 1587) (the Taxonomy) [6]. The taxonomy contains qualitative and quantitative criteria and requirements for verifying ESG investment projects. In general, the independent legal regulation of financial direction is not broad; most of the changes are made by amending the already existing acts. The normative part is more developed due to the need for clarifying and methodological explanations. Collectively, these documents form the foundation for the operation of the consortium of ESG funding institutions (Fig. 1).

The central ESG institution is the Ministry of Economic Development, empowered by the Government of the Russian Federation with the coordinating function of implementing the financial support of ESG investment.

Table 1 The main normative and legal acts in the field of ESG financing in the Russian Federation

Level	List of regulations
Orders and Decrees of the Government	Decree of the Government of the Russian Federation "On the coordinating role of the Ministry of Economic Development of Russia on the development of investment activities and attraction of extra-budgetary funds in sustainable (including green) development projects in the Russian Federation" (November 18, 2020 No. 3024-r) [5] Decree of the Government of the Russian Federation "On approval of the goals and main directions of sustainable (including green) development of the Russian Federation" (July 14, 2021 No. 1912-r) [7] Decree of the Government of the Russian Federation "On approval of the criteria for development projects in the Russian Federation and methodological guidelines aimed at achieving the goals and main directions of sustainable (including green) development in the Russian Federation" (September 21, 2021 No. 1587) [6].
Documents of the Central Bank of the Russian Federation	Regulation of the Bank of Russia "On standards of securities issue" (December 19, 2019 No. 706-p) [1]; Information letter "On recommendations for the implementation of the principles of responsible investment" (July 15, 2020 No. IN-06-28/111) [2] Information letter "On recommendations for disclosure by public joint stock companies of non-financial information related to the activities of such companies" (July 12, 2021 No. IN-06-28/49) [3] Information letter "On accounting of climatic risks in the activities of certain financial market participants" (August 17, 2021 No. IN-015-38/64) [4].
Exchange documents	Listing Rules of PJSC "Moscow Exchange" [17]
VEB.RF documents	IRIIS Impact and responsible investing for infrastructure sustainability [21]

Source Compiled by the authors based on the report "ESG and green finance in Russia 2018–2022" [10]

The methodological center for organizing the functioning of green financial instruments is the Corporation for Economic Development VEB.RF, which develops normative and legal acts, collects and analyzes statistical data, and exercises the powers of investors and co-investor in the collective financing of ESG projects.

An Interdepartmental Working Group (IWG) focused on the development of investment activities and attracting extra-budgetary funds for ESG projects was established through the Ministry of Economic Development to coordinate the actions of ministries, agencies, sector regulators, and other market participants.

As the primary financial market regulator, the Central Bank of Russia plays a key role in financing ESG projects through lending and other financial instruments. It also oversees its subordinate institutions.

Within the areas supervised by the Central Bank of Russia, there are working groups with thematic subgroups of narrower specialization. In terms of organizing the financing of ESG projects, the Central Bank has introduced new standards for issuing green and social bonds [1]. Furthermore, it has issued recommendations for responsible investment directed at institutional investors [2, 4]. The Bank of Russia

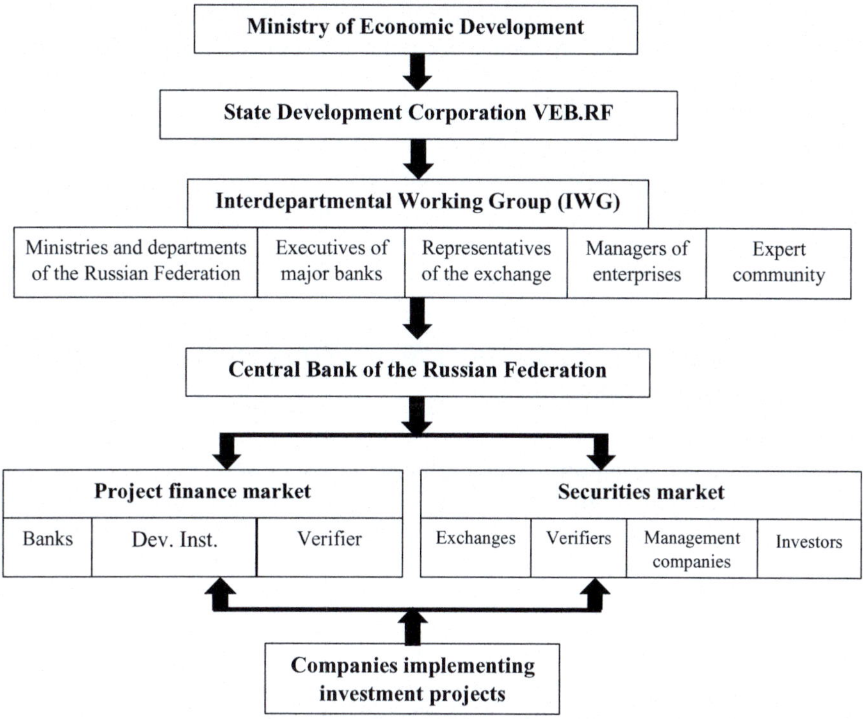

Fig. 1 Institutional structure for financing ESG investments. (*Source* Developed by the authors)

has also introduced new provisions for the disclosure of information by issuers regarding the publication of data on ESG projects [3].

The ESG project financing market operates in two ways. The first is project financing, encompassing traditional and syndicated bank lending. In this regard, the primary platform is the VEB.RF Project Finance Factory, which accounts for more than 10% of implemented syndicated lending in Russia in 2020 [22]. Additionally, there are approximately 30 more private banking platforms. Besides participating in project financing, banks are developing their own business lending systems based on ESG assessments (e.g., Sberbank, VTB, Raiffeisenbank, and Gazprombank). Such lending is frequently offered at preferential rates, often linked to specific initial KPIs. Presently, reporting remains a significant challenge in project finance and lending. While Central Bank has regulated reporting disclosure requirements, concerns about the reliability of data provided by borrowers concerning the implementation of planned ESG projects persist, leading to a high level of investor distrust.

The second approach involves attracting funds from investors using securities market instruments. In this context, the stock exchange plays a crucial role as the central conduit for channeling capital into ESG projects. With the support of the Central Bank, the Moscow Exchange, Russia's primary trading platform, has established a sustainable development section comprising green bonds, social bonds, sustainable development bonds, and a national and adaptation projects segment. The platform enables both private and institutional investors to invest in ESG financing instruments.

However, for a financial instrument to receive ESG status, it must undergo verification by a specialized organization. The Methodological Center of VEB.RF maintains a list of such organizations. As of 2022, this list includes four rating agencies (ACRA, NRA, NCR, and Expert RA), one auditing company (FBK), and other legal entities.

3.2 Instruments of Financing ESG Investments in the Russian Securities Market

Investors may purchase instruments for financing ESG investments on the Russian market independently or through the financial intermediation of management companies. Instruments for financing ESG investments in the Russian financial market include bonds and responsible investment funds.

3.2.1 Bonds

The regulation of the Bank of Russia "On standards of securities issue" (706-p) [1] outlines the issuance of the following types of bonds:

1. Bonds related to the financing or refinancing of projects aimed at preserving and protecting the environment (green bonds).
2. Bonds related to the financing or refinancing of adaptation projects (adaptation bonds).
3. Bonds related to the financing or refinancing of projects aimed at the development of public life (social bonds).
4. Bonds related to the financing or refinancing projects for the creation or reconstruction of the property.

From 2018 to 2021, there have been 33 issues of sustainability bonds in Russia, totaling approximately 418 billion rubles [10]. Funds were raised through bond placements on the Moscow Exchange, foreign exchanges, and over-the–counter markets. Comparing the list of allowed types of ESG bonds under the law with the actual issued instruments, it becomes evident that bonds related to property reconstruction are not in high demand. Additionally, the segment of adaptation bonds is currently represented by only one issue. The limited issuer activity in these areas may be attributed to the time required for the project preparation and implementation, as this bond type only emerged in Russia in 2021. Nevertheless, the introduction of adaptation bonds is significant for the market as it allows a broader range of companies to secure concessional financing for gradually transitioning to ESG standards.

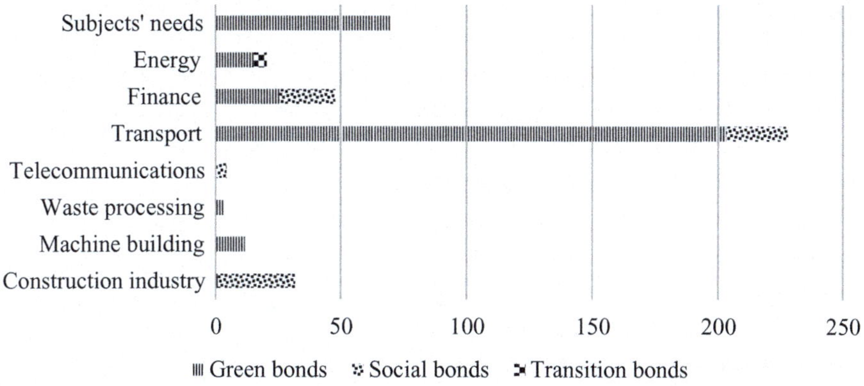

Fig. 2 Volume of ESG bond issues in Russia from 2018 to 2021 by industry, billion rubles. (*Source* Compiled by the authors based on the report "ESG and green finance in Russia 2018–2022" [10])

Green bonds constitute the largest volume of issued bonds, with a significant portion related to transportation (Fig. 2).

This distribution can be attributed to the increasing adaption of urban electric transport. It is worth noting the entrance of the city of Moscow into the financial market in 2021 with an environmentally focused bonded loan of 70 billion rubles. The construction industry has also shown substantial activity in developing and implementing real estate projects serving the public interest.

3.2.2 Responsible Investment Funds

Responsible investment funds began to appear in Russia in early 2020. By the end of 2021, there were six funds managed by five management companies (MCs). In 2021, an additional three funds were introduced. As of December 30, 2021, the total net asset value (NAV) of responsible investment funds operating in Russia, including open-end and exchange-traded funds, amounted to about 16 billion rubles [10]. According to the "Investing in sustainable recovery" report [20], in 2020, foreign companies managed 1738 million dollars in assets, with the majority in the EU (1373 million dollars). The strained geopolitical situation significantly worsened this gap. The NAV of Russian funds decreased by nearly half, reducing their attractiveness to foreign investors (Table 2).

The study of return on ESG investments in Russia from the position of analysis of the dynamics of the index "RSPP Sustainability Vector of the Development of Russian Issuers" relative to the MICEX general index (IMOEX) showed an almost complete identity of movement; the calculated correlation coefficient is 0.987, which indicates the absence of increased returns on investment in ESG securities of Russian securities market issuers.

Table 2 NAV of the main investment funds of the ESG sector at the end of 2022

	Management company	Name	Type	NAV (rubles)
1	VIM investments (formerly VTB Capital)	"Equity Fund. Responsible investments"	Open-end mutual fund	500,371,079.07 (as of November 25, 2022)
2		"Bond Fund. Responsible investments"	Open-end mutual fund	247,113,241.31 (as of November 25, 2022)
3		"Fund for sustainable development of Russian companies"	Exchange-traded mutual fund	104,694,185.06 (as of November 25, 2022)
4	RSHB Asset Management	"RSHB – MosBeX Index – RSPP Sustainability Vector, Total Gross Return"	Exchange-traded mutual fund	97,468,277.36 (as of November 25, 2022)
5		"RSHB – Russian Corporate Eurobonds, vector E.S.G."	Exchange-traded mutual fund	91,127,744.21 (as of November 25, 2022)
6	Tinkoff Capital	"Tinkoff Green Technology Index"	Exchange-traded mutual fund	568,509,180 (as of November 25, 2022)
7	Sberbank Asset Management	"Sber – Responsible Investment"	Exchange-traded mutual fund	935,043,183.10 (as of November 25, 2022)
8	Raiffeisen Capital	"Raiffeisen – USA Fund"	Open-end mutual fund	6,306,239,125.61 (as of November 25, 2022)
9	TRINFICO	"TRINFICO ECO FINANCE"	Open-end mutual fund	2,532,681.45 (as of November 25, 2022)
Total:				**8,853,098,697.17**

Source Developed by the authors

4 Conclusion

In view of the global trend, as well as the raw material orientation of the Russian economy, the significant number of industrial facilities, the vast territory, and the low quality of life, the market for sustainable investments in the Russian Federation has great development potential. However, the current level of implementation of the ESG approach is in its infancy. The volume of investment lags many times behind Europe, and the number of financial instruments is insufficient and does not cover all opportunities provided by the regulatory framework. Further development is complicated by geopolitical conditions and the reorientation of the Russian economy from European to Asian markets. It is possible that this transition will not reduce the need to develop ESG financing. China is no less active in the transition to sustainable development and will require further synchronization of industries and processes. However, there are no predictions at this time.

There are also several problems in the functioning of the financing of the Russian market of ESG investment. Currently, the achievements within the corporate ESG processes are declared only by the companies. These achievements are not subject to verification. Accordingly, conclusions about the effectiveness of investments are highly subjective. The introduction of an external independent audit of the results of ESG projects will improve the institutional structure of the market and increase confidence, thereby expanding the range of potential investors.

The volume of investment in the ESG sector is clearly insufficient. The value of funds under management has almost halved due to high geopolitical tensions. It is possible to increase the amount of funding for ESG projects by prioritizing the placement of public investment as the primary source of liquidity in the short term. This can provide an additional positive effect in the form of an inflow of funds to socially responsible companies. The potential positive impact of the government is not limited to direct cash infusions. A positive effect can be achieved by subsidizing coupon rates on ESG bonds, providing tax incentives (including professional securities market participants), and including ESG sector securities in the Lombard list of the central bank to increase their liquidity and demand. Crowdfunding platforms are also a promising avenue for raising funds from private investors for small investment projects.

Further development and improvement of ESG projects can increase the investment attractiveness of the Russian Federation by improving the rating positions and, accordingly, the volume and value of funds raised. In 2021, Russia ranked 46th in the Sustainable Development Report 2021 [19], standing between Bulgaria and Bosnia and Herzegovina.

References

1. Central Bank of the Russian Federation (2019) Regulation of the Bank of Russia "On standards of securities issue" (December 19, 2019 No. 706-p). Bank of Russia, Moscow
2. Central Bank of the Russian Federation (2020) Information letter of the Bank of Russia "On recommendations for the implementation of the principles of responsible investment" (July 15, 2020 No. IN-06-28/111). Bank of Russia, Moscow
3. Central Bank of the Russian Federation (2021) Information letter of the Bank of Russia "On recommendations for disclosure by public joint stock companies of non-financial information related to the activities of such companies" (July 12, 2021 No. IN-06-28/49). Bank of Russia, Moscow
4. Central Bank of the Russian Federation (2021) Information letter of the Bank of Russia "On accounting of climatic risks in the activities of certain financial market participants" (August 17, 2021 No. IN-015-38/64). Bank of Russia, Moscow
5. Government of the Russian Federation (2020) Decree "On the coordinating role of the Ministry of Economic Development of Russia on the development of investment activities and attraction of extra-budgetary funds in sustainable (including green) development projects in the Russian Federation" (November 18, 2020 No. 3024-r), Moscow
6. Government of the Russian Federation (2021) Decree "On approval of the criteria for development projects in the Russian Federation and methodological guidelines aimed at achieving

the goals and main directions of sustainable (including green) development in the Russian Federation" (September 21, 2021 No. 1587), Moscow
7. Government of the Russian Federation (2021) Decree "On approval of the goals and main directions of sustainable (including green) development of the Russian Federation" (July 14, 2021 No. 1912-r), Moscow
8. Belokrylova OS, Situkho AN (2020) Prospects for green investment in technologies of the fourth industrial revolution. In: Project and investment management in the digital economy: proceedings of the national scientific and practical conference, Krasnodar, pp 30–36
9. EKG Rating (2021) Assessment of the "health" of companies aimed at determining the level of their reliability, social and environmental responsibility. Retrieved from https://экг-рейтинг.рф/. Accessed 22 Nov 2022
10. ESG Consulting (2022) Report "ESG and green finance in Russia in 2018–2022", Moscow. Retrieved from https://esg-consulting.ru/wp-content/uploads/2022/03/infragreen_green_finance_esg_in_russia_2018-2022.pdf?ysclid=lbaus5flq4653023824. Accessed 14 Nov 2022
11. Gamilovskaya AA (2020) Transformation of global risks: causes and consequences. In: Modernization of Russian society and education: new economic benchmarks, management strategies, law enforcement, and training issues: proceedings of the XXI National Scientific Conference (with international participation). Taganrog Institute of Management and Economics, Taganrog, pp 99–102
12. Ivanova NI, Levchenko LV (2017) "Green" economy: the essence, principles, and prospects. Herald of Omsk University. Ser: Econ 2(58):19–28
13. Kabir LS (2019) State support for "green" investments and market "green" financing: foreign experience. Innov Expert Exam 1(26):97–108
14. Kabir LS, Sigova MV (eds) (2020) Green finance: the process of development and prospects for transformation. International Banking Institute named after Anatoliy Sobchak, St. Petersburg
15. Klyutchnikov OI (2019) "Green" finance as a production function of a sustainable environment. Proc Int Bank Inst 3(29):52–64
16. Makarov IN (2018) Formation of the project financing system for infrastructure network development: theory, methodology, and practice (Dissertation of Doctor of Economic Sciences). International Banking Institute named after Anatoliy Sobchak, St. Petersburg
17. Moscow Exchange (2021) Listing rules of PJSC "Moscow Exchange." Retrieved from https://rcgroup.pro/tpl/pdf/listing-rules.pdf?ysclid=lb3gsdiqua845959545. Accessed 28 Nov 2022
18. Porfirev BN (2016) Green trends in the global financial system. World Econ Int Relat 60(9):5–16. https://doi.org/10.20542/0131-2227-2016-60-9-5-16
19. Sustainable development report 2021: rankings. Retrieved from https://dashboards.sdgindex.org/rankings. Accessed 28 Nov 2022
20. UNCTAD (2021) World investment report 2021: investing in sustainable recovery. Retrieved from https://unctad.org/system/files/official-document/wir2021_en.pdf. Accessed 28 Nov 2022
21. VEB.RF (n.d.) IRIIS Impact and responsible investing for infrastructure sustainability. Retrieved from https://veb.ru/downloads/iriis_draft_methodology.pdf. Accessed 28 Nov 2022
22. VEB.RF (n.d.) Project finance factory. Retrieved from https://вэб.рф/biznesu/fabrika-proektnogo-finansirovaniya/?ysclid=lb0udwulu516662150. Accessed 22 Nov 2022
23. World Economic Forum (2020) The global risks report 2020, 15 Jan 2020. Retrieved from https://www.weforum.org/reports/the-global-risks-report-2020. Accessed 14 Nov 2022
24. World Economic Forum (2021) The global risks report 2021, 19 Jan 2021. Retrieved from https://www.weforum.org/reports/the-global-risks-report-2021/. Accessed 14 Nov 2022

Digitalization of the Nuclear Industry for Sustainable Development

Anastasia V. Sheveleva and Marina A. Jordanovski

1 Introduction

The observed global scale of natural resource use, as well as major human-made impacts leading to further increases in resource use, has given rise to justified concerns about the future of the environment and the planet at large. The search for a solution to prevent a socio-environmental collapse has involved governments, international organizations, and businesses.

Despite the global geopolitical situation, sustainable development remains an issue of great relevance. Its importance has been reaffirmed at the highest levels. For instance, in 2015, the UN adopted a new program titled "Transforming our World: The 2030 Agenda for Sustainable Development." In addition to achieving more sustainable economic development and fighting poverty, this program seeks to remedy the situation brought about by the irreversible effects of climate change.

Under the aegis of the United Nations, as well as international political and economic unions, efforts are being made toward the achievement of sustainable development goals, which aim to solve the problems mentioned above at the global and regional levels. To ensure the effective implementation of a number of these goals, it is imperative to forge tools for sustainable development and look for new ways of solving global issues at the level of countries and individual entities (regions, cities, and businesses). One of the most promising tools at the micro level is the use of digitalization as a driver of sustainable development. The contemporary economy is based on continuous improvement, i.e., enhancing, updating, and inventing new management models, products, and systems. In recent years, this mode of development has been directly linked to the concept of digitalization. Many authors mistakenly assume that digitalization is a separate process for company-wide automation linked solely to increased labor productivity and production modernization.

A. V. Sheveleva (✉) · M. A. Jordanovski
MGIMO University, Moscow, Russia
e-mail: a_sheveleva@rambler.ru; m.a.jordanovski@inno.mgimo.ru

E. G. Popkova (ed.), *Smart Green Innovations in Industry 4.0*, Springer Climate,
https://doi.org/10.1007/978-3-031-45830-9_24

However, the current approach requires every process and phenomenon to be studied comprehensively, which is why it is essential to give a theoretical and practical substantiation outlining the effects of digitalization on improving a company's economic indicators, reducing its environmental impacts and advancing its stakeholder relationships as the tools for assessing the effects of digitalization on sustainable development become increasingly relevant.

In this regard, the nuclear industry is no exception. This is despite the fact that the market is quite narrow due to the small number of businesses offering a full range of services for construction and subsequent maintenance of nuclear power plants. Nevertheless, competitiveness in this market is a major factor in finding potential customers; adherence to sustainable development paradigms gives businesses a distinct advantage.

In today's world, many experts view nuclear power as an eco-friendly energy source producing the smallest possible amounts of harmful emissions into the environment. However, this is not quite the case: uranium is a non-renewable resource, nuclear power plants (NPPs) produce waste, and the disasters that took place at the Chornobyl NPP in 1986 and the Fukushima NPP in 2011 have caused harm not only to the environment by polluting the air and water resources with nuclear substances but also to human lives and health (Table 1).

This is why the risk of error at NPPs must be reduced to zero. The authors believe that digitalization, which has a special role in implementing global sustainable development goals specifically, could be helpful in this regard.

Therefore, it seems relevant to analyze the practice of application of digital technology by nuclear companies in Russia and abroad to understand the experience accumulated in this field, future prospects, and options for NPPs that only start to integrate digital technologies in their activities.

2 Methodology

The authors have canvassed numerous papers, reviewing and analyzing the existing opinions on the digitalization of the nuclear industry for sustainable development, and have concluded that the digitalization of the nuclear industry significantly improves operating safety at NPPs, makes business processes more efficient, and enables sustainable development.

Issues related to digitalization in Russia and other countries alike have been discussed by Russian and foreign authors [1, 7, 8, 14, 16, 18].

A number of papers have been written on using digital technologies to achieve sustainable development goals [4, 5, 11, 13, 15, 17, 19, 20].

Based on the aforementioned papers, it appears that integrating the use of digital technologies in the nuclear industry will allow Russia and other countries in the long term to reach the objectives of improving the peaceful use of the atom and facilitate the implementation of sustainable development goals.

In analyzing these materials, the authors have used various methods, such as compilation, systemic approach, synthesis, and comparative analysis, to draw the conclusions outlined below.

Table 1 Consequences of accidents at Chornobyl and Fukushima NPPs

	Chornobyl	Fukushima
Level on the International Nuclear and Radiological Event Scale (INES)	Level 7. Major accident (highest level)	Level 7. Major accident (highest level)
Causes of accident	Human error and violation of procedures	Earthquake and tsunami
Highest level of radiation detected	300 sieverts/h soon after the explosion near the reactor core	530 sieverts/h inside the Unit 2 containment vessel in 2017, according to the *Japan Times*
Radioactivity released	According to the IAEA, the release totaled 14 EBq (14,000 PBq), including 5.2 EBq (5200 PBq) in iodine-131 equivalent	Experts estimated that as of 2014, the total amount released was 340–780 PBq, with 80% in the Pacific Ocean; radiation continues to seep from the groundwaters
Affected area	According to the UN, the contaminated area stretches to 500 km (310 mi)	According to officials, radiation levels exceeding the annual limits are observed within 60 km (37 mi) to the northwest and 40 km (25 mi) to the southwest
Direct victims of the accident	2 trauma deaths; 28 deaths from acute radiation syndrome; 4 deaths in an industrial accident (helicopter crash); 15 deaths from radiation-induced genetic thyroid cancer; 4000–9000 cancer-related deaths	One confirmed death from cancer, deaths from radiation expected in the long term
Current status	All reactors were shut down by the year 2000. The damaged reactor was encased in a steel-and-concrete sarcophagus. In November 2016, a new safe structure was installed to facilitate the decontamination and decommissioning of the plant	A cold shutdown was announced on December 16, 2011. The decommissioning process might take 30–40 years. All fuel rods in the Unit 4 cooling pool were removed. Mitigation efforts continue

Source Compiled by the authors

3 Results

According to the World Nuclear Industry Status Report 2022 [9], the countries with the largest number of nuclear reactors in the world are the USA, France, China, and Russia (Table 2).

The countries with the largest number of nuclear reactors are primarily interested in developing technologies conducive to preventing incidents and reducing costs.

In the USA, outdated reactors have been undergoing digital modernization since 2012, which is indicative of a major shift in climate change action. Digital systems can warn reactor operators about faulty components, automatically intervene to eliminate problems and help solve many issues by updating software instead of searching for old, hard-to-access, and expensive components. In 2019, Reactor Number One (PUR-1) of Purdue University in the state of Indiana replaced all

Table 2 Top 10 countries with the most nuclear reactors in operation

Country	Number of reactors		Share of power produced by NPPs, %
	2011	2022	
USA	103	92	19.6
France	58	56	69
China	12	55	5
Russia	32	38	20
South Korea	21	24	28
India	20	19	3.2
Canada	18	17	14.3
Ukraine	19	15	55
UK	15	11	14.8
Japan	48	10	7.2

Source Compiled by the authors based on the World Nuclear Industry Status Report 2022 [9] and World Nuclear Energy [6]

obsolete analog controls with state-of-the-art digital ones. In 2021, the U.S. Energy Department allocated $27 million to support digital twin projects in the nuclear industry. In October 2022, researchers at Idaho National Laboratory performed the first digital twin test for a simulated microreactor. This technology will make it possible to carry out remote monitoring, autonomous control, and forecasting at NPPs, thereby reducing the operating costs of microreactors and making them safer.

In France, many existing nuclear facilities use risk-analysis software. In 2020, France's Framatome, together with the EDF Group, the Alternative Energies and Atomic Energy Commission, and other partners, launched an ambitious initiative. In the course of 4 years, more than 100 experts will be creating virtual clones of the reactors operating in France. It is planned to use these clones as training simulators for operators and as simulation environments for engineering experiments and research.

Chinese authorities emphasized the safe development of the nuclear industry. Wang Shoujun, President of the Chinese Nuclear Society, has put forward a proposal to create a special fund for the digital transformation of the nuclear industry, adopt relevant standards, devise a policy fostering localization efforts, and actively promote digital products [16].

In 2021, a wireless Nu-WiFi system by the Chinese company UltraPower was installed at Units 5 and 6 of the Tianwan NPP, one of the largest NPPs in China. The system withstands a high level of radiation and keeps data secure. Additionally, Nu-WiFi supports mobile communication, precise positioning, broadband data transmission, and access to the IoT. This technology helps obtain full and accurate information about equipment, facilitating the operational stability of the facility.

In Russia, the digitalization process is spearheaded by, among others, State Atomic Energy Corporation Rosatom. In February 2021, the company commenced an IT project titled "Digital NPP Operation Template." The project offers a unified industry-specific digital solution for the efficient operation of nuclear facilities in Russia and abroad, which ensures that the processes at NPPs function in an optimal way from the design stage to decommissioning [12].

Several versions of digital twins depending on their field of application have been developed within Rosatom. The first version, intended for developing NPPs, is called the research version. This digital twin helps engineers optimize design and engineering and technological solutions and save time on looking for the best version of an NPP project. The twin plays an important role in implementing new technologies that have never been used before. Moreover, it can help address flaws in the technologies that have already been implemented or improve their quality, as NPP construction usually follows the principle of going from the known to the unknown because there must be absolutely no margin for error [2].

The second one is the design version. This digital twin facilitates various scenarios of NPP operation. It applies first and foremost to use a scenario in normal conditions to get as much data as possible about every element at play. The design version also has practical applications and is used during emergencies to develop the most useful solutions with minimal consequences [2].

The VR/AR model offers full immersion into a virtual reality through which one can address various tasks pertaining to business and production processes in the nuclear industry. The model helps improve the quality of employee training and increase the safety of operations involving high-risk technological facilities. The uses of the virtual/digital NPP include identification of possible accidents, emergency response measures, and personnel training during accidents because these events cannot be simulated at a real NPP [3].

On the whole, digitalization makes the nuclear industry safer and helps achieve sustainable development goals (Fig. 1).

For instance, big data technologies help reduce the risk of error when analyzing important indicators by 25%: installation of sensors at the core units of a nuclear power plant makes it possible to control equipment conditions and monitor the system's life cycle [10].

Digital twins can be used as training simulators for operators and as simulation environments for engineering experiments and research. Virtual NPPs will be useful in simulating any mode of operation at power units, from normal operations to complex emergencies, thereby preventing incidents and reducing costs.

Virtual reality technologies are indispensable in training engineers and specialists whose qualifications and actions are crucial to NPP safety. VR models offer an opportunity to "visit" an NPP, take a look at the equipment, and learn to work with turbines, which is conducive to making an informed decision if an emergency occurs in reality.

Predictive analytics enable the collation and program analysis of all incoming data to generate predictions regarding equipment operation and possible malfunctions. This information can be used to perform preventive maintenance and anticipate abnormal situations.

Predictive analysis helps compile and study large amounts of information to detect and diagnose existing and future defects.

In October 2022, another product, the Digital Heat Supply software and hardware package, emerged on the energy market. This package aims to enhance the efficiency of power generation by enterprises via automation and analysis of certain business processes. Digital Heat Supply has three areas of application.

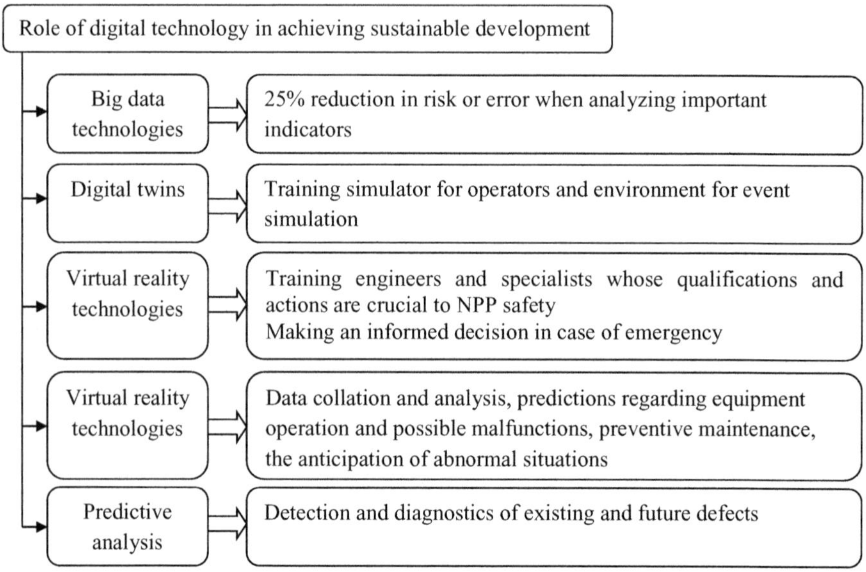

Fig. 1 Role of digital technologies in the sustainable development of the nuclear industry. (*Source* Compiled by the authors)

The first area is productivity enhancement and automation of a given utility company.

The second area concerns increasing the efficiency of power consumption by such consumers as kindergartens, hospitals, and houses. Since this package allows users to manage heat consumption and save resources properly, it is planned to reduce heat consumption by up to 40% while the number of emergencies is to decrease by up to 90%.

The third area was developed to assist the authorities in keeping oversight of utility companies by using business analytics technologies to respond to malfunctions in a timely manner and control the quality of preparations for the heating season.

These focus areas facilitate the achievement of a whole range of sustainable development goals, including industry, innovation, and infrastructure; sustainable cities and human settlements; climate action and environmental protection.

4 Conclusions

Digitalization is essential to the development of the nuclear industry. It applies to all processes, from direct control and management to business planning and documentation with the use of digital data and digital infrastructure. On the one hand, digital transformation is in its early stages. On the other hand, digitalization leads to substantial advances in the nuclear industry. Companies that use digital technologies enjoy a significant advantage over their competitors by nearly every indicator.

Digital transformation influences all metrics at a given company, i.e., production efficiency, safety, reliability, and compliance with legal acts in the field of environmental protection and sustainable development.

Russia pays particular attention to doctrine and theory, which are continuously put into practice. A good example is Rosatom and the digital technologies it has developed, namely, digital twins, big data technology, virtual reality technology, predictive analytics, and predictive analysis. These helped put forward ideas facilitating sustainable development of the nuclear industry. Continuous improvement of the existing technologies and adoption of new ones make it possible to simulate the processes of a nuclear power plant without conducting actual experiments. Therefore, it can be concluded that Russian digitalization looks to the future. These methods are conducive to environmental conservation and emergency response and mitigation at NPPs.

The authors believe that the Russian technologies being used are more efficient than their Western European and Asian counterparts. Foreign technologies are mainly focused on the real-time operation or, in other words, on making nuclear power plants operate more efficiently. Very little consideration is given to theory or adopting new management methods and increasing power capacity. Experience has shown that foreign companies start implementing new technologies after Russian companies have already proven their effectiveness in practice. It can be said that Russian technologies shape the trends in digitalization and sustainable development of the nuclear industry at the global level.

Therefore, the digitalization of the nuclear industry improves NPP safety, efficiency, and performance; helps monitor and control operations in real-time; enhances engineering, operating and production processes providing clients with various environmental and social benefits; and facilitates the implementation of sustainable development goals.

Acknowledgments The authors would like to thank the conference organizers for the opportunity to take part in this event. They would like to share the results of their research and hear the views of other conference participants on relevant and timely topics.

Additionally, the authors would like to express their gratitude to the reviewers for the attention they devoted to this research.

References

1. Arenkova IA, Lezina TA, Tsenzharik MK, Chernova EG (2019) Business management in the digital economy: challenges and solutions. St. Petersburg State University, St. Petersburg
2. Atomprom Bulletin (2021) Look into the future. A virtual digital nuclear power plant helps to predict the behavior of its real prototype, 22 Dec 2021. Retrieved from https://atomvestnik.ru/2021/12/22/zagljanut-v-budushhee/. Accessed 25 Nov 2022
3. Atomprom Bulletin (2021) Scaling new digital heights. Atomprom Bulletin 10:12–24. Retrieved from https://atomvestnik.ru/wp-content/uploads/2021/12/VA10_Book_spread.pdf. Accessed 25 Nov 2022

4. Baykoucheva S (2022) Driving science information discovery in the digital age: Chandos information professional series. Chandos Publishing, Cambridge. https://doi.org/10.1016/B978-0-12-823723-6.20001-3
5. Cassenti D, Scataglini S, Rajulu S, Wright J (eds) (2021) Advances in simulation and digital human modeling: proceedings of the AHFE 2020 virtual conferences on human factors and simulation, and digital human modeling and applied optimization, July 16–20, 2020. Springer, Cham. https://doi.org/10.1007/978-3-030-51064-0
6. Econs.online (2022) World nuclear energy, 3 Nov 2022. Retrieved from https://econs.online/articles/details/atomnaya-energiya-mira/. Accessed 25 Nov 2022
7. Gavrikova E, Volkova I, Burda Y (2022) Implementing asset data management in power companies. Int J Qual Reliab Manage 39(2):588–611. https://doi.org/10.1108/IJQRM-10-2020-0346
8. Lenchuk EB, Vlaskin GA (2018) Formation of the digital economy in Russia: problems, risks, prospects. Bull Inst Econ Russ Acad Sci 5:9–21. https://doi.org/10.24411/2073-6487-2018-00001
9. Mycle Schneider Consulting (2022) World nuclear industry status report 2022. Retrieved from https://www.worldnuclearreport.org/IMG/pdf/wnisr2022-lr.pdf. Accessed 25 Nov 2022
10. Nuclear Power 2.0 Research Portal (2020) How does the Internet of Things affect the nuclear industry? 15 Oct 2020. Retrieved from https://www.atomic-energy.ru/articles/2020/10/15/107953. Accessed 25 Nov 2022
11. Popkova EG, Zavyalova E (eds) (2021) New institutions for socio-economic development: The change of paradigm from rationality and stability to responsibility and dynamism. De Gruyter, Berlin/Boston. https://doi.org/10.1515/9783110699869
12. Rosenergoatom (n.d.) Digital products. Digital energy of the future. Retrieved from https://www.rosenergoatom.ru/stations_projects/opornyy-tsod/press-tsentr/novosti/DataCenter%200305%20rus.compressed.pdf. Accessed 25 Nov 2022
13. Sheveleva AV, Cherevik MV (2022) Digital technologies in the oil and gas sector and their contribution to UN Climate Action Goal. In: Zavyalova EB, Popkova EG (eds) Industry 4.0: fighting climate change in the economy of the future. Palgrave Macmillan, Cham. https://doi.org/10.1007/978-3-030-79496-5_28
14. Sheveleva AV, Tyaglov SG, Khaiter P (2021) Digital transformation strategies of oil and gas companies: preparing for the fourth industrial revolution. In: Konina N (ed) Digital strategies in a global market: navigating the fourth industrial revolution. Palgrave Macmillan, Cham, pp 157–171. https://doi.org/10.1007/978-3-030-58267-8_12
15. Soares I, Sousa RB, Petry M, Paulo M (2021) Accuracy and repeatability tests on HoloLens 2 and HTC Vive. Multimodal Technol Interact 5(8):47. https://doi.org/10.3390/mti5080047
16. Surwillo I (2020) Energy security logics in Europe: Threat, risk or emancipation? 1st edn. Routledge, London
17. Tyaglov SG, Sheveleva AV, Rodionova ND, Guseva TB (2021) Contribution of Russian oil and gas companies to the implementation of the Sustainable Development Goal of combating climate change. IOP Conf Ser: Earth Environ Sci 666:022007. https://doi.org/10.1088/1755-1315/666/2/022007
18. Vasileva L, Polyanskaya O, Khartanovich K, Milenkiy A, Tereshchenko S (2021) Digital technologies in the economy: opportunities and threats. IOP Conf Ser: Earth Environ Sci 806:012036. https://doi.org/10.1088/1755-1315/806/1/012036
19. Zavyalova EB, Krotova TG (2021) Methods for achieving SDG 1, poverty eradication. In: Popkova EG, Sergi BS (eds) Modern global economic system: evolutional development vs. revolutionary leap. Springer, Cham, pp 1987–2001. https://doi.org/10.1007/978-3-030-69415-9_218
20. Zhang X, Deng Z, Ma Q, Ming-Ming L (2021) Design and implementation of translation-based virtual DCS based on Simulink. In: Xu Y, Sun Y, Liu Y, Gao F, Gu P, Liu Z (eds) Nuclear power plants: innovative technologies for instrumentation and control systems. Springer, Singapore, pp 292–300. https://doi.org/10.1007/978-981-16-3456-7_29

Environmental Factors' Moderating Effect on Intangible Organizational Resources and Performance of Insurance Brokers in Zambia

Maikisa Ilukena, Lubinda Haabazoka, and Taonaziso Chowa

1 Introduction

As businesses operate, their activities are influenced by or spiked by the environment in which they are established [4]. The business environment encompasses various elements, such as institutions, individuals, groups, forces, policies, laws, and actions, that directly or indirectly creates opportunities or hinders the effective functioning of business organizations [4]. The business environment involves interactions between the business organization and its stakeholders at both macro and micro levels. The macro level includes factors such as Act-of-God, legal, international, political, economic, sociocultural, technological, and demographic aspects, while the micro level comprises the board of directors, employees, customers, suppliers, and owners [4].

Environmental factors significantly impact a firm's performance. These factors encompass political influences stemming from regulatory bodies, statutory bodies, and government policies [22, 33], as well as legal influences originating from national constitutions and economic factors, such as tax effects and inflation [41].

In Zambia, all insurance-related businesses, including insurers and insurance intermediaries, are regulated and supervised by the Pensions and Insurance Authority. The Pensions and Insurance Authority formulates policies, guidelines, and regulations for the effective supervision, regulation, and development of the sector, which all regulated entities are required to adhere to. Additionally, regulated entities must comply with other laws administered by various regulatory and statutory bodies, such as tax laws, competition regulations, and consumer protection

M. Ilukena (✉) · L. Haabazoka
Graduate School of Business, The University of Zambia, Lusaka, Zambia

T. Chowa
Graduate School of Business/School of Natural Sciences, The University of Zambia, Lusaka, Zambia

E. G. Popkova (ed.), *Smart Green Innovations in Industry 4.0*, Springer Climate,
https://doi.org/10.1007/978-3-031-45830-9_25

laws. Consequently, it is crucial to investigate whether these laws, regulations, and guidelines provide certain insurance brokers with an advantage, leading to their superior performance compared to others.

2 Theoretical Literature Review

2.1 Resource-Based View

The resource-based view is a theory that explains why some organizations perform better than others [12, 32]. This theory posits that resources possessing specific characteristics are critical determinants of firm success [5]. It emphasizes that possessing and controlling valuable, rare, inimitable, and non-substitutable resources will lead to a competitive advantage and, consequently, superior performance [5, 6].

A firm's resources include both tangible and intangible assets [43]. Research has demonstrated that intangible resources contribute more to a competitive advantage than tangible ones [5, 7, 18, 21, 42]. Thus, they are considered the main reasons as to why firms differ in performance. Intangible resources can be classified as follows:

1. Relational capital: Encompasses external relationships of the firm with investors, customers, suppliers, and competitors [10, 30, 36].
2. Structural capital: Represents the knowledge retained within the organization, including routines, software, databases, work procedures, organizational culture, etc. [8, 13, 16, 31, 34].
3. Human capital: Encompasses the knowledge, skills, experiences, and abilities of people [31].

3 Empirical Literature Review

3.1 Environmental Factors and Performance of Firms

The study conducted by Xavier [44] found that pricing decisions of firm products and services are influenced by both internal company factors and external environmental factors in which the firm operates. The study further found that external environmental factors such as taxation and inflation were more complicated factors to be handled by the firm due to their uncontrollable nature than internal factors. Lazzarin [28] observed that government industry policies affect the firm performance, either hindering performance potential or reducing performance [28]. Adeoye and Elegunde [2] found that the external business environment impacts organizational effectiveness, efficiency, sales, and achievement of corporate goals. Koumparoulis [26] observed that exploring environmental factors helps firms in devising competitive strategies that may lead to superior performance. Adeoye [3] found that government plays a crucial role in creating a favorable business environment.

Aka [4] found that inflation rate can affect the output of business operations with a 1% rise in the inflation rate leading to reduced business output. However, increasing interest rates can improve the contribution of business operations to the overall output of goods and services, along with reducing the unemployment rate. Chawdhury [11] found that an unstable political environment is detrimental to the speed of economic development for any country. The researcher also found that political factors such as wars, coups d'état, and transitions to and from democracy are significant barriers to international and domestic trade, affecting firms.

These studies highlight the critical importance of examining environmental factors, as they can either facilitate or hinder the performance of firms. The studies further show that the performance of firms does not wholly rely on internal factors only, also relying on external environmental factors. This research focuses on environmental factors. Environmental factors considered in this research comprised political, legal, and economic factors.

4 Methodology

In terms of research methodological approaches [39], the research adopted objectivism as the ontology [9], positivism as the epistemology [17], and a quantitative research strategy [27] as the methodology. This research employed an explanatory approach [25] and adopted deduction as the research approach process [14]. The research used inferential statistics, specifically regression analysis. The software for data analysis was Stata version 14. In terms of time horizon, the research is cross-sectional because data were collected from the population at one specific point in time [38]. The target population of interest in the research was licensed insurance brokers as of December 31, 2021. The total number of licensed insurance brokers was 61 as of December 31, 2020 [35], making the research sample essentially the total population of licensed insurance brokers, with a sample size of 61. The research used mainly primary data collected via a closed-ended questionnaire from insurance brokers, while secondary data were obtained from the Pensions and Insurance Authority.

Organizational concepts are measured at the organizational level [37]. Given that organizations cannot fill out a questionnaire by themselves or take part in a survey, data were obtained from a single participant, in this case, the Chief Executive Officer, who filled in the questionnaire on behalf of the insurance broker. The use of Chief Executive Officers in organizational research is common because organizations are a reflection of their top management [15, 22–24]. Furthermore, researchers of the past resource-based approach have used Chief Executive Officers as single informants for organizational research [1, 19, 21, 40].

4.1 Empirical Model

To establish whether environmental factors moderates the relationship between intangible organizational resources and the performance of insurance brokers in Zambia, the authors estimated two linear models, as shown below. Initially, model (1) is estimated to determine the nature of the relationship between dependent and independent variables [20, 33]. Next, the authors estimated model (2), which determines the effect of the moderator, in this case, environmental factors on the relationship [33].

$$Y = B0 + B1\mathrm{IOR} + B3\mathrm{E} + \varepsilon \tag{1}$$

$$Y = B0 + B1\mathrm{IOR} + B2\mathrm{E} + B3\mathrm{E.IOR} + e \tag{2}$$

where:

Y – performance
$B0$, $B1$, $B2$, and $B3$ – coefficients
IOR – intangible organizational resources
E – environmental factors
E.IOR – product of environmental factors X intangible Organizational Resources

To decide on whether environmental factors moderate the relationship between intangible organizational resources and the performance of insurance brokers in Zambia, the researchers tested two conditions, as recommended by Mackinnon et al. [29], Gakenia [20], and Njoroge [33]. First, the researchers tested whether environmental factors are significant when introduced into model (1) because this is the first condition of moderation. Second, the researchers tested whether environmental factors are significant where products of environmental factors and intangible organizational resources are introduced in model (2). To prove moderation, the coefficients for environmental factors in model (1) and model (2) have to be significant [41].

5 Results and Discussion

In terms of the response rate, 41 out of 61 respondents participated in the research, representing a response rate of 67.2%.

The research tested a hypothesis that sought to investigate whether environmental factors have a moderating effect on the relationship between intangible organizational resources and the performance of insurance brokers in Zambia. A null hypothesis (H0) was stated as follows: "Environmental factors have no moderating effect on the relationship between intangible organizational resources and performance of insurance brokers in Zambia." Table 1 shows the results of moderating effect of environmental factors on independent and dependent variables.

Table 1 Step 1. Results of moderating effect of environmental factors on independent and dependent variables

Goodness of fit	Test Statistic	*P*-Value	
Adjusted *R*-squared	0.3625		
F-statistic (8.32)	3.84	0.0029	
Dependent variable = Performance	Linear Regression Results		
	Coefficients	*t*-statistic	*P*-value
Intangible organizational resources	0.576	0.55	0.589
Environmental factors	−8.711	−1.40	0.170
Dummy 1: Majority shareholders			
Private individuals (foreign)	0.9162	0.60	0.552
Foreign company	7.070	2.48	0.019
Other shareholders	1.939	1.11	0.277
Dummy 2: Foreign company relations			
Subsidiary of foreign company	−2.887	−108	0.288
Associate of a foreign company	−2.584	−1.37	0.179
Other foreign company relations	3.5	3.01	0.005
Constant	27.095	2.62	0.013

Source Developed by the authors

$$Y = 27.095 + 0.5764\text{OR} - 8.711E + \varepsilon \quad (3)$$

The regression results in Table 1 indicate that the adjusted *R*-squared is 0.3625, meaning that the model explains 36.25% of the variation in insurance broker performance, with the remaining percentage explained by variables not included in the model. From Table 1, the *F*-statistic is 3.84, and the *p*-value is 0.0029, indicating that independent variables are collectively significant in explaining variation in insurance broker performance. Table 1 shows that the coefficient of environmental factors is −8.711, the *t*-statistic is −1.40, and the *p*-value is 0.170. Since the *p*-value is above 0.05, environmental factors are insignificant when introduced into model (1). Table 2 shows the results of the moderating effect of the product of environmental factors and intangible organizational resources.

$$Y = -20.78125 + 18.2804\text{OR} + 15.4474\text{E} - 8.8822\text{E.OR} + \quad (4)$$

Table 2 shows the results where products of environmental factors and intangible organizational resources were used to estimate the moderating effects. From Table 2, the coefficient for environmental factors is 15.44, with *t* equal to 1.05 and the *p*-value equal to 0.302. Intangible organizational resources have a coefficient of 18.2, with *t* equal to 1.85 and the *p*-value equal to 0.07. Table 2 shows that the product of environmental factors and organizational resources has a coefficient of 8.8822, with *t* equal to −1.80 and the *p*-value equal to 0.082. In this regard, they are insignificant.

Table 2 Step 2. Results of the moderating effect of the product of environmental factors and intangible organizational resources

Goodness of fit	Test statistic	*P*-value	
Adjusted *R*-squared	0.4040		
F-statistic (9.31)	4.01	0.0018	
Dependent variable = Performance	Linear Regression Results		
	Coefficients	*t*-statistic	*P*-Value
Intangible organizational resources	18.2804	1.85	0.074
Environmental factors	15.4474	1.05	0.302
Product of environmental factors and intangible Organizational Resources	−8.8822	−1.80	0.082
Dummy 1: Majority shareholders			
Private individuals (foreign)	0.758	0.51	0.611
Foreign company	6.599	2.38	0.023
Other shareholders	1.946	1.15	0.259
Dummy 2: Foreign company relations			
Subsidiary of foreign company	−2.976	−1.15	0.258
Associate of a foreign company	−3.334	−1.79	0.084
Other foreign company relations	3.5	2.74	0.010
Constant	−20.78125	−0.73	0.471

Source Developed by the authors

Since environmental factors in model 3 and model 4 are insignificant, we fail to reject the null hypothesis and state that environmental factors do not moderate the relationship between intangible organizational resources and insurance broker performance in Zambia. The findings are in tandem with the findings of Njoroge [33], who also found that environmental factors do not moderate the relationship between organizational resources and the performance of mobile telecommunications companies in Kenya.

6 Conclusion and Recommendations

The findings showed that environmental factors have no moderating effect on the relationship between intangible organizational resources and the performance of insurance brokers in Zambia, as they are not significant. As such, laws, regulations, guidelines, etc., are not formulated to give an advantage to certain insurance brokers relative to other insurance brokers.

The research recommends that insurance brokers should not disregard the aspect of the fluctuating environment of political, legal, and economic factors because they have the potential to positively or negatively affect the performance of insurance brokers in Zambia.

References

1. Aaker DA (1989) Managing assets and skills: the key to a sustainable competitive advantage. Calif Manag Rev 31(2):91–106. https://doi.org/10.2307/41166561
2. Adeoye AO, Elegunde AF (2012) Impacts of external business environment on organisational performance in the food and beverage industry in Nigeria. Br J Arts Soc Sci 6(2):194–201. Retrieved from http://www.bjournal.co.uk/paper/BJASS_6_2/BJASS_06_02_05.pdf. Accessed 31 Jan 2023
3. Adeoye MO (2013) The impact of business environment on entrepreneurship performance in Nigeria. Comput Inf Syst Dev Inf Allied Res 4(4):59–66. Retrieved from https://www.iiste.org/Journals/index.php/CIS/article/view/10176. Accessed 31 Jan 2023
4. Aka U (2015) Environmental factors and business operations in Nigeria. Int J Bus Soc Sci 6(11):144–154. Retrieved from https://www.ijbssnet.com/journals/Vol_6_No_11_November_2015/18.pdf. Accessed 31 Jan 2023
5. Barney J (1991) Firm resources and sustained competitive advantage. J Manag 17(1):99–120
6. Barney JB (2002) Gaining and sustaining competitive advantage. Pearson Prentice, Upper Saddle River
7. Barney JB, Hesterly WS (2010) Strategic management and competitive advantage concept. Pearson Prentice Hall, Upper Saddle River
8. Bontis N, Serenko A (2009) A causal model of human capital antecedents and consequents in the financial service industry. J Hum Cap 10(1):53–69. https://doi.org/10.1108/14691930910922897
9. Bryman A (2012) Social research methods, 4th edn. Oxford University Press, New York
10. Carson E, Ranzijn R, Winefield A, Marsden H (2004) Intellectual capital: mapping employee and work group attributes. J Intellect Cap 5(3):443–463. https://doi.org/10.1108/14691930410550390
11. Chawdbury J (2016) Political instability a major obstacle to economic growth in Bangladesh. A (Thesis of Degree Programme Business Management). Central University of Applied Sciences, Kokkola
12. Clulow V, Gerstman J, Barry C (2003) The resource-based view and sustainable competitive advantage: the case of a financial service firm. J Eur Ind Train 27(5):220–232. https://doi.org/10.1108/03090590310469605
13. Collis DJ, Montgomery CA (1995) Competing on resources: strategy in the 1990s. Harvard Business Review. Retrieved from https://hbr.org/1995/07/competing-on-resources-strategy-in-the-1990s. Accessed 31 Jan 2023
14. Creswell JW, Clark PVL (2007) Designing and conducting mixed methods research. Sage Publication, Thousand Oaks
15. Doving E (1996) In the image of man: organizational actions, competence and learning. In: Grant D, Oswick C (eds) Metaphors and organizations. Sage Publication, London
16. Edvinsson L, Sullivan P (1996) Developing a model for managing intellectual capital. Eur Manag J 14(4):356–364. https://doi.org/10.1016/0263-2373(96)00022-9
17. Elliot J, Lukes D (2008) Epistemology as ethics in research and policy: the case of case studies. University of East Anglia, Norwich
18. Eustace C (2000) The intangible economy: impact and policy issues. Report of the European high level expert group on the intangible economy. European Commission, Brussels
19. Fahy J (2002) A resource-based analysis of sustainable competitive advantage in a global environment. Int Bus Rev 11(1):57–77. https://doi.org/10.1016/S0969-5931(01)00047-6
20. Gakenia NJ (2015) Organizational resources and performance of mobile companies in Kenya. (Thesis of Doctor of Philosophy). Kenyatta University, Nairobi
21. Hall R (1992) The strategic analysis of intangible resources. Strateg Manag J 13(2):135–144. https://doi.org/10.1002/smj.4250130205
22. Hambrick D (1981) Environment, strategy and power within top management teams. Adm Sci Q 26(2):253–275. https://doi.org/10.2307/2392472

23. Hambrick D, Mason P (1984) Upper Echelons: the organization as a reflection of its top executives. Acad Manag Rev 9(2):193–207. https://doi.org/10.2307/258434
24. Huber GP, Power DJ (1985) Retrospective reports of strategic-level managers: guidelines for increasing their accuracy. Strateg Manag J 6(2):171–180. https://doi.org/10.1002/smj.4250060206
25. Kothari CR (2014) Research methodology: methods and techniques, 3rd edn. New Age International (P) Limited, New Delhi
26. Koumparoulis M (2013) Environment, competitive strategies and performance. Strateg Manag J 25:890–910
27. Layder D (1993) New strategies in social research: an introduction and guide. Polity, Cambridge
28. Lazzarin W (2012) Strategizing by the government: industrial policy and sustainable competitive advantage. Insper working paper no. 289/2012. Insper Institute of Education and Research, São Paulo
29. MacKinnon DP, Fairchild AJ, Fritz MS (2007) Mediation analysis. Annu Rev Psychol 58(1):593–614. https://doi.org/10.1146/annurev.psych.58.110405.085542
30. Marr B (2004) Measuring and benchmarking intellectual capital. BIJ 11(6):559–570. https://doi.org/10.1108/14635770410566474
31. Meritum (2002) Guidelines for managing and reporting on intangibles (Intellectual Capital Report), TSER Programme. Retrieved from https://cristinachaminade.files.wordpress.com/2018/07/meritum-guidelines.pdf. Accessed 31 Jan 2023
32. Newbert SL (2007) Empirical research on the resource-based view of the firm: an assessment and suggestions for future research. Strateg Manag J 28(2):121–146. https://doi.org/10.1002/smj.573
33. Njoroge JG (2020) Moderating role of environment on organizational resources and performance in telecommunication industry in Kenya. Int J Hum Soc Stud 8(5):313–322. https://doi.org/10.24940/theijhss/2020/v8/i5/HS2005-042
34. Ordonez de Pablos P (2004) Measuring and reporting structural capital: lessons from European learning firms. J Intellect Cap 5(4):629–647. https://doi.org/10.1108/14691930410567059
35. Pensions and Insurance Authority (2021) PIA annual report, 2021. Lusaka. Retrieved from https://www.pia.org.zm/?wpdmpro=pia-annual-report-2021. Accessed 31 Jan 2023
36. Roos G, Roos J (1997) Measuring your company's intellectual performance. Long Range Plan 30(3):413–426. https://doi.org/10.1016/S0024-6301(97)90260-0
37. Rousseau D (1985) Issues of level in organizational research: multi-level and cross-level perspectives. In: Cummings LL, Shaw BM (eds) Research in organizational behavior: an annual series of analytical essays and critical reviews. JAI Press, Greenwich, pp 1–37
38. Saunders M, Lewis P, Thornhill A (2007) Research methods for business students, 4th edn. Prentice Hall Financial Times, Harlow
39. Saunders M, Lewis P, Thornhill A (2009) Research methods for business students, 5th edn. Pearson Education Limited Edinburgh Gate, Essex
40. Spanos YE, Lioukas S (2001) An examination of the causal logic of rent generation: contrasting Porter's competitive strategy framework and the resource-based perspective. Strateg Manag J 22(10):907–934. https://doi.org/10.1002/smj.174
41. Srivastava P, Frankwick GL (2011) Environment, management attitude, and strategy tripod. Acad Manag Perspect 23(3):63–81
42. Teece DJ (1998) Capturing value from knowledge assets: the new economy, markets for know-how, and intangible assets. Calif Manag Rev 40(3):55–79. Retrieved from http://capitalintelectual.egc.ufsc.br/wp-content/uploads/2016/05/9-teece-capturing-value-from-knowledge-assets.pdf. Accessed 31 Jan 2023
43. Wernerfelt B (1984) A resource-based view of the firm. Strateg Manag J 5(2):171–180. https://doi.org/10.1002/smj.4250050207
44. Xavier K, Patrick R (2011) Behavioural economics and customer complaints in communication markets. Australian Communications and Media Authority, Melbourne

Integration of ESG Principles in the Practice of Managing Enterprises in the Agro-industrial Complex

Tatiana N. Litvinova, Natalia N. Balashova, Olga M. Zemskova, and Anna A. Karpova

1 Introduction

Entrepreneurship in Russia operates in a fundamentally new business environment, which was formed in early 2022 and will remain for the coming years: from the medium term (until 2025) to the long term (until 2030 and beyond). The emergence of this environment is due to the general deglobalization of the world economy in general and the aggravation of the international sanctions crisis in particular. In this connection, the problem of adapting Russian entrepreneurship to the current business environment is relevant, which requires the improvement of management practices.

In the business environment of the early twenty-first century, characterized by intense globalization and free trading, the practices of corporate social and environmental responsibility have gained wide popularity worldwide and in Russia. Russia supported the global initiative to implement the Sustainable Development Goals (SDGs) in the Decade of Action. This was an effort to strengthen the reputation and position of Russian entrepreneurship in world markets. The need for this was no longer necessary when the business environment changed.

Russian entrepreneurship is returning to traditional management practices, which involve striving to maximize economic efficiency by increasing profits and reducing costs. Corporate social and environmental responsibility contradicts this practice because it is associated with higher costs and, therefore, underestimates the economic efficiency of entrepreneurship. Simultaneously, the international practice sees increased popularity of ESG-based management.

T. N. Litvinova (✉) · N. N. Balashova · O. M. Zemskova · A. A. Karpova
Volgograd State Agricultural University, Volgograd, Russia
e-mail: balashova_nat@volgau.com; zemskova_olga@volgau.com; karpova_anna@volgau.com

E. G. Popkova (ed.), *Smart Green Innovations in Industry 4.0*, Springer Climate,
https://doi.org/10.1007/978-3-031-45830-9_26

The peculiarity of this management is related to the fact that corporate environmental (E) and social (S) responsibility is supplemented by commercial responsibility (G) to the owners of enterprises, including shareholders and investors. Thanks to this, the interests of stakeholders are balanced in the management of the enterprise, which contributes to maximizing the effectiveness of its activities. Therefore, the question arises as to whether ESG principles can be useful for improving the management practices of Russian enterprises, particularly their successful adaptation to the new business environment.

This research seeks to answer the question posed. Particular attention is paid to entrepreneurship in the agro-industrial complex (AIC) because its high efficiency and the disclosure of its development potential are crucial to the food security of Russia. The research aims to investigate the best Russian practices and prospects for improving the practice of managing entrepreneurship (using the AIC as an example) through the integration of ESG principles in the new business environment of Russia (2022–2023).

2 Literature Review

The research is grounded in the concept of entrepreneurship management and examines the benefits of ESG principles for improving the quality of life and supporting social adaptation in the evolving business environment [4]. This is due to the fact that compliance with ESG principles ensures an increase in the quality of business products and its corporate social and environmental responsibility [6]. Thanks to this, ESG principles can help support social adaptation to the new business environment in Russia: prevent product shortages, curb inflation, support employment and income growth of the working population, and develop a green economy [2, 5, 15].

Abuzyarova [1] and Popkova et al. [11] proposed a purely commercial approach as the preferred managerial approach to ensure the greatest sustainability of market positions of Russian entrepreneurship in the new business environment. Based on the experience of large businesses, this approach, associated with the restriction and relegation of corporate social and environmental responsibility (e.g., the production of eco-class "Euro 0" cars), allows maintaining the break-even business in crisis conditions and saving jobs with a small amount of solvent demand [10].

Literature review showed that the issues of entrepreneurship management had been extensively studied in the available literature. The peculiarities of entrepreneurship management in the AIC are disclosed in the works of Borisova et al. [3], Litvinova et al. [7], and Troyanskaya et al. [14]. The essence of ESG principles and the international practice of relying on them in business management is reflected in the publications of Shevyakova et al. [13]. Nevertheless, the contribution of ESG management to the sustainability of market positions of Russian entrepreneurship in the new business environment (2022–2023) is virtually unstudied and unknown, which is a gap in the literature.

This raises the following research question (RQ): “How does ESG management contribute to the sustainability of market positions (capitalization) of Russian enterprises in the new business environment (2022–2023)?” The research hypothesis is that ESG management increases the sustainability of market positions (capitalization) of Russian entrepreneurship in the new business environment (2022–2023). To test the hypothesis, the author uses factor analysis to identify the benefits of managing entrepreneurship based on ESG principles using the example of small and medium-capitalization companies in Russia in the new business environment (2022–2023).

3 Materials and Methods

To test the hypothesis and determine the benefits of entrepreneurship management based on ESG principles, the author applies the method of regression analysis. Using this method, the author carries out econometric modeling of the contribution of ESG management to the sustainability of market positions of Russian enterprises in the new business environment. The research was conducted between February 24, 2022, and January 24, 2023.

The Moscow Exchange Small and Medium Capitalization Index (MCXSM) [8] serves as an indicator of the stability of the market position of Russian enterprises. The Moscow Exchange RSPP Responsibility and Transparency Index (MRRT) [9] serves as an indicator of the ESG governance of Russian enterprises. The evidence base for the research is shown in Fig. 1 (211 observations).

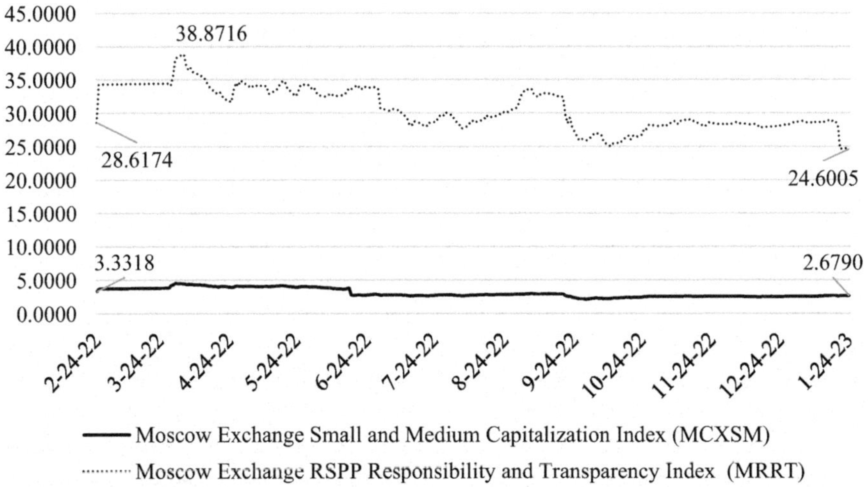

Fig. 1 Dynamics of capitalization of Russian companies in the MCXSM and MRRT indices from February 24, 2022, to January 24, 2023, trillion rubles. (*Source* Developed by the author based on the materials of the Moscow Exchange [8, 9])

According to Fig. 1, the value of MCXSM [8] decreased by 19.5912% by January 24, 2023: from 28.6174 trillion rubles (February 24, 2022) to 24.6005 trillion rubles. Similarly, the value of the MRRT [9] decreased by 14.0368%: from 3.3318 trillion rubles (February 24, 2022) to 2.6790 trillion rubles. The research model is as follows:

$$\text{MCXSM} = a + b * \text{MRRT} \tag{1}$$

The proof of the hypothesis is the positive value of the regression coefficient (b) in the research model (1). The reliability of the econometric model is assessed using the correlation coefficient, Fisher's F-test, and Student's t-test.

4 Results

4.1 Benefits of Managing Entrepreneurship Based on the ESG Principles: The Case of Small- and Medium-Capitalization Companies in Russia

The results of econometric modeling demonstrate a positive relationship between ESG management and the sustainability of market positions of Russian enterprises in the new business environment from February 24, 2022, to January 24, 2023, provided the following equation of simple (paired) linear regression:

$$\text{MCXSM} = -2.7227 + 0.1894 * \text{MRRT} \tag{2}$$

According to Eq. (2), if the capitalization of Russian enterprises in the MRRT [9] increases by 1 trillion rubles, the capitalization of Russian enterprises in the MCXSM [8] increases by 0.1894 trillion rubles. The regression analysis of the dependence of capitalization of Russian entrepreneurship on ESG management in 2022–2023 (Table 1) reveals the relationship between these indicators in more detail.

According to Table 1, the sustainability of market positions (capitalization) of Russian enterprises is explained by ESG management by 86.50% (correlation coefficient) in 2022–2023. The significance of F was $1.66 * 10^{-64}$; thus, model (2) corresponds to the highest level of significance: 0.001. In this case, tabular F is 11.1401, and observed F is 620.9092, which exceeds the tabular value. Thus, Fisher's F-test is passed. Tabular t is 3.29, and observed t is 24.9180, which exceeds the tabular value. Consequently, the Student's t-test is passed. Thus, model (2) is reliable at a significance level of 0.001 (its error is less than 0.1%).

Thus, the regression coefficient (b) in model (2) took a positive value (0.1894), which proves the research hypothesis and confirms that ESG management increases the stability of market positions (capitalization) of Russian enterprises in the new business environment (2022–2023). Based on model (2), the author established that the recovery of MCXSM [8] to the level of February 24, 2022, (28.62 trillion rubles)

Table 1 Regression analysis of the dependence of capitalization of Russian enterprises on ESG management in 2022–2023

Regression statistics						
Multiple R	0.8650					
R^2	0.7482					
Normalized R^2	0.7470					
Standard error	0.3319					
Observations	211					
Variance analysis						
	df	SS	MS	F	Significance of F	Level of significance
Regression	1	68.4	68.4000	620.9092	$1.66 * 10^{-64}$	0.001
Residual	209	23.02366	0.1102			
Total	210	91.42365				
Parameters of the regression model						
	Coefficients	Standard error	t-statistics	P-value	Lower 95%	Upper 95%
Y-intercept	−2.7227	0.2318	−11.7469	$8.52 * 10^{-25}$	−3.17965	−2.2658
Capitalization, trillion rubles	0.1894	0.0076	24.9180	$1.66 * 10^{-64}$	0.1744	0.2044

Source Calculated and compiled by the author

in 2023 could be achieved through increasing the MRRT [9] by 29.88% (to 31.95 trillion rubles).

4.2 *Improving the Practices of Entrepreneurship Management in the Agro-industrial Complex of Russia Based on ESG Principles*

The obtained results indicate the feasibility of expanding the practice of ESG management in Russian enterprises. The ESG ranking of Russian companies for December 2022 by RAEX-Europe [12] shows few companies in the agro-industrial complex. This ranking of 160 companies includes the following agribusinesses:

- Manufacturers of agrochemicals – "Phosagro" Group, "Uralkali," "Eurochem" Group, "Akron" Group, "Togliattiazot," "Uralchem" United Chemical Company, "Miniudobrenia," "August" Company, "Azot" SDS, and "Shchyokinazot"
- Agricultural producers – "STEP" Agroholding and "Rusagro" Group of Companies

This indicates the need to improve the practice of entrepreneurship management in the AIC of Russia based on ESG principles. For this purpose, the author proposes the following recommendations for the integration of ESG principles in the practice of business management in the AIC (using agricultural enterprises as an example):

- Increasing corporate environmental responsibility (E) through climate-resilient innovations that reduce the dependence of agricultural production and food quality on climatic factors and contribute to a better climate.
- Increasing corporate social responsibility (S) through the creation of year-round, high-performance, and knowledge-intensive green jobs for the digital agricultural workforce.
- Improving corporate governance (G) through increased use of smart technology to automate phytomonitoring and plant care.

For the systemic implementation of the set of author's recommendations, it is recommended to create a network of smart vertical farms in Russia, the organization and management of which is based on ESG principles. The activities of smart vertical farms are proposed to be harmonized with the strategic vision of the green economy of Russia (in particular, with the strategy of socio-economic development of Russia with low greenhouse gas emissions until 2050 and with the network of carbonic ranges of Russia), as well as food security (in particular, with the Doctrine of Food Security of the Russian Federation).

5 Discussion

The contribution of the research to the literature consists of the development of scientific provisions of the concept of entrepreneurship management through clarification of the contribution of ESG management in the sustainability of market positions (capitalization) of Russian entrepreneurship in the new business environment (2022–2023). The new results obtained in the research are compared with the existing literature in Table 2.

Table 2 Comparison of new results obtained in the research with the available literature

Comparison criterion	Available literature: Scientific provisions of the available literature	Available literature: References	New results obtained in this research
The usefulness of ESG principles in the new business environment	Supporting social adaptation and increasing the quality of life	Ali et al. [2], Denisov et al. [4], Liang et al. [5], Lisin et al. [6], and Zhang and Jin [15]	Supporting the adaptation of entrepreneurship and the growth of its market capitalization
Management approach, ensuring the sustainability of market positions of Russian entrepreneurship in the new business environment	Pure commercial approach (from the experience of big business)	Abuzyarova [1], Pies and Schultz [10], and Popkova et al. [11]	ESG approach, flexibly combining elements of commercial and non-commercial management (for small and medium-sized businesses, including AIC)

Source Developed by the author

As shown in Table 2, in contrast to Ali et al. [2], Denisov et al. [4], Liang et al. [5], Lisin et al. [6], and Zhang and Jin [15], this research substantiates that the usefulness of ESG principles in the new business environment is not only to support social adaptation and increase the quality of life but also to support the adaptation of entrepreneurship and increase its market capitalization.

In contrast to Abuzyarova [1], Pies and Schultz [10], and Popkova et al. [11], the research proves that the stability of market positions of Russian enterprises in the new business environment is best provided not by a purely commercial approach (more suitable for large businesses) but by ESG approach, which flexibly combines elements of commercial and non-commercial management and is suitable for small and medium businesses, including AIC enterprises.

6 Conclusion

In conclusion, this research filled a gap in the literature and answered the research question, proving that ESG management significantly contributes to the sustainability of market positions (capitalization) of Russian enterprises in the new business environment (2022–2023). Using the experience of small- and medium-capitalization companies in Russia from February 24, 2022, to January 24, 2023, the author proves that the stability of market positions (capitalization) of Russian enterprises is by 86.50% (correlation coefficient) explained by ESG management. On this basis, the author proposed integrating ESG principles in the practice of business management in the AIC, for which a set of author's recommendations is proposed (using agricultural entrepreneurship as an example).

The theoretical significance of these results is that they reveal a new angle of the study of ESG management of entrepreneurship – from the perspective of the interests of strengthening the position in domestic markets. This perspective is most promising in the new business environment, in which the former perspective (from the perspective of strengthening the position in foreign markets) is much less relevant. The practical significance of the author's conclusions and recommendations is related to the fact that they allow Russian enterprises to better adapt to the new business environment and increase the sustainability of their market positions.

References

1. Abuzyarova MI (2017) Methodological approaches to ensure the competitiveness of organizations. J Bus Retail Manag Res 11(2):125–132. Retrieved from https://jbrmr.com/cdn/article_file/i-26_c-261.pdf. Accessed 25 Jan 2023
2. Ali Q, Salman A, Parveen S (2022) Evaluating the effects of environmental management practices on environmental and financial performance of firms in Malaysia: the mediating role of ESG disclosure. Heliyon 8(12):e12486. https://doi.org/10.1016/j.heliyon.2022.e12486

3. Borisova O, Abramova L, Zageeva L, Popkova E, Morozova I, Litvinova T (2015) Role of agricultural clusters in provision of food security. Eur Res Stud J 18(3):287–298. https://doi.org/10.35808/ersj/472
4. Denisov IV, Khachaturyan MV, Umnova MG (2018) Corporate social responsibility in Russian companies: introduction of social audit as assurance of quality. Qual – Access Success 19(164):63–73
5. Liang Y, Lee MJ, Jung JS (2022) Dynamic capabilities and an ESG strategy for sustainable management performance. Front Psychol 13:887776. https://doi.org/10.3389/fpsyg.2022.887776
6. Lisin A, Kushnir A, Koryakov AG, Fomenko N, Shchukina T (2022) Financial stability in companies with high ESG scores: evidence from North America using the Ohlson O-Score. Sustainability 14(1):479. https://doi.org/10.3390/su14010479
7. Litvinova TN, Kulikova ES, Kuznetsov VP, Taranov PM (2017) Marketing as a determinant of the agricultural machinery market development. In: Popkova E (ed) Overcoming uncertainty of institutional environment as a tool of global crisis management. Springer, Cham, pp 465–471. https://doi.org/10.1007/978-3-319-60696-5_59
8. Moscow Exchange (2023) Moscow Exchange RSPP Responsibility and Transparency Index (MRRT) (MRRT) in rubles: archive of values by days from February 24, 2022, to January 24, 2023. Retrieved from https://www.moex.com/ru/index/MRRT/archive/?from=2022-02-24&till=2023-01-24&sort=TRADEDATE&order=desc. Accessed 25 Jan 2023
9. Moscow Exchange (2023) Moscow Exchange Small and Medium Capitalization Index (MCXSM) in rubles: archive of values by days from February 24, 2022, to January 24, 2023. Retrieved from https://www.moex.com/ru/index/MCXSM/archive/?from=2022-02-24&till=2023-01-24&sort=TRADEDATE&order=desc. Accessed 25 Jan 2023
10. Pies I, Schultz FC (2023) The governance of sustainable business model innovation—an ordonomic approach. Scand J Manag 39(1):101246. https://doi.org/10.1016/j.scaman.2022.101246
11. Popkova EG, Bogoviz AV, Pozdnyakova UA, Przhedetskaya NV (2018) Specifics of economic growth of developing countries. In: Endovitsky D, Popkova E (eds) Management of changes in socio-economic systems. Springer, Cham, pp 139–146. https://doi.org/10.1007/978-3-319-72613-7_12
12. RAEX-Europe (2023) ESG ranking of Russian companies (December 2022). Retrieved from https://raex-rr.com/esg/ESG_rating. Accessed 25 Jan 2023
13. Shevyakova A, Petrenko E, Vechkinzova Y, Koroleva A (2019) Features of the development of female entrepreneurship in Kazakhstan. In: Proceedings of the IBIMA 2019: 33rd International Business Information Management Association Conference "Education Excellence and Innovation Management through Vision 2020", Granada, pp 6587–6595
14. Troyanskaya MA, Ostrovskiy VI, Litvinova TN, Matkovskaya YS, Bogoviz AV (2017) Possibilities and perspectives for activation of sales in the agricultural machinery market within sectorial development of Russian and European economies. In: Popkova E (ed) Overcoming uncertainty of institutional environment as a tool of global crisis management. Springer, Cham, pp 473–480. https://doi.org/10.1007/978-3-319-60696-5_60
15. Zhang C, Jin S (2022) What drives sustainable development of enterprises? Focusing on ESG management and green technology innovation. Sustainability 14(18):11695. https://doi.org/10.3390/su141811695

The Role of Public-Private Partnerships for the Development of Green Building

Ellina A. Shamanina

1 Introduction

The contemporary organization of socio-economic life within the framework of the sustainable development concept, which implies economic growth and the solution of social contradictions without harming the environment, reinforces the need to create a green economy and activates the process of its formation.

The concept of the green economy is not new. It appeared in the late twentieth–early twenty-first century. According to the definition of the United Nations Environment Program (UNEP), a green economy is about improving the well-being of society and achieving social justice by reducing threats to the environment, particularly the risks caused by climate change, and increasing the efficiency of the use of energy and other resources to save stability of functioning and development of the ecosystem of our planet [16].

The transition to a green economy should be accompanied by the formation of appropriate conditions, which include the creation of a legal framework and the organization of support programs, and financial support. Green economy requires the strengthening of organizational structures and the development of the country's sustainable infrastructure, which is necessary to ensure social welfare and economic growth. In this case, sustainability implies a long-term process that guarantees stability in the development of the natural environment respecting the interests of society and keeping economic growth in balance. Investing in such infrastructure contributes to this balance, which, in turn, directly affects the achievement of all sustainable development goals (SDGs).

Infrastructure projects play a vital role in driving economic growth and restructuring the country's economy. The aggravation of the environmental situation, including those associated with climate change and natural disasters, leads to

E. A. Shamanina (✉)
MGIMO University, Moscow, Russia

E. G. Popkova (ed.), *Smart Green Innovations in Industry 4.0*, Springer Climate,
https://doi.org/10.1007/978-3-031-45830-9_27

damage and deterioration of infrastructure systems, which causes disruptions in the provision of services needed by the population [17].

Construction is undeniably one of the most polluting industries, contributing to increased greenhouse gas emissions and construction waste during infrastructure projects. Hence, there is a growing need for infrastructure construction that aligns with environmental standards while considering economic, social, and environmental parameters.

Among the most important criteria are the requirements for creating a comfortable indoor and outdoor environment, environmental safety, as well as increasing the energy efficiency of the operated facility. Moreover, according to UN forecasts, by 2030, the number of people living in cities will constitute more than 60% of the total world population, the urban population in developing countries will double, and the territory occupied by cities will triple. By that time, urban areas are expected to generate 80% of the world GDP and consume three-fourths of the world's energy, which will inevitably lead to the expansion of economic, energy, environmental, and social problems caused by increased government and consumer spending and population growth [11].

With the rising cost of energy resources, the issue of energy efficiency becomes extremely urgent. Construction is the most energy-intensive sector of the economy. According to the Global Alliance for Buildings and Constructions (GABS), more than half of global energy demand and about 30% of global CO_2 emissions associated with energy generation come from the construction sector [3]. In this regard, technologies and principles for energy efficiency and energy profitability of housing and other infrastructure are being actively developed worldwide, including in CIS countries. The legal framework for their application is being formed and improved [2].

To implement the energy transition of the world economy by 2030, it is necessary to reduce the energy intensity of GDP by 4.2% annually. Nevertheless, the global rate of reduction of energy intensity of global GDP for 2015–2022 was 1.6%, 2% for the EU countries, and 0.6% for Russia [12].

Simultaneously, the construction industry has significant potential for the rational use of energy resources in the production of building materials, as well as at the stages of the creation and operation of the final infrastructure facility. According to the same report by the Global Alliance for Buildings and Constructions (GABS), energy savings in the construction sector could reach more than 50% by 2050.

The main criterion for success in changing the energy intensity of implemented infrastructure facilities is the introduction of sustainable or green building technologies, which ensure economic efficiency while maintaining social and environmental parameters.

According to the World Green Building Council, green infrastructure has the following set of characteristics:

- Efficiency in the use of natural (energy, water) and other resources
- The use of renewable energy sources

- The use of environmentally friendly materials
- "Flexible" design of the infrastructure facility that makes it possible to adapt to changing environmental conditions

Thus, any object of industrial (road, bridge, power plant, etc.) and social (school, hospital, etc.) infrastructure can be sustainable and green if all the above criteria are observed [7].

The trend toward developing green buildings is driven by several advantages, which can be categorized into three areas:

1. Environmental. Green infrastructure makes it possible to reduce the burden on the environment through the efficient use of natural resources. Particularly, by using innovative green technologies, it reduces greenhouse gas emissions into the atmosphere. According to UNEP, green buildings can reduce greenhouse gas emissions by 84 gigatons by 2050. In particular, green infrastructure will save more than 50% of energy consumption. Reducing the volume of waste from construction activities prevents air, water, and soil pollution [8].
2. Economic. Green infrastructure has some economic and financial advantages for all stakeholders. These advantages include savings on utility bills due to efficient use of resources, which can save up to 25% of the energy consumed and reduce water consumption by 30%. This type of construction improves financial and economic indicators. On the one hand, profitability increases. According to the Green Building Council of Canada, the implementation of green infrastructure projects provided an increase in GDP of $23.45 billion in 2014. On the other hand, the costs of operating the infrastructure facility are reduced. For example, during the first year of operating, there are savings of 10%; over 5 years, savings increase up to 16%. Additionally, the asset value of green building facilities increases by 7% compared to the facilities implemented by the traditional construction scheme [4].
3. Social. An undeniable advantage of green building is its positive impact on public health and well-being. For example, according to a report of the Harvard T. H. Chan School of Public Health, workers in green buildings have increased cognitive performance, that is, brain function. Along with this criterion, according to the World Green Building Council, more than 50% of those surveyed reported that green buildings increase employee productivity. The increase in productivity is provided by up to 8%. For example, in developing countries such as Colombia, South Africa, India, Mexico, and Saudi Arabia, this indicator is critical for the creation of infrastructure with green technology [8].

Therefore, the development of green infrastructure benefits the economy, the environment, and societal well-being.

2 Materials and Methodology

This research employs a methodology based on theoretical and statistical analysis to explore the infrastructure development necessary for implementing green building concepts.

The comprehensiveness of the conclusions and reliability of the results of the study are provided by the use of general scientific methods of knowledge, such as the method of grouping, generalization, classification, analogy, and comparison. Descriptive methods used in the research are based on the collection and study of international practice of using public-private partnership (PPP) tools in the implementation of green building projects.

To understand the development of the green building, its main advantages, and trends, including projects for the development of green infrastructure on the PPP principles, the author analyzed foreign sources, particularly research articles, reports of consulting companies, PPP development centers, and the Council on Green Building (World Green Building Council). This analysis served as the basis for the disclosure of the essence of green building, its key features, as well as opportunities for the integration of PPP mechanisms in the implementation of projects to create sustainable infrastructure.

Statistical data presented in the research were collected and analyzed using international and Russian databases, particularly materials from the World Bank, consulting agency Deloitte, McKinsey, Dodge Data & Analytics, UNEP, Global Center on Adaptation (GCA), Asian Development Bank (ADB), African Development Bank, Global Data Agency, and ANO "National Center for PPP Development." The methodological framework of the work was also formed by regulatory documents, international standards, and guidelines related to the regulation of the development of the green building, quality infrastructure, and the development of PPP mechanisms.

3 Results

Currently, the global market for green infrastructure is experiencing rapid growth, with a significant portion of green infrastructure initiatives occurring in developing markets. According to Global Data, as of 2021, the leading countries in the green building include China, the USA, India, Indonesia, and Mexico.

The leading market for green building, with a value of $178.1 million, is China, followed by the USA, with $83.1 billion. They are followed by India ($21 million), Mexico ($19 million), and Indonesia ($18.5 million) [6]. Residential projects comprise the bulk of the entire green infrastructure market (41%), while office building initiatives account for 23%. Promising industries for the future are as follows:

1. Commercial and non-commercial construction
2. Implementation of projects in the field of construction of low- and high-rise blocks
3. Reconstruction of already existing infrastructure

For example, the first and second groups of future industries include India, Cameroon, Saudi Arabia, and China. The USA, Canada, and Brazil are focused on modernizing existing infrastructure facilities.

More than 14,000 projects, or more than 40% of the total green building projects portfolio, are implemented in the UK. France and the Netherlands are in second and third position – 11% and 8%, respectively. Simultaneously, the share of the portfolio of project initiatives in Russia is still only 0.7%, or 223 buildings as of 2021 [10].

The reasons for the intensified investment growth in sustainable infrastructure projects are the savings on further facilitation of operations. For example, according to the World Bank, the annual costs of households from the use of infrastructure built on traditional construction technologies (due to infrastructure deterioration) amount to about $390 billion. Simultaneously, the net benefit of investing in green infrastructure and its further operation, especially in low-income countries, is $4.2 trillion [10].

Nevertheless, extra financing is needed for the further development of the green building. The global deficit in infrastructure is estimated at more than $4.5 trillion per year. Moreover, increasing investment from 9% to 27% is important to comply with sustainable development principles.

In this regard, the underfunding is an incentive to attract private financing on the PPP principles. The basis for the attractiveness of PPP tools in implementing green building projects is its traditional advantages.

The first advantage is the return on investment based on particular project specifications. The private partner can reimburse the funds spent on the project through user-pay (e.g., road tolls) or through availability payments (government-pay), where the government makes compensation to the private sector. Such compensation is usually based on the achievement of appropriate performance indicators under the project. In this way, the application of green building principles can be included as a criterion, the fulfillment of which determines the project's payback.

The second advantage is duration. One of the key characteristics of PPP is the long duration of project initiatives. Thus, after the completion of the facility construction, the private partner usually undertakes obligations for its further maintenance and operation within the period set out in the agreement. Due to the fact that the private contractor can operate the facility for several decades, it is more interested in its long-term trouble-free service so that it is resistant to changing climatic conditions, including in order to reduce the cost of its operation. In this regard, during the project design, the private partner can implement green principles.

The third advantage is risk sharing. This characteristic of partnership is also relevant in the case of the implementation of green building principles. For example, if the design, construction, and maintenance risks are allocated to the private partner, then in the event of problems with an infrastructure facility caused by the climate factor, it can absorb and manage these risks more quickly and effectively than the public partner, particularly to ensure the infrastructure rehabilitation. Digital technology makes it possible to increase the stress resistance of infrastructure facilities at the design level, reduce or eliminate their operational risks, and better account for the allocation of partners' responsibilities [9].

In turn, the implementation of green building projects through PPP corresponds to the UN initiative adopted in 2018 to revitalize the partnership with its reorientation to ensure the "value for money" and the "value for people" [15]. The adopted model, termed as people-first PPP, is based on the following key principles:

- Ensuring equal access to infrastructure and public services for all social groups
- Achieving economic efficiency
- Investing in sustainable infrastructure projects
- Ensuring the engagement of a wide range of stakeholders

Thus, the current PPP model is being transformed, becoming a more effective tool for addressing the UN 2030 agenda.

International practice demonstrates that climate-resilient infrastructure can be easily built through the PPP model. One of the most successful examples is the "As-Samra Wastewater Treatment Plant project in Jordan" [5].

The As-Samra Wastewater Treatment Plant was built to upgrade the city's aging water infrastructure and increase the available water supply. The project was constructed through the BOT ("build-operate-transfer") model with blended financing. The project company "Samra Plant Company" received a 25-year concessional loan. The project investment was split among the Government of Jordan, the US Agency for International Development (USAID), The Millennium Challenge Corporation, and a consortium of banks led by Arab Bank and Samra Plant.

The As-Samra Wastewater Treatment Plant supplies enough water to irrigate about 4000 farms on 10,000 acres. The plant generates 80% of its electricity from on-site renewable sources, which saves about $14 million. It saves about $14 million a year in operating costs.

Another successful project delivered through PPP is the "New flood proof district in Bilbao, Spain." The project was structured as a mixed-enterprise project, for which a Special Purpose Vehicle (SPV) "Comisión Gestora de Zorrotzaurre" was created, consisting of the Zorrotzaurre landowners and the public authorities. The participants contributed financially in proportion to land ownership (51% public, 49% private investment) [5].

The project included the construction of an artificial island in the Deusto area of Bilbao. To protect it from flooding, the island was raised 1.5 m above ground level, and a protective wall was built. The district is equipped with affordable housing, 50% of which is social housing, and two-thirds of the area is for public use, which includes the creation of educational, health, sports, and cultural facilities necessary for citizens. To improve the territory, green open spaces have been created, such as a 20 m wide park along the river bank (7.5 km long), a 40,000 sq meters park in the island's center, and more than 5000 sq meters of green spaces inside the island.

Creating large open green spaces and socially friendly areas positively affect public health, thereby reducing the risk of illness and stress and promoting psychological well-being and social cohesion. Likewise, urban green spaces reduce air pollution while cleaning the air. The economic benefits include a boost to the local budget through domestic and inbound tourism.

This project is a good example of how PPP principles can be integrated into green infrastructure projects, particularly to create sustainable urban spaces.

Thus, among the abovementioned advantages of PPP in the development of infrastructure, we can also include the widespread use of PPP tools. The presented examples demonstrate that the governments of economically developed countries have considerable experience in implementing partnerships to develop infrastructure complexes. Emerging economies also integrate sustainability principles into their PPP practice [14]. Moreover, PPP may become a relevant tool in solving the problems of green building in countries with rapidly developing economies, such as Africa and Asia. For example, the African Development Bank estimates that Africa's infrastructure financing gap ranges from \$68 billion to \$108 billion [1]. The Asian Development Bank estimates this figure in Asia at \$907 billion [13].

4 Conclusion

The value of the green building has recently increased significantly in many countries. It has become one of the basic tools aimed at preserving the environment and improving the quality of people's life.

The widespread implementation of sustainable development principles to create climate-resilient infrastructure allows the green building market to expand, introduce new forms of cooperation (e.g., PPP), create commercial real estate and social infrastructure, and build up entire neighborhoods (e.g., the experience of Spain).

Proper implementation of green standards makes it possible to build durable, energy-efficient, sustainable infrastructure facilities and reduce future operating costs. As a result, green infrastructure facilities require fewer repairs and maintenance costs in the long term.

To overcome the barrier of the high cost of using green building technologies that prevent attracting investors and expanding the project portfolio, the practice of implementing PPP tools is becoming increasingly widespread. Due to its specific features, PPP can be an effective tool for optimizing funding and closing the sustainable infrastructure gap. The global practice demonstrates that PPP tools can be easily and effectively integrated in sustainable infrastructure projects delivering.

Undoubtedly, further introduction of PPP forms will require solving several issues, including the following:

- Whether the project complies with the principles of partnership
- Whether the private partner can manage climate risks

- Whether the current payment mechanisms can encourage the private partner to implement green technologies to mitigate climate risks, etc.

Nevertheless, international practice shows that the improvement of industrial and social infrastructure requires the attraction of additional resources, including extra investments. In this case, the PPP mechanisms are a proven practice to accelerate infrastructure development, including green building.

Due to the investment stability, as well as the already accumulated global experience in infrastructure projects, delivered through PPP, most countries in the world can implement their green building initiatives through the toolkit of partnership.

References

1. African Development Bank (2018) Africa's infrastructure: great potential but little impact on inclusive growth. In: African economic outlook 2018, Abidjan, pp 63–94. Retrieved from https://www.afdb.org/fileadmin/uploads/afdb/Documents/Publications/2018AEO/African_Economic_Outlook_2018_-_EN_Chapter3.pdf. Accessed 12 Jan 2023
2. Assylbayev AB, Safronchuk MV, Niiazalieva KN, Brovko NA (2023) Green transformation and the concept of energy efficiency in the housing sector. In: Lazareva EI, Murzin AD, Rivza BA, Ostrovskaya VN (eds) Innovative trends in international business and sustainable management. Springer, Singapore, pp 567–577. https://doi.org/10.1007/978-981-19-4005-7_61
3. Dean B, Dulac J, Petrichenko K, Graham P (2016) Global status report: towards zero-emission efficient and resilient buildings. Global Alliance for Buildings and Construction (GABC), Paris. Retrieved from https://backend.orbit.dtu.dk/ws/portalfiles/portal/127199228/GABC_Global_StatusReport_V09_november.pdf. Accessed 17 Dec 2022
4. First Green Consulting (2022) Why are green buildings so important? 22 Feb 2022. Retrieved from https://www.firstgreen.co/why-green-buildings/. Accessed 20 Dec 2022
5. Global Center on Adaptation (2021) Knowledge module on PPPs for climate-resilient infrastructure: handbook. GCA, Rotterdam. Retrieved from https://gca.org/wp-content/uploads/2021/08/GCA-Handbook-V2.0-13-September-2021-2.pdf. Accessed 25 Dec 2022
6. GlobalData (2021) Top five green building markets in 2021. Retrieved from https://www.globaldata.com/data-insights/construction/the-top-five-green-building-markets-in-2021/. Accessed 25 Dec 2022
7. Green Build Council the Russian Federation (GBCRU) (n.d.) What is green building? Retrieved from https://gbcru.org/about/index.php. Accessed 22 Dec 2022
8. Green Build Council the Russian Federation (GBCRU) (n.d.) What are the benefits of Green Building? Retrieved from https://gbcru.org/about/plyusy.php. Accessed 22 Dec 2022
9. Ivanitskaya NV, Baybulov AK, Safronchuk MV (2020) Modelling of the stress-strain state of a transport tunnel under load as a measure to reduce operational risks to transportation facilities. J Phys Conf Ser 1703:012024. https://doi.org/10.1088/1742-6596/1703/1/012024
10. Jones SA (ed) (2021) Smart market report: world green building trends 2021. Dodge Construction Network, Bedford. Retrieved from https://www.corporate.carrier.com/Images/Corporate-World-Green-Building-Trends-2021-1121_tcm558-149468.pdf. Accessed 17 Dec 2022
11. Lederer EM (2016) UN report: by 2030 two-thirds of world will live in cities. Associated Press, 19 May 2016. Retrieved from https://apnews.com/article/40b530ac84ab4931874e1f7efb4f1a22. Accessed 19 Dec 2022

12. Ministry of the Economic Development of the Russian Federation (2021) State report “On energy saving and increasing energy efficiency in the Russian Federation in 2020”, Moscow. Retrieved from https://www.economy.gov.ru/material/file/45f28379991124fa0098b17f21f169ed/Energyefficiency2021.pdf. Accessed 20 Dec 2022
13. Ra S, Li Z (2018) Closing the financing gap in Asian infrastructure. ADB South Asia Working Paper Series No. 57. Retrieved from https://www.adb.org/sites/default/files/publication/431261/swp-057-financing-gap-asian-infrastructure.pdf. Accessed 12 Jan 2023
14. Starikova EA (2020) Public–Private Partnership as an instrument to implement the Sustainable Development Goals in the developing countries. In: Ivanov OV, Inshakova AO (eds) Public-private partnerships in Russia: institutional frameworks and best practices. Springer, Cham, pp 293–305. https://doi.org/10.1007/978-3-030-56352-3_16
15. UNECE (2018) Guiding principles on People-First Public-Private Partnerships (PPPs) for the United Nations Sustainable Development Goals (UN SDGs). Retrieved from https://unece.org/fileadmin/DAM/ceci/documents/2018/PPP/Forum/Documents/The_8_Guiding_Principles_for_People-first_PPPs_in_support_of_the_UN_SDGs-Part_II.pdf. Accessed 10 Jan 2023
16. UNEP (n.d.) What does green economy matter? Retrieved from https://www.unep.org/explore-topics/green-economy/why-does-green-economy-matter. Accessed 17 Dec 2022
17. Zavyalova EB, Popkova EG (eds) (2021) Industry 4.0: exploring the consequences of climate change. Palgrave Macmillan, Cham. https://doi.org/10.1007/978-3-030-75405-1

Improving the Quality of Corporate Governance with Reliance on ESG-Based HR Management

Zhanna V. Gornostaeva and Yulia S. Chernysheva

1 Introduction

With Industry 4.0 and intense digital competition, the focus is placed on the quantitative results of corporate governance. Consequently, large-scale automation is deployed to increase production capacity, extract economies of scale, and reduce the influence of the human factor on entrepreneurial processes. In this regard, corporate management is reoriented from human resource management (HR management) to technology and innovation management.

The problem is that human potential is not fully unlocked due to the lack of attention to HR management. Issues related to workplace organization in the contemporary economy are much more studied from the standpoint of technical equipment than from the standpoint of comfort for workers. The beginning of the Fifth Industrial Revolution indicates that in the present, near future, and in long-term entrepreneurship, automation will not lead to the full autonomy of smart enterprises despite the undoubted growth of its scale.

Smart technology does not completely replace human resources, complementing them and making more rational use of them by outsourcing routine processes to automation. In this way, there remain the most complex business processes, the knowledge intensity of which increases as the technical equipment of jobs becomes more complex [1]. These complex business processes are assigned to human resources.

This determines the relevance of the study of the prospects for improving the quality of corporate governance, which involves optimizing job structures to attract and retain top talent while harnessing their human potential. This research proposes ESG-based HR management as a promising practical solution to the problem. The

Z. V. Gornostaeva · Y. S. Chernysheva (✉)
Don State Technical University, Rostov-on-Don, Russia

E. G. Popkova (ed.), *Smart Green Innovations in Industry 4.0*, Springer Climate,
https://doi.org/10.1007/978-3-031-45830-9_28

research aims to investigate the current international experience and prospects of improving the quality of corporate governance through ESG-based HR management.

2 Literature Review

The fundamental basis of the research conducted in this research was formed by the concept of HR management. According to this concept, the main characteristics of the quality of corporate governance are as follows:

- The level of development of entrepreneurial culture
- Attracting young personnel as the most creative one
- Engagement of stakeholders
- Pay flexibility
- Professionalism of management
- Pay equity
- Staff retention [2].

The established vision of corporate governance in the existing literature, including the works of Bilek-Steindl and Url [3], Borrego et al. [4], Khalique et al. [5], Kreinin and Aigner [6], and Makarenko et al. [7], suggests the improvement of the highlighted characteristics of corporate governance separately. Corporate HR management falls within the realm of a Sustainable Development Goal (SDG) such as SDG 8.

ESG-based HR management is considered one of the most progressive practices of HR management [8]. Its potential benefits for the quality of corporate governance are based on a systemic combination of the social orientation of jobs (e.g., providing a wide range of social and labor guarantees and opportunities for creative work and development, including through training, and the discovery of human potential) and their environmental friendliness (e.g., high energy efficiency, economy and reuse, that is, the circular use of raw materials and supplies, minimizing waste production, and using eco-packaging of products), on the one hand, and high production efficiency, on the other [9–11].

The implementation of ESG-based HR management in modern business creates green jobs, embodying all features noted: social orientation, environmental friendliness, and high economic efficiency [12–14]. ESG-based HR management contributes to the practical implementation of SDG 12. Amos and Lydgate [15], Beccarello and Di Foggia [16], Ferreira-Lopes et al. [17], and Palakshappa and Dodds [18] note that SDG 8 and SDG 12 are implemented separately. Numerous publications are devoted to the issues of ESG management, which indicates a high degree of their elaboration.

Nevertheless, the impact of ESG-based HR management on the quality of corporate governance is insufficiently studied and unknown, which is a gap in the literature. This research seeks to fill an identified gap by posing and answering the following research question (RQ): "What impact does ESG-based HR management have on the quality of corporate governance?" Based on the works of Amon et al. [19] and Velte [20], which noted the benefits of ESG-based HR management and the green jobs created through it, this research puts forward hypothesis H that ESG-based HR management contributes to the quality of corporate governance.

3 Materials and Methods

The methodological apparatus of this research relies on the method of regression analysis. Using the names of the indicators from official statistics, the hypothesis is formulated as follows: as green jobs proliferate (x, Global Green Growth Institute [21]; figure calculated for 2021), the following is observed:

- The level of development of entrepreneurial culture increases (y_1, Institute of Scientific Communications [22]; the original figure is calculated by WIPO for 2021).
- Youth unemployment rate decreases (y_2, Institute of Scientific Communications [22]; the original figure calculated by ILO for 2021).
- Stakeholder engagement increases (y_3, Institute of Scientific Communications [22]; the original figure calculated by WEF for 2019).
- Pay flexibility increases (y_4, Institute of Scientific Communications [22]; the original figure calculated by WEF for 2019).
- Professionalism of management increases (y_5, Institute of Scientific Communications [22]; the original figure calculated by WEF for 2019).
- Pay equity increases (y_6, Institute of Scientific Communications [22]; the original figure calculated by WEF for 2019).
- Staff retention improves (y_7, Institute of Scientific Communications [22]; the original figure calculated by WEF for 2019).

The sample of countries was formed according to the criterion of the sufficiency of statistics for all indicators (absence of gaps in the data). The empirical basis of the research is shown in Table 1.

Based on the statistics from Table 1, the authors compiled a model of the impact of ESG-based HR management on the quality of corporate governance. To check the quality, statistical rigor, and the model's reliability, the authors conducted a heteroscedasticity test (R^2 and Fisher's F-test) and Student's t-test. Prospects for improving the quality of corporate governance through ESG-based HR management are determined using the substitution of the maximum possible value of green jobs (100 points) in the regression model.

Table 1 Statistics on ESG-based HR management and quality of corporate governance

Country	Green employment, score 0–100	Entrepreneurship policies and culture, score 0–100	Youth unemployment rate, %	Multi-stakeholder collaboration, score 0–100	Flexibility of wage determination, score 0–100	Reliance on professional management, score 0–100	Pay and productivity, score 0–100	Hiring and firing practices, score 0–100
	x	y_1	y_2	y_3	y_4	y_5	y_6	y_7
Belarus	49.75	0.77	6.6	n/d	n/d	n/d	n/d.	n/d
Brazil	23.48	12.80	22.62	44.3	54.8	58.6	40.4	29.3
Greece	32.25	17.13	28.38	34.9	55.8	49.2	41.4	46
Dominican Republic	16.20	39.13	14.56	43.3	60.7	52.6	47.1	44.2
Canada	69.67	64.30	11.08	63.8	72.4	75.3	65	60.9
Mexico	54.56	50.61	6.96	45.3	70	53.9	46.4	40.5
Panama	15.54	34.61	20.03	43.3	59.9	51.6	37.6	36.3
Republic of Korea (South Korea)	73.93	69.82	7.89	56.9	63.1	58.6	63.8	42.4
Slovakia	80.26	8.56	13.05	43.4	65.9	57.7	53.6	37.9
Switzerland	100.00	79.99	7.53	72.1	79	81.2	74.6	80.4

Source: Compiled by the authors based on the materials of the Global Green Growth Institute [21] and the Institute of Scientific Communications [22]

4 Results

4.1 *Model of the Impact of ESG-Based HR Management on the Quality of Corporate Governance*

To determine the impact of ESG-based HR management on the quality of corporate governance, the authors conducted a regression analysis of the data from Table 1 (Table 2).

As shown in Table 2, all obtained equations of simple (paired) linear regression are valid, reliable, statistically accurate, and rigorous. The correlation of the indicators in all cases was high; Fisher's F-test was passed. Thus, there is no heteroscedasticity. The t-test was also performed in all cases. This allows us, based on Table 2, to create a model (system of equations) of the impact of ESG-based HR management on the quality of corporate governance:

$$\begin{cases} y_1 = 13.2257 + 0.4761x; \\ y_2 = 22.5176 - 0.1677x; \\ y_3 = 34.9295 + 0.2853x; \\ y_4 = 53.8145 + 0.2088x; \\ y_5 = 46.4333 + 0.2593x; \\ y_6 = 33.1235 + 0.3687x; \\ y_7 = 29.7433 + 0.3224x. \end{cases} \quad (1)$$

Table 2 Regression statistics

	Characteristics of a simple (paired) linear regression model						
Regression statistics	y_1	y_2	y_3	y_4	y_5	y_6	y_7
Correlation, R^2	0.5013	0.6521	0.7418	0.8105	0.7307	0.8879	0.6517
Significance of F	0.1399	0.0410	0.0221	0.0081	0.0253	0.0014	0.0572
Significance level	0.15*	0.05*	0.05**	0.01**	0.05**	0.01**	0.10**
Tabular F	2.6854	5.3177	5.5914	12.2464	5.5914	12.2464	3.5894
Observed F	2.5352	5.9201	8.5663	13.3984	8.0172	26.0719	5.1674
F-test	Passed	Passed	Passed	Passed	Passed	Passed	Passed
Constant	13.2257	22.5176	34.9295	53.8145	46.4333	33.1235	29.7433
Regression coefficient	0.4761	−0.1677	0.2853	0.2088	0.2593	0.3687	0.3224
Standard error at x	0.2905	0.0689	0.0975	0.0570	0.0916	0.0722	0.1418
Tabular t	1.62***	2.26***	2.31****	3.36****	2.31****	3.36****	1.86****
Observed t	1.6387	−2.4331	2.9268	3.6604	2.8315	5.1061	2.2732
t-test	Passed	Passed	Passed	Passed	Passed	Passed	Passed

Note: *at $k_1 = m = 1$; $k_2 = n-m-1 = 10-1-1 = 8$; **at $k_1 = m = 1$; $k_2 = n-m-1 = 9-1-1 = 7$; ***number of degrees of freedom = $n-1 = 10-1 = 9$; ****number of degrees of freedom = $m-1 = 9-1 = 8$

Source: Calculated and compiled by the authors

According to the econometric model (1), for every 1–point increase in the prevalence of green jobs, the following changes are observed:

- The level of development of entrepreneurial culture increases by 0.4761 points.
- Youth unemployment declines by 0.1677%.
- Stakeholder engagement increases by 0.2853 points.
- The flexibility of pay increases by 0.2088 points.
- Management professionalism increases by 0.2593 points.
- The fairness of pay increases by 0.3687 points.
- Staff retention improves by 0.3224 points.

This proves the proposed hypothesis H.

4.2 Prospects for Improving the Quality of Corporate Governance with Reliance on ESG-Based HR Management

Prospects for improving the quality of corporate governance through ESG-based HR management are identified based on model (1) and shown in Fig. 1.

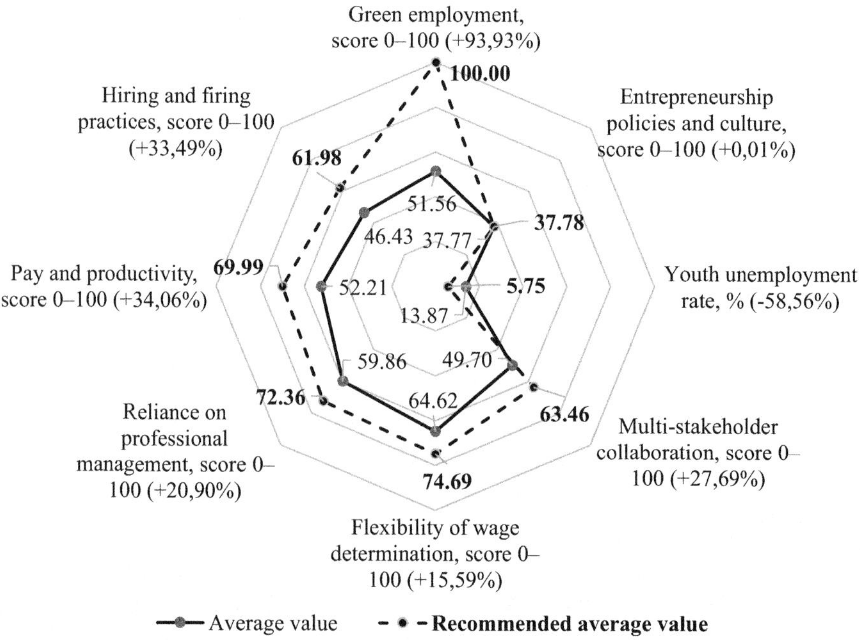

Fig. 1 Prospects for improving the quality of corporate governance through ESG-based HR management. (Source: Calculated and compiled by the authors)

According to Fig. 1, in the long term, due to the increase in the prevalence of green jobs to 100 points (+93.93%), the level of entrepreneurial culture will increase by 0.01%, youth unemployment will decrease by 58.56%, stakeholder engagement will increase by 27.69%, pay flexibility will increase by 15.59%, management professionalism will increase by 20.90%, pay equity will increase by 34.06%, and retention will improve by 33.49%.

5 Discussion

This research contributes to the literature through the development of scientific provisions of the concept of HR management by substantiating the significant contribution of ESG-based HR management to improve the quality of corporate governance. In contrast to Amos and Lydgate [15], Beccarello and Di Foggia [16], Ferreira-Lopes et al. [17], and Palakshappa and Dodds [18], this research substantiates that SDG 8 and SDG 12 can and should be implemented jointly, which is facilitated by the creation of green workplaces through ESG-based HR management.

The econometric model of the impact of ESG-based HR management on the quality of corporate governance, compiled in the research, demonstrated that, in contrast to Bilek-Steindl and Url [3], Borrego et al. [4], Khalique et al. [5], Kreinin and Aigner [6], and Makarenko et al. [7], the characteristics of corporate governance do not necessarily have to be improved individually – it is also possible to improve them comprehensively through the creation of green workplaces through ESG-based HR management.

6 Conclusion

Thus, the research revealed the prospects of improving the quality of corporate governance with the support of ESG-based HR management. The econometric model of ESG-based HR management on the quality of corporate governance, based on international best practices and authoritative statistics of Global Green Growth Institute, WEF, ILO, and WIPO 2019–2021, proved that the creation of green jobs through ESG-based HR management explains the level of entrepreneurial culture development by 80.13%, youth unemployment rate – by 65.21%, stakeholder engagement – by 74.18%, pay flexibility – by 81.05%, management professionalism – by 73.07%, pay equity – by 88.79%, and staff retention – by 65.17%.

The key conclusion from the research is that the prospects for improving the quality of corporate governance through ESG-based HR management are associated with the creation of green jobs through ESG-based HR management. The theoretical significance of the research results is that they redefine the essence of corporate governance in the Fifth Industrial Revolution, which should not be limited to the introduction of smart technology, but also involve the improvement of the organization of jobs to unlock human potential.

The practical significance of these results is that they make it possible to improve the practice of HR management through the creation of green jobs, the high attractiveness of which for business is proved in the research from the perspective of a large contribution to improving the quality of corporate governance. The social significance of the author's conclusions and recommendations is due to the fact that they opened the possibility and formed the scientific and methodological support for the joint implementation of SDG 8 and SDG 12.

References

1. Sergi BS, Popkova EG (2022) Towards a 'wide' role for venture capital in OECD countries' industry 4.0. Heliyon 8(1):e08700. https://doi.org/10.1016/j.heliyon.2021.e08700
2. Popkova E (2021) The social management of human capital: basic principles and methodological approaches. Int J Sociol Soc Policy 41(1–2):24–36. https://doi.org/10.1108/IJSSP-03-2020-0062
3. Bilek-Steindl S, Url T (2022) Nowcasting and monitoring SDG 8. Empirica 49(2):313–345. https://doi.org/10.1007/s10663-022-09533-0
4. Borrego AC, Carreira FA, Pardal P, Abreu R (2022) Social responsibility and SDG 8 during the first wave of the COVID-19 pandemic: the role of chartered accountants in Portugal. Sustainability 14(14):8625. https://doi.org/10.3390/su14148625
5. Khalique F, Madan P, Puri G, Parimoo D (2021) Incorporating SDG 8 for decent work practices: a study of MNC subsidiaries in India. Aust Account Bus Financ J 15(5SI):99–114. https://doi.org/10.14453/aabfj.v15i5.7
6. Kreinin H, Aigner E (2022) From "decent work and economic growth" to "sustainable work and economic degrowth": a new framework for SDG 8. Empirica 49(2):281–311. https://doi.org/10.1007/s10663-021-09526-5
7. Makarenko I, Plastun A, Petrushenko Y, Vorontsova A, Alwasiak S (2021) SDG 4 and SDG 8 in the knowledge economy: a meta-analysis in the context of post-COVID-19 recovery. Knowl Perform Manag 5(1):50–67. https://doi.org/10.21511/kpm.05(1).2021.05
8. Liang Y, Lee MJ, Jung JS (2022) Dynamic capabilities and an ESG strategy for sustainable management performance. Front Psychol 13:887776. https://doi.org/10.3389/fpsyg.2022.887776
9. Chouaibi Y, Zouari G (2022) The effect of corporate social responsibility practices on real earnings management: evidence from a European ESG data. Int J Discl Gov 19(1):11–30. https://doi.org/10.1057/s41310-021-00125-1
10. Niu S, Park BI, Jung JS (2022) The effects of digital leadership and ESG management on organizational innovation and sustainability. Sustainability 14(23):15639. https://doi.org/10.3390/su142315639
11. Schleich MV (2022) What are the human resources policies and practices most used by companies with the best ESG ratios in Brazil? RAE Revista de Administracao de Empresas 62(6):e02021–e00370. https://doi.org/10.1590/S0034-759020220511x
12. Aldieri L, Barra C, Ruggiero N, Vinci CP (2021) Green energies, employment, and institutional quality: some evidence for the OECD. Sustainability 13(6):3252. https://doi.org/10.3390/su13063252
13. Bohnenberger K (2022) Is it a green or brown job? A taxonomy of sustainable employment. Ecol Econ 200:107469. https://doi.org/10.1016/j.ecolecon.2022.107469

14. Sharpe SA, Martinez-Fernandez CM (2021) The implications of green employment: making a just transition in ASEAN. Sustainability 13(13):7389. https://doi.org/10.3390/su13137389
15. Amos R, Lydgate E (2020) Trade, transboundary impacts and the implementation of SDG 12. Sustain Sci 15(6):1699–1710. https://doi.org/10.1007/s11625-019-00713-9
16. Beccarello M, Di Foggia G (2022) Sustainable development goals data-driven local policy: focus on SDG 11 and SDG 12. Adm Sci 12(4):167. https://doi.org/10.3390/admsci12040167
17. Ferreira-Lopes L, Van Rompay-Bartels I, Bezanilla MJ, Elexpuru-Albizuri I (2022) Integrating SDG 12 into business studies through intercultural virtual collaboration. Sustainability 14(15):9024. https://doi.org/10.3390/su14159024
18. Palakshappa N, Dodds S (2021) Mobilising SDG 12: co-creating sustainability through brands. Mark Intell Plan 39(2):265–283. https://doi.org/10.1108/MIP-08-2018-0360
19. Amon J, Rammerstorfer M, Weinmayer K (2021) Passive ESG portfolio management – the benchmark strategy for socially responsible investors. Sustainability 13(16):9388. https://doi.org/10.3390/su13169388
20. Velte P (2019) The bidirectional relationship between ESG performance and earnings management – empirical evidence from Germany. J Glob Responsib 10(4):322–338. https://doi.org/10.1108/JGR-01-2019-0001
21. Global Green Growth Institute (2023) Green Growth index. Retrieved from https://gggi-simtool-demo.herokuapp.com/SimulationDashBoard/country-profile. Accessed 1 Feb 2023
22. Institute of Scientific Communications (2023) Dataset “Social entrepreneurship in the global economy: the path from virtual evaluations to big data – 2022.” Retrieved from https://datasets-isc.ru/data2/787-data-set-sotsialnoe-predprinimatelstvo-v-mirovoj-ekonomike-put-ot-virtualnykh-otsenok-k-bolshim-dannym-2020. Accessed 1 Feb 2023

Digital Technologies in the Development of Hydrogen Energy

Anastasia V. Sheveleva and Vladislav I. Solomos

1 Introduction

The climate agenda plays a significant part in economic development. One of its significant items is hydrogen energy, which is considered one of the promising areas for implementing decarbonization programs and achieving carbon neutrality by many countries worldwide.

The role of hydrogen in the global energy balance will increase due to the process of decarbonization and the transition to carbon-free energy, from the embrace of the West, even despite the energy crisis, and due to the need to balance the system associated with emissions of energy sources.

According to the moderate scenario of the specialists of the EnergyNet Center, adopted as the baseline, the global hydrogen energy market should increase its volume by $26 billion by 2025. During this period, in 2025–2040, hydrogen fuel prices should decrease from $4000 to $2000 per ton. According to Persistence Market Research forecasts, the average annual growth of the world market is estimated at 6.1% in 2020–2025; by the end of 2025, its impact volume will be $200 billion. In the APS (Announced Commitment Scenario) of the IEA, the amount of annual production assumed in accordance with hydrogen strategies should increase to an average of 150 million tons by 2035 and 250 million tons by 2050. Simultaneously, according to the forecast of the Ministry of Energy of the Russian Federation, the global consumption of pure hydrogen and hydrogen in the mixture will amount to 156 million tons by 2030, which is 37 million tons more than the current volume, estimated at 119 million tons.

According to the American consulting company ADI Analytics, the demand for hydrogen will more than double in 2020–2050 (Fig. 1).

A. V. Sheveleva (✉) · V. I. Solomos
MGIMO University, Russia, Moscow
e-mail: a_sheveleva@rambler.ru

E. G. Popkova (ed.), *Smart Green Innovations in Industry 4.0*, Springer Climate,
https://doi.org/10.1007/978-3-031-45830-9_29

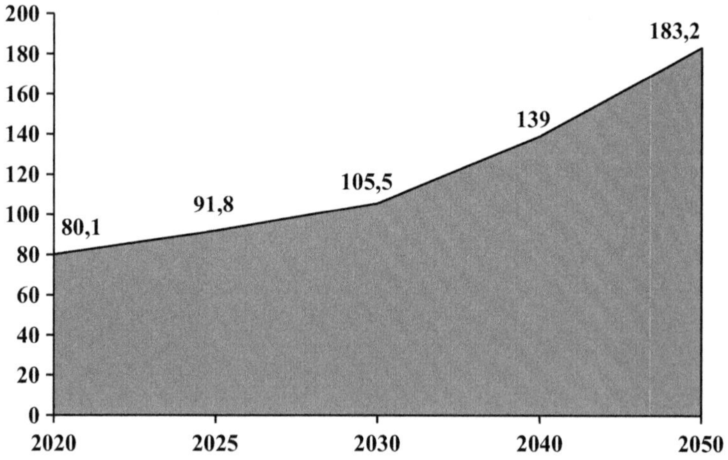

Fig. 1 Forecast of global hydrogen demand, million tons. (Source: Compiled by the authors based on [1])

Currently, several existing companies are engaged in hydrogen technologies in Russia, including Gazprom, Rosatom, Rosnano, Novatek, H2 Clean Energy, and others. These companies invest in developing technologies for producing, storing, and transporting objects and using CO_2. They also work on using technologies in various industries, energy, and transport.

In 2023, hydrogen test sites should appear in Yamal and Sakhalin, within which more than ten pilot projects for the production, storage, delivery, and presentation of the structure will be launched.

However, in connection with the introduction of sanctions against Russia, the Ministry of Energy of the Russian Federation worsened the forecast for the export of infrastructure of the Russian Federation in its draft Comprehensive Program for the Development of Hydrogen Energy until 2030. Russia has actually lost the location of hydrogen importers (Germany, Japan, and South Korea). The sales market, in the end, can shrink only to China. Exports from the Russian Federation may decrease from the previously forecasted 2.2 million tons to 1.4 million tons per year by 2030. Investments in the production of derivative metals in Russia amount to $21.1 billion, and export earnings are currently estimated at $12.8 billion.

The interest in hydrogen is due to the fact that it has high energy characteristics and is an environmentally friendly energy carrier. As a result, hydrogen is a valuable resource that can help companies achieve carbon emissions reductions and cut energy costs. Nowadays, hydrogen is used in various sectors of the economy, including petrochemistry, mineral fertilizers, metallurgy, and a cooler in energy blocks (Fig. 2).

Simultaneously, the use of hydrogen is associated with problems of explosion hazard and hydrogen-induced cracking of metals, resulting in possible accidents

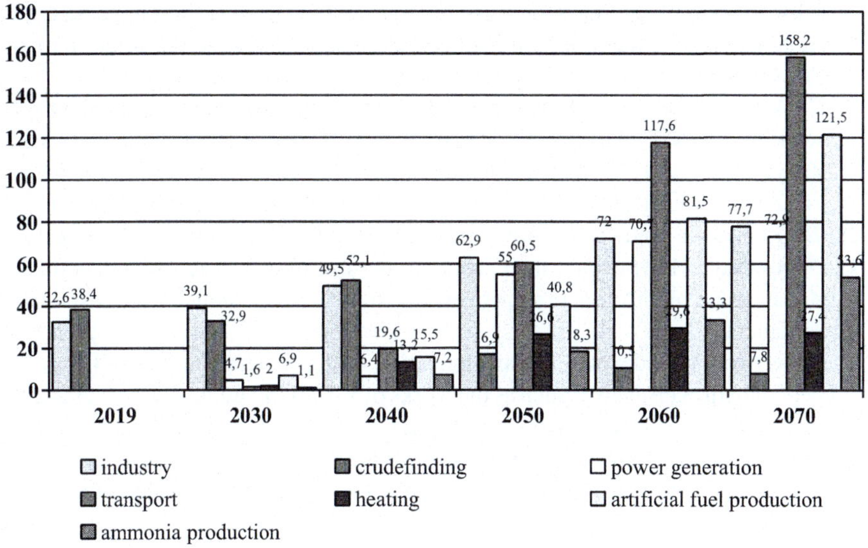

Fig. 2 Forecast of hydrogen consumption by segment, million metric tons per year. (Source: Compiled by the authors based on [2])

during transportation, storage, production, and consumption of hydrogen, accompanied by ecological backlash consequences that exacerbate climate change.

Digital technologies play an essential role in dealing with climate change at the present stage, among which it is necessary to highlight artificial intelligence, big data, and the Internet of Things (IoT).

These technologies should reduce climate risks, systematize large amounts of data, develop climate models that can more accurately predict climate parameters, monitor the state of the environment and predict its changes, consider the carbon footprint, reduce CO_2 emissions, and achieve carbon neutrality.

Digitalization also promotes the development of hydrogen energy, helping to design equipment and develop practical solutions for hydrogen technologies more accurately. Digitalization enables more complex energy systems to be managed, opens up access to future markets, and significantly improves the performance of the hydrogen industry.

In this regard, the analysis of digital technologies in the development of hydrogen energy is quite relevant.

2 Methodology

The authors studied the works of Russian and foreign authors on the development of hydrogen energy and the role of digital technologies in this process. The conducted critical analysis and generalization of separate results of these works made it possible to form the author's view on this issue and prove that digital technologies promote the development of hydrogen energy and, accordingly, the solution to climate problems.

C. Abreu Lopes and M. Erridge [3]; A. Dementieva and E. Zavyalova [4]; G. Kyriakopoulos, E. Kuzmin, V. Akberdina, and V. Kumar [5]; E. G. Popkova and E. Zavyalova [6]; A. V. Sheveleva and M. V. Cherevik [7]; A. V. Sheveleva and N. S. Zagrebelnaya [8]; A. V. Sheveleva, S. G. Tyaglov, and P. Khaiter [9]; E. B. Zavyalova and T. G. Krotova [10]; and E. Zavyalova, J. S. Lee, and E. Ostrovskaya [11] devoted their works to the use of digital technologies for the implementation of the climate agenda and the transition of companies to sustainable development.

Fewer works are devoted to the application of digital technologies in the development of hydrogen energy, such as the works of V. L. Bondarenko, D. N. Ilyinskaya, A. A. Kazakova, P. S. Kozlovtsev, and N. A. Lavrov [12, 13]; B. Gerard, E. Carrera, O. Bernard, and D. Lun [14]; and D. Walwyn [15].

The analysis of the works mentioned above has shown that the role of digitalization of hydrogen energy in terms of the implementation of the Sustainable Development Goals (SDGs) in general and the climate agenda in particular has not been studied. For this purpose, the authors of this research suggest that digitalization of the production, transportation, and storage of hydrogen, considered as a source of clean energy, despite certain difficulties, can ensure the environmental safety and economic efficiency of these processes.

The use of generalization and synthesis as part of the research, as well as a systematic approach, allowed the authors to come to certain conclusions.

3 Results

Energy consumption, plant productivity, cleanliness, and storage are among the key performance indicators (KPIs) for hydrogen generation that demand transparency to ensure efficient production. The Internet of Things can offer rapid anomaly detection using intelligent alarms, sensors on assets to monitor key performance indicators and asset status, and cloud-based remote monitoring outside of dispatcher services. Providing real-time monitoring of plant operation and asset status in combination with remote management can reduce costs by 10–20% by reducing energy consumption and optimizing the workforce. The use of monitoring models using digital twins allows investors to view how they evaluate the business project and take actions to reduce possible losses.

Reducing downtime is an ongoing problem for industrial companies, and this is no exception for hydrogen production enterprises. Early warning and diagnostics of equipment performance are necessary to ensure that enterprises can operate at full capacity when needed, preventing mechanical or technological failures. With predictive analytics based on artificial intelligence, asset-intensive companies can

Table 1 Digital technologies in hydrogen value chain optimization

The element of the hydrogen value chain	Digital technologies	Results of the application of digital technologies
Accelerate hydrogen production innovation	Process modeling	Accelerate research to construction pipeline
		Improve conversion processes
		Exhaustively evaluate alternatives
		Improve fuel cell economics
		De-risk storage and transport
Blue hydrogen and ammonia asset optimization	Digital twin, energy modeling, advanced process control	Optimize energy use
		Optimize production
		Carbon capture
De-risk and optimize the hydrogen value chain	Systems risk modeling, planning and scheduling	Built the right value chain
		Optimize integration with natural gas networks
Rapid repeatable design and operation	Process modeling	Need to scale
		Operate complex plants with minimum staff

Source: Compiled by the authors based on [16]

reduce equipment downtime and improve operational reliability while cutting operating and maintenance costs.

Advanced analytics can transform data into business analytics with practical conclusions. Energy losses can be staved off by predicting failures and optimizing equipment uptime, which increases revenue and reduces operating costs.

Digital technology will be essential in delivering the hydrogen economy, accelerating and de-risking innovation, de-risking adoption, and enabling faster and better scale-up and optimization of the hydrogen value chain. It will be fundamental in overcoming many value chain obstacles, maximizing commercialization, design, and supply chains, and boosting production and economics (Table 1).

Digital technologies for hydrogen energy act as activators because models that optimize its value chains become cheaper. Developing hydrogen as an energy source involves the complete value chain from production to end-use – and encompasses the entire commercialization lifecycle, from innovation to reliable operation at scale. Innovations in asset optimization software now span design, operations, supply chain, and maintenance; they are uniquely suited to address these challenges.

All countries worldwide strive for the sustainable development of their energy industry. Large-scale fuel crises, climate changes, and the situation in 2022 have shown the failure of hydrocarbon energy to achieve this goal, while hydrogen energy has such potential.

For example, EU countries expect to achieve zero net emissions by 2050. For this purpose, it is necessary to transfer production from coking coal to hydrogen and renewable energy. For example, green hydrogen will become an important technology in decarbonizing heavy industrial processes, such as steel production. Therefore, the importance of digital technologies for designing hydrogen energy equipment is increasing. In 2020, the Swedish company H2 Green Steel developed strategic software to optimize the configuration of a hydrogen unit that provides complete decarbonization of the steel industry [17].

The goal is to create environmentally friendly hydrogen units that would replace traditional non-ecological steel production technologies with energy solutions contributing to achieving climate goals. The software will recommend the optimal setting depending on the factors. Instead of spending 6 months on the hydrogen plant design process, using this technology, one can get a ready-made concept of an enterprise to produce green hydrogen in about 2 weeks.

The Russian Federation is also developing hydrogen energy to reduce greenhouse gas emissions. In August 2021, the Government of the Russian Federation approved the "Concept for the development of hydrogen energy in the Russian Federation" for the medium-term period until 2024, the long-term period until 2035, as well as the main focus for the future until 2050, where one of the directions indicates the development of high-tech solutions, including digital technologies [18].

In 2021, at the National Research Tomsk Polytechnic University, scientists created a digital twin of the power system with alternative and hydrogen elements [19]. It allows one to develop and research a management strategy for systems that include elements of alternative energy – wind generators, solar panels, and hydrogen energy storage. The digital twin allows one to simulate regular situations and emergencies, including emergency ones.

Scientists of Tomsk Polytechnic University will also improve hydrogen storage systems using digital modeling and 3D prototyping methods as part of the support of the Russian Science Foundation. The goal is to develop efficient and safe hydrogen storage systems.

However, several risks arise when introducing hydrogen energy based on Industry 4.0. The first group includes production and technological risks because no necessary infrastructure exists. Second, there are financial and economic risks associated with the increase in the costs of financing digitalization projects. The third group is organizational and administrative risks because the introduction of digital technologies requires the restructuring of the existing organizational structure of management, as well as planning and control processes. However, despite all advantages of hydrogen energy, the prospects for the global market are still unclear.

4 Conclusions

In general, the following conclusions can be drawn.

The course toward implementing the Sustainable Development Goals, the climate agenda, the decarbonization of the economy, and the "achievement of zero emissions" pushes countries to develop hydrogen energy.

Digital technologies will become an important component in providing hydrogen energy, accelerating innovation, reducing the risks of implementation, and ensuring faster and better scaling and optimization of the hydrogen value chain.

Digitalization is a powerful incentive to optimize the large-scale production of green hydrogen, ensuring the profitable production of everything from green steel to ammonia and methanol.

Acknowledgments The authors express their gratitude to the conference organizers for the opportunity to participate in the conference, present their results, and get acquainted with the opinions of other participants on such an urgent problem. The authors would especially like to note the professionalism in preparing and holding this event, as well as the timely manner of informing the participants.

The authors also thank the reviewers who took the time to familiarize themselves with the research results and give valuable comments and advice.

References

1. Gupta U, Turaga U (2019) Hydrogen is back in vogue. ADI Analytics. Retrieved from https://adi-analytics.com/2019/11/12/hydrogen-is-back-in-vogue/. Accessed 21 Dec 2022
2. IEA (2022) Hydrogen. Retrieved from https://www.iea.org/fuels-and-technologies/hydrogen. Accessed 21 Dec 2022
3. Abreu Lopes C, Erridge M (2021) Digital technologies for sustainable development. In: Andrew J, Bernard F (eds) Human rights responsibilities in the digital age: states, companies and individuals. Hart Publishing, Oxford, pp 205–220. https://doi.org/10.5040/9781509938865.ch-012
4. Dementieva A, Zavyalova E (eds) (2020) Corporate governance in Russia: quo vadis? De Gruyter, Berlin/Boston. https://doi.org/10.1515/9783110695816
5. Kyriakopoulos G, Kuzmin E, Akberdina V, Kumar V (eds) (2022) Research topic: sustainability of digital transformation for the environment. Retrieved from https://www.frontiersin.org/research-topics/48140/sustainability-of-digital-transformation-for-the-environment. Accessed 21 Dec 2022
6. Popkova EG, Zavyalova E (eds) (2021) New institutions for socio-economic development: the change of paradigm from rationality and stability to responsibility and dynamism, Berlin/Boston. https://doi.org/10.1515/9783110699869
7. Sheveleva AV, Cherevik MV (2022) Digital technologies in the oil and gas sector and their contribution to UN climate action goal. In: Zavyalova EB, Popkova EG (eds) Industry 4.0: fighting climate change in the economy of the future. Palgrave Macmillan, Cham, pp 307–315. https://doi.org/10.1007/978-3-030-79496-5_28
8. Sheveleva AV, Zagrebelnaya NS (2020) The practice of introducing digital technologies in the oil and gas industry. In: Isachenko TM (ed) Modern international economic relations in the post-bipolar era. MGIMO University, Moscow, pp 244–225
9. Sheveleva AV, Tyaglov SG, Khaiter P (2021) Digital transformation strategies of oil and gas companies: preparing for the fourth industrial revolution. In: Konina N (ed) Digital strategies in a global market: navigating the fourth industrial revolution. Palgrave Macmillan, Cham, pp 157–171. https://doi.org/10.1007/978-3-030-58267-8_12
10. Zavyalova EB, Krotova TG (2021) Methods for achicving SDG 1, poverty eradication. In: Popkova EG, Sergi BS (eds) Modern global economic system: evolutional development vs. revolutionary leap. Springer, Cham, pp 1987–2001. https://doi.org/10.1007/978-3-030-69415-9_218

11. Zavyalova E, Lee JS, Ostrovskaya E (2020) Corporate social responsibility. In: Dementieva A, Zavyalova E (eds) Corporate governance in Russia: quo vadis? De Gruyter, Berlin/Boston, pp 173–187. https://doi.org/10.1515/9783110695816-016
12. Bondarenko VL, Ilyinskaya DN, Kazakova AA, Kozlovtsev PS, Lavrov NA (2022) Introduction to digital hydrogen energy. Chem Pet Eng 58:42–46. https://doi.org/10.1007/s10556-022-01052-w
13. Bondarenko VL, Ilyinskaya DN, Kazakova AA, Kozlovtsev PS, Lavrov NA (2022) Prospects and current problems of hydrogen energy digitalization. Chem Pet Eng 58:63–67. https://doi.org/10.1007/s10556-022-01056-6
14. Gerard B, Carrera E, Bernard O, Lun D (2022) Smart design of green hydrogen facilities: a digital twin-driven approach. E3S Web Conf 334:02001. https://doi.org/10.1051/e3sconf/202233402001
15. Walwyn D (2017) Building the hydrogen economy through niche experimentation and digitalization. In: Proceedings of the IAMOT 2017: 26th annual conference of the International Association for Management of Technology, Austria, Vienna. https://doi.org/10.13140/RG.2.2.24322.68801
16. Beck R (n.d.) Accelerating the hydrogen economy through digitalization. Retrieved from https://www.aspentech.com/en/resources/white-papers/accelerating-the-hydrogen-economy-through-digitalization. Accessed 21 Dec 2022
17. H2 Green Steel (2022) Green hydrogen enables the decarbonization of steel production. Retrieved from https://www.h2greensteel.com/articles/green-hydrogen-enables-the-decarbonization-of-steel-production. Accessed 21 Dec 2022
18. Government of the Russian Federation (2021) Order "On approval of the concept of hydrogen energy development in The Russian Federation". August 5, 2021 No. 2162. Moscow, Russia
19. Graf N (2021) Tomsk Polytechnic University has created a digital twin of the electric power system. Rossiyskaya Gazeta. Retrieved from https://rg.ru/2021/12/08/reg-sibfo/v-tomskom-politehe-sozdali-cifrovoj-dvojnik-elektroenergeticheskoj-sistemy.html. Accessed 22 Dec 2022

Central Bank Mandate in the Age of Climate Change: Global and National Perspective

Vasily N. Tkachev, Igor B. Turuev, and Anastasia A. Zhukova

1 Introduction

Climate change and global warming tend to be considered primarily through the lens of environmental degradation and the increased frequency of natural calamities. Meanwhile, the implications of the process are reaching far beyond these trends, affecting the global financial system as well. Although many established financial institutions have committed to reaching net-zero financed emissions by the mid-century, the path to a climate-neutral business model and green asset portfolios is fraught with risks. Financial authorities are to play a pivotal role in spearheading future trends in the green finance market and averting the manifestation of climate-related risks. Likewise, private actors could contribute to the green agenda by channeling resources to eligible financial instruments and projects.

During the UN Climate Change Conference (COP27) held in Sharm el-Sheikh, Egypt, in November 2022, the international community reiterated its commitment to bring down global temperatures. The cornerstone of the debates during COP27 was the issue of financing inclusive and sustainable transition pathways, with the necessary investments estimated at roughly 4–6 trillion dollars per year [13]. This emphasizes the need to streamline sustainable finance initiatives currently underway, including creating a conducive environment for raising green funds.

Therefore, the primary objective consists of coordinating the efforts of financial authorities, governments, and private actors with a view to mobilizing the necessary financial resources without any detriment to financial stability. With regard to central banks and their mandates, which, as a rule, center around price stability [4], this means revisiting their operational goals and striking a balance between conventional areas of regulation and emerging ones, including climate-related financial

V. N. Tkachev (✉) · I. B. Turuev · A. A. Zhukova
MGIMO University, Moscow, Russia
e-mail: i.turuev@inno.mgimo.ru

E. G. Popkova (ed.), *Smart Green Innovations in Industry 4.0*, Springer Climate,
https://doi.org/10.1007/978-3-031-45830-9_30

risks and the application of Industry 4.0 technologies in the financial sector (in some cases, these two spheres may interact, for example, when AI is used for shaping a climate-neutral asset portfolio).

2 Methodology

This research is based on the analysis of reports and data provided by specialized international organizations and networks, academic literature, official statistical data, and information from official websites of financial regulators developing regulation in the sphere of climate-related risks and financial institutions involved in the agenda through their lending and investment activities.

The impact of the green agenda on the following areas of central banks' activities has been regarded:

1. Mitigation and supervision of climate-related risks.
2. Supporting private initiatives and creating sustainable finance market infrastructure.
3. International cooperation. Although there is no uniform high-level definition of climate risks, for the purpose of the research, climate-related risks refer to potential risks ensuing from the direct consequences of climate change or measures to curb it [1].

These risks are typically divided into two broad categories – physical and transition risks. Additionally, there is a need to distinguish between the concepts of green, sustainable, and transition finance. Green finance comprises financing activities that can create a positive environmental effect. More broadly, sustainable finance is regarded as a process of taking due account of environmental, social, and governance (ESG) aspects, as well as the UN Sustainable Development Goals (SDGs), when making investment decisions. Finally, transition finance focuses primarily on financing activities that enable the low-carbon transition (including those promoting environmentally friendly practices among carbon-intensive companies).

3 Results

3.1 Climate Change as a Source of Financial Instability

Nowadays, it is possible to draw a parallel between climate change and tail risk events referred to as black swans. Since the transition to a greener economy is itself associated with risks characterized by deep uncertainty and imbalances, climate change can be regarded as a green swan suggesting that its development is hard to predict and its chances of manifestation are anything but reflected in the past data [2].

Climate-related risks are divided into two categories. First, physical risks originate from extreme climate-related events such as storms, floods, and wildfires that cause financial losses and physical damage. In their turn, transition risks are related to an unclear financial effect that might arise from a quick and chaotic transition to a low-carbon economy; they may include governmental changes, reputational consequences, technological advances, and shifts in consumers' preferences. Such risks combined are anticipated to wipe out up to 18% of the global GDP by 2050 and leave billions of stranded assets subject to premature devaluation.

Climate change might translate into financial risks for banks through the following important components:

1. Climate risk drivers (climate-related changes entailing financial hazards)
2. Transmission channels (causal linkages explaining how climate risk drivers affect banks directly and indirectly through their counterparties, assets, and the economy they operate in)
3. Variance sources (the geographical location of a particular bank, its assets or exposures that might exacerbate or mitigate effects)
4. Financial risks (credit, market, liquidity, etc.)

However, it is pivotal to note that the unique features of climate-related risks may hamper the proper use of the existing framework, which is largely backward-looking and based on historical loss experience [5]. Besides, capital requirements particularly focus on risks that will materialize over a relatively short-term horizon, whereas climate risks will take time to demonstrate their impact.

3.2 Global Regulatory Response to Climate-Related Financial Risks

With regard to the scope for regulatory and supervisory response to climate-related financial risks, central banks' mandate in this sphere appears to be quite multifaceted, ranging from bridging data gaps and stress testing to propping up green financial markets and targeted prudential measures. The activities of foreign regulators are primarily focused on risk assessment exercises (scenario analyses and stress tests) and integration of risks into the operational framework (long-term strategy, management of own portfolio, internal control), whereas regulatory measures are still under development.

Overall, ongoing efforts in setting supervisory expectations for climate-related and broader environmental risk management have greatly improved (Fig. 1). According to the Network for Greening the Financial System, a group of central banks and supervisors that exchange best practices in climate risk management, most central banks are developing climate-related (83%) risks supervisory

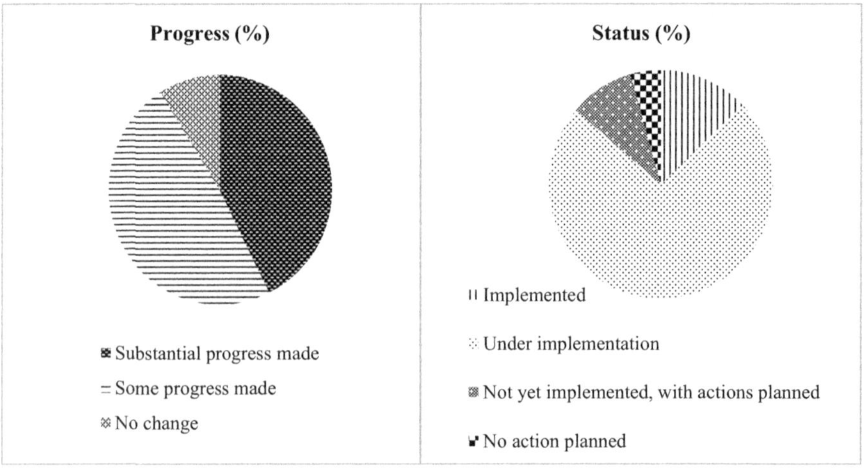

Fig. 1 The approaches of NGFS members to setting climate-related supervisory expectations in accordance with NGFS recommendations. (*Source* Compiled by the authors based on NGFS Progress Report on the Guide for Supervisors [8])

expectations. Nevertheless, some members (24%) only consider actions related to climate risks. Meanwhile, supervisors have made comparatively less progress in incorporating climate-related risks into their supervisory toolbox, with devising a set of formal and binding rules appearing to be the next step for most.

The most common analytical techniques are climate scenario analysis and stress tests designed to give supervisors an understanding of the extent to which supervised firms are exposed to climate-related financial risks. According to a study conducted by the Financial Stability Board, stress tests undertaken by central banks and regulators tend to capture both types of risks, albeit across varying timeframes. Moreover, in terms of sectoral coverage, supervisors are more likely to concentrate on the banking and insurance sectors, whereas a minority of the surveyed financial organizations examine the exposure of asset managers and pension funds to climate-related risks.

Progress is more palpable in relation to disclosure requirements, which are crucial to understanding banks' exposure to climate risks. The framework for environmental disclosure is also important as part of the green finance market infrastructure since it enhances investors' awareness of available options. To narrow the persistent data gaps, which are hampering the proper financing of climate-friendly bankable projects, the EU market participants are required to disclose the environmental and social impacts of investment decisions. Besides, in early 2022, the European Commission enacted Corporate Sustainability Reporting Directive that requires listed companies, banks, and insurance companies to reveal information about sustainability factors in their business. The proposals also urge financial firms and advisers to include sustainability in the investment advice they deliver to clients. Likewise, Singapore- and Hong Kong-based ESG-labeled investment funds and asset managers are to unveil their focus, investment strategy, asset allocation, risks, and reference benchmarks.

Moreover, one of the salient features of the contemporary paradigm of socio-economic development is dynamism; hence, the wide proliferation of Industry 4.0 technologies in financial regulation, including climate-oriented ones [10]. Since one of the challenges the regulators worldwide face is persisting data gaps that hinder the effective implementation of regulatory measures, it is only natural to assume that advancements in data processing should help mitigate the problem. To alleviate possible repercussions, a number of financial market participants have already begun to use their own solutions in this sphere. For instance, the People's Bank of China, which has made strides in regulating FinTech [12], is testing a cloud-based platform for sharing green credit reporting between banks and regulators based on big data and artificial intelligence. The platform is a repository with statistical information on green lending, relevant analytics, and reviews.

It is crucial to note that embedding climate considerations into the regulatory framework is a double-edged sword; that is, it has benefits and drawbacks. With regard to potential positive effects, we can list the enhanced resilience of the country's financial system along with fostering redistribution of capital in favor of eligible green projects. Nevertheless, there are still some backlashes to consider. For example, a poorly designed climate-oriented prudential regulation may precipitate adverse shocks in the financial system and leave carbon-intensive assets stranded (i.e., exposed to premature and eminent devaluation). Besides, there is still no sufficient evidence to prove that green assets are less risky.

3.3 Supporting Private Climate-Related Initiatives

Finance is crucial in response to economic shocks. However, mobilizing capital for green investments can be difficult because of market failures that include the lack of data on risks and opportunities of a green transition and the underpricing of externalities. Other barriers facing green financial markets include a lack of necessary knowledge of environmental issues and green investment skills in the financial sector, weak oversight mechanisms, short-termism (i.e., the pursuit of short-term goals disregarding longer-term ones), and information asymmetries.

In recent years, private actors have embraced the climate agenda at an accelerating pace. The most prominent example of this trend is the issuance of green bonds. Green bonds are standard bonds with proceeds going to green and climate-related assets, projects, and activities. They are issued in compliance with the Green Bond Principles developed by ICMA, which incorporate the following four core components:

1. The use of proceeds
2. Project evaluation and selection
3. Management of proceeds
4. Reporting (Fig. 2)

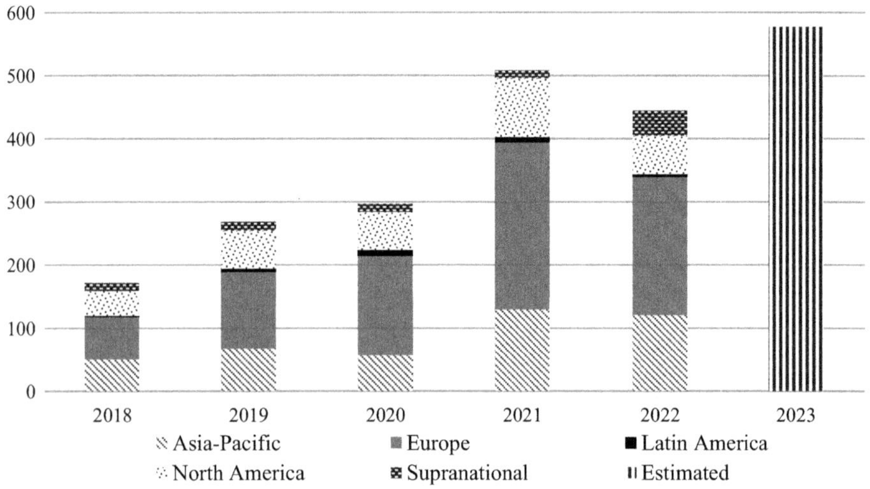

Fig. 2 Regional structure of the green bond issuance in 2018–2022 and expected issuance in 2023 ($ billion). (*Source* Compiled by the authors based on Climate Bonds Initiative (CBI) Market Data [3])

In 2022, amid rising downside risks, investors have grown wary of green bonds, whereas corporates have scaled down on new issuances. The decline amounted to $64 billion (12.8%). Europe has reasserted itself as a leading region in terms of green bond issuance, which can be explained by its role as a pioneer of the climate agenda and the state of development of its financial markets and infrastructure [9]. Nevertheless, the expectations of major players for 2023 are more sanguine: the volume of issuance is expected to return to 2021 levels subject to softer regulations and incentives. However, shrinking issuance volumes in response to crisis events underscore the necessity for financial authorities to spearhead the development of green financial instruments.

With regard to the concept of greening central banks' mandate, the regulator may contemplate on adjusting collateral requirements for refinancing operations to embed climate considerations in its policy. Nevertheless, this measure may not be expedient for asset purchase programs (APPs). On the one hand, green financial instruments are in no way a substitute for more secure sovereign bonds. On the other hand, large-scale APPs are not in retreat ceding place to the hawkish stance of most central bankers amid entrenched high inflation. Another aspect of a green mandate is the management of reserves: central banks may consider either explicitly including climate considerations into their FX reserve framework as a main goal or implicitly – as sideline aspects of conventional policy goals [6]. To this end, the regulators should strike a balance between liquidity and sustainability; then, trade off safety and return. Alternatively, for institutions where sustainability is deemed a first-order consideration, it could precede the whole process: for example, by setting liquidity, safety, and return requirements only using instruments whose environmental impact or footprint has been assessed beforehand.

In a move to enhance clarity for investors, there was a steady proliferation of detailed guidance in the form of taxonomies [7]. The latter can be defined as a classification system of activities or assets delivering on key environmental objectives. Many jurisdictions worldwide are either drafting green taxonomy regulations or already have them in place, including China, the EU, Japan, Malaysia, and Russia. Additionally, a growing number of countries are investigating transition taxonomies that define and identify actions that are compatible with a transition toward green aims. Such taxonomies are distinguished by a higher emphasis on entity-level transition and business model redesign.

3.4 Current Situation in Russia

Given the potential vulnerability of Russia to climate-related risks due to a traditionally high share of carbon-intensive industries, the green agenda is pivotal. Besides, against the backdrop of geopolitical challenges, Russia is undergoing a gradual structural economic transformation – the key to decent economic performance and bridging the technological gap are green investments [14]. Therefore, such steps as the development of a taxonomy of green projects, the increase in green financial instruments issuance, and the spread of responsible investment practices are worth considering.

With regard to the history of green finance in Russia, the first green bonds were issued in December 2018. Until now, there have been 13 subsequent green bonds and four social bond issuances demonstrating an appetite from financiers for sustainable investment solutions. In 2021, the sustainable finance market reached $3.82 billion in value terms, with green bonds accounting for 80% of the market and prioritizing investments in clean transportation and renewable energy. Nevertheless, the development of the considered market cannot gain momentum without lifting some underlying barriers, namely its relatively low attractiveness for investors seeking generous returns. Investments in financial instruments for environmental projects are so far perceived as riskier and less lucrative since companies raising such funds generally require a higher cost of capital for their projects. Against the backdrop of sanctions, the main restraining factor has become the reluctance of Russian corporates to issue green bonds: in 2022, only VEB.RF raised 50 billion rubles of new green debt financing at the rate of 10.2%.

During the last 2 years, the Bank of Russia has accelerated the integration of climate considerations into its activities by establishing a special Working Group on Financing for Sustainable Development. In December 2022, the Bank of Russia also issued a consultation paper on the consideration of transition and physical risks by banks and insurance companies under changing external circumstances. In the longer run, it may also be advisable to consider some adjustments in the regulatory framework aimed at creating a favorable environment for the green transformation of carbon-intensive companies. Among the options worth considering are direct quantitative limits on financing the so-called brown economic activities and reducing capital requirements in relation to loans provided for projects in line with the national environmental objectives.

To a great extent, the current geopolitical tensions have shifted the vector of the development of the Russian ESG market. Nevertheless, sustainability issues remain among the priorities. Export-oriented ESG targets may lose their importance as the sanctions have revealed the fault lines of the country's financial system, including its high dependence on foreign infrastructure [11]. For example, Russian corporates have been excluded from ESG rankings compiled by foreign rating agencies, which creates barriers to aligning with international sustainability standards. However, transitional and social projects corresponding to local needs will continue to develop at a higher pace.

4 Conclusion and Recommendations

Climate risks are generally recognized by the civilized world and carry economic threats that can potentially affect the financial component of the global economy. The transition to a green economy potentially disrupts established economic systems and affects the prospects for previously made investments, particularly in carbon energy and industrial production. Banks are at risk in terms of credits and possible loss of liquidity and market reputation.

Regulatory financial oversight and tools at the global and national levels are currently insufficient, being limited mainly to work on climate risk assessment but not their management. There are gaps in the regulatory legal framework, a lack of information about the "green" economy and finance, and a lack of knowledge and skills to work in the new circumstances. This hinders the development of the green financial market and investment prospects.

Developing the green concept of the economy and finance is also important for Russia, which has formulated its environmental agenda and is developing green financial instruments and programs, improving the legal framework and regulatory control by the Bank of Russia and stimulating the green transition. Creating a set of bankable projects may revitalize capital investments that are vital to establishing a new model of economic development and harnessing the potential of the country's resources.

The effects of climate-oriented financial regulation are a mixed blessing. Its potential benefits include the enhanced resilience of a country's financial system, as well as the reallocation of capital toward green projects. Simultaneously, there are potential downsides, particularly if the stimulus measures are driven solely by political reasons. In such a case, inefficient capital reallocation is possible, worsening the financial standing of brown borrowers due to a significant tightening of financial conditions and ensuing credit squeeze.

On the contrary, it is more important at this stage to focus on transforming the business models of carbon-intensive companies and financing their gradual transition to more environmentally friendly practices rather than abruptly withdrawing access to finance. Reduced capital requirements for banks on loans and bond investments to finance green and adaptation projects should be considered the main incentive tool. The incentives should be differentiated according to the project's contribution to achieving national carbon neutrality goals and the borrower's specific risk mitigation factors (e.g., a sound financial position or a low debt burden). Moreover, in the longer run, the Bank of Russia could consider applying direct quantitative limits within its macroprudential framework. In particular, this measure implies reducing financial institutions' exposure to carbon-intensive industries by limiting the maximum amount of loans issued for certain types of borrowers.

References

1. Bank of Russia (2022) Climate-related risks in changing economic conditions: consultation paper. Moscow, Russia. Retrieved from https://www.cbr.ru/Content/Document/File/143643/Consultation_Paper_21122022.pdf. Accessed 25 Jan 2023
2. Bolton P, Despres M, Pereira da Silva LA, Samama F, Svartzman R (2020) The green swan: central banking and financial stability in the age of climate change. BIS, Basel. Retrieved from https://www.bis.org/publ/othp31.pdf. Accessed 24 Jan 2023
3. Climate Bonds Initiative (n.d.) Market data. Retrieved from https://www.climatebonds.net/market/data/. Accessed 10 Feb 2023
4. Dikau S, Volz U (2019) Central bank mandates, sustainability objectives and the promotion of green finance. Ecol Econ 184:107022. https://doi.org/10.1016/j.ecolecon.2021.107022
5. Fabris N (2020) Financial stability and climate change. J Cent Bank Theory Pract 9(3):27–43
6. Fender I, McMorrow M, Sahakyan V, Zulaica O (2020) Reserve management and sustainability: the case for green bonds? BIS Working Papers No 849. BIS, Basel. Retrieved from https://www.bis.org/publ/work849.pdf. Accessed 27 Jan 2023
7. G20 SFWG (2021) 2021 synthesis report. Retrieved from https://g20sfwg.org/wp-content/uploads/2021/11/Synth_G20_Final.pdf. Accessed 27 Jan 2023
8. NGFS (2021) Progress report on the Guide for Supervisors. Retrieved from https://www.ngfs.net/sites/default/files/progress_report_on_the_guide_for_supervisors_0.pdf. Accessed 26 Jan 2023
9. Petrov MV, Tkachev VN, Turuev IB (2023) Development of national systems of green finance in the context of Industry 4.0. In: Popkova EG, Sergi BS (eds) Current problems of the global environmental economy under the conditions of climate change and the perspectives of sustainable development. Springer, Cham, pp 131–141. https://doi.org/10.1007/978-3-031-19979-0_15
10. Popkova EG, Zavyalova EB (eds) (2021) New institutions for socio-economic development: the change of paradigm from rationality and stability to responsibility and dynamism. De Gruyter, Berlin; Boston. https://doi.org/10.1515/9783110699869
11. Tkachev V, Grishchenko V, Summanen K (2022) Chapter 40: the Russian financial sector: opportunities in an unstable environment and sanctions. In: Nguyen DK (ed) Handbook of banking and finance in emerging markets. Edward Elgar Publishing, Cheltenham
12. Turuev I, Shashkina E (2022) Transformation of the banking sector under the influence of FinTech: China's experience vs Russia's experience. Finance and Credit 28(2):372–411. https://doi.org/10.24891/fc.28.2.372
13. UNFCCC (2022) Sharm el-Sheikh implementation plan. Retrieved from https://unfccc.int/documents/624444. Accessed 24 Jan 2023
14. Zavyalova EB, Studenikin NV (2019) Green investment in Russia as a new economic stimulus. In: Sergi BS (ed) Modelling economic growth in contemporary Russia. Emerald Publishing Limited, Bingley, pp 273–296. https://doi.org/10.1108/978-1-78973-265-820191011

Integrated Reporting as an Implementation Tool of ESG Strategies and Anti-inflationary Effect

Gulbaira B. Usmanalieva, Marina V. Safronchuk, and Natalia A. Brovko

1 Introduction

Globalization and the complication of the economic and institutional environment for companies and markets impose new requirements on the corporate reporting system that could satisfy a wide range of interests of its users.

The transparency of information on the firms, regions, and country activities is very important. In world practice, firms are not limited in providing information; in financial statements, they disclose non-financial information concerning climate change, environmental pollution, the prolonged COVID-19 pandemic, and the fight against corruption, inflation, and inequality. However, these problems are huge challenges for the entire global community and economy. It confirms the problem of insufficient information in financial reports and the need to implement public reporting with disclosure of responsible investment factors (ESG). In the annual report for 2022, the IMF declared to support the governments in their policies concerning climate change and ecology [17, p. 76]. In 2023, the UN Global Compact is introducing a new Communication on Progress (CoP) system and a dedicated reporting platform. "Corporate transparency and fair reporting are becoming increasingly important to advance the ten principles of the Global Compact, the Sustainable Development Goals, the Paris Climate Agreement," said Sanda Ojiambo, Executive Director of the UN Global Compact [9]. The Progress Report is the main mechanism for companies participating in the Global Compact to demonstrate commitment to the Principles and Guidelines of Sustainable Development. Harmonization of financial and non-financial reporting integrates ESG factors and

G. B. Usmanalieva · N. A. Brovko
Kyrgyz-Russian Slavic University named after B. Yeltsin, Bishkek, Kyrgyzstan

M. V. Safronchuk (✉)
MGIMO University, Moscow, Russia

E. G. Popkova (ed.), *Smart Green Innovations in Industry 4.0*, Springer Climate,
https://doi.org/10.1007/978-3-031-45830-9_31

financial performance. The ESG is a development strategy that provides for transparency in management and concern for the environment and social responsibility. ESG reports are annual and detail the actions, initiatives, plans, goals, and achievements of companies in ESG efforts. They comply with GRI, SASB, TCFD, CDP, GRESB, and SDG standards [8, p. 4].

2 Methodology

This research employed general scientific methods of cognition, including analysis, synthesis, hypothesis, and a systemic approach to the studied problems. The theoretical and methodological provisions for constructing financial and integrated statements have been generalized. A comparative analysis of their concepts has been carried out. Based on the International Framework for Integrated Reporting (IFR) and International Financial Reporting Standards (IFRS), the authors disclosed and analyzed the aspects of a systemic approach to forming integrated reporting. A hypothesis has been formed about the need to develop an innovative approach in the system of public reporting – integrated reporting, which is an institutional factor in implementing the ESG strategy and investment attractiveness in the conditions of Kyrgyzstan and other post-Soviet countries. The hypothesis that ESG transformation creates anti-inflationary effects on the scale of the economy was built based on institutional and economic analysis, including the method of extrapolating microeconomic results to the macro-level. The hypothesis is confirmed by the study of the dynamics of macroeconomic indicators in Kyrgyzstan and their comparison with some other post-Soviet countries for over the past 20 years. The authors also used macroeconomic indicators of the Trading Economics statistical information website and the Center for Economics and Business Research (CEBR) study, conducted for the global audit company Moore Global and published in 2022.

3 Discussion

Financial statements originate from the accounting system. Accounting is an orderly system for collecting, measuring, recording, and summarizing information about the assets, liabilities, equity, income, and expenses of an economic entity. Based on this information, financial reports are compiled for making informed decisions [18, p. 500]. Accounting methods record and reflect retrospective information. Nowadays, users are interested in forward-looking information about the strategy and the creation of value in different time horizons, considering ESG factors. Such information is provided by integrated reporting, which is a tool for implementing the company's ESG strategy.

In the research, external users are segmented according to the degree of their financial interest:

- users with direct financial interest (shareholders, potential and future investors, creditors);
- users with indirect financial interest – state authorities and administrations, tax authorities, stock market participants;
- users without financial interest (audit companies, statistics agencies, etc.).

The owners and administrative and management personnel of the company need analytical information to assess the organization's activities, make reasonable managerial decisions, and determine the company's strategy.

The interests of users of financial statements are closely interrelated with the objectives of the company's financial statements in accordance with IFRS. They can be ranked as general purpose (reflected in CFFR), individual purpose (in accordance with IAS 1), and consolidated purpose (in accordance with IAS 1) [10–12].

The goals, principles, and concepts of IR are reflected in the International Framework for Integrated Reporting (IFR). It shows how the company creates, retains, or loses value over different time horizons in accordance with the specifications of production and economic situation. This information is vital for all types of investors, employers, customers, suppliers, partners, local communities, legislators, regulators, and policymakers to create value [16, p. 6, 53]. It shows that the key positions of the ESG transformation contain systematic changes of the firm toward responsible business conduct in the context of social and environmental safety. The company's ESG strategy should include a new business model for achieving sustainable development goals.

ESG principles were first formulated by the former UN Secretary-General Kofi Annan. He suggested that the managers of large global companies include these principles in their strategies to combat climate change [20]. The combined structure of integrated reporting forms their concepts, which have a common foundation (Fig. 1).

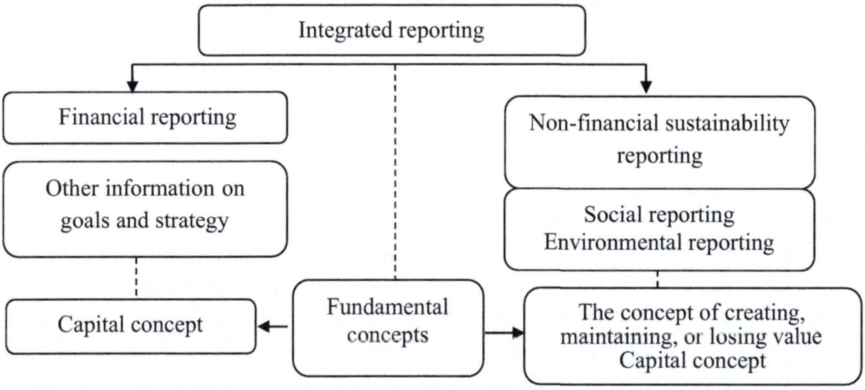

Fig. 1 Concepts and types of corporate reporting in the structure of integrated reporting. (*Source* Developed and compiled by the authors based on [16, 21])

Figure 1 shows that integrated reporting requires information from financial and non-financial reporting. According to Fig. 1, integrated reporting is a comprehensive source. In the accounting system, one can get information about the goals and strategies for achieving them. This information is contained mainly in managerial reporting, which is non-public and unavailable to external users [19, p. 17]. The formation of financial statements is based on the basic concept of capital (financial or physical). The concept of capital, used in forming integrated reporting, reveals information about financial, production, human, natural, intellectual, social, and reputational capital through the prism of their interaction with each other. The concept of creating, maintaining, or losing value is closely related to the concept of capital.

Fundamental principles of financial and integrated reporting are compared in Fig. 2.

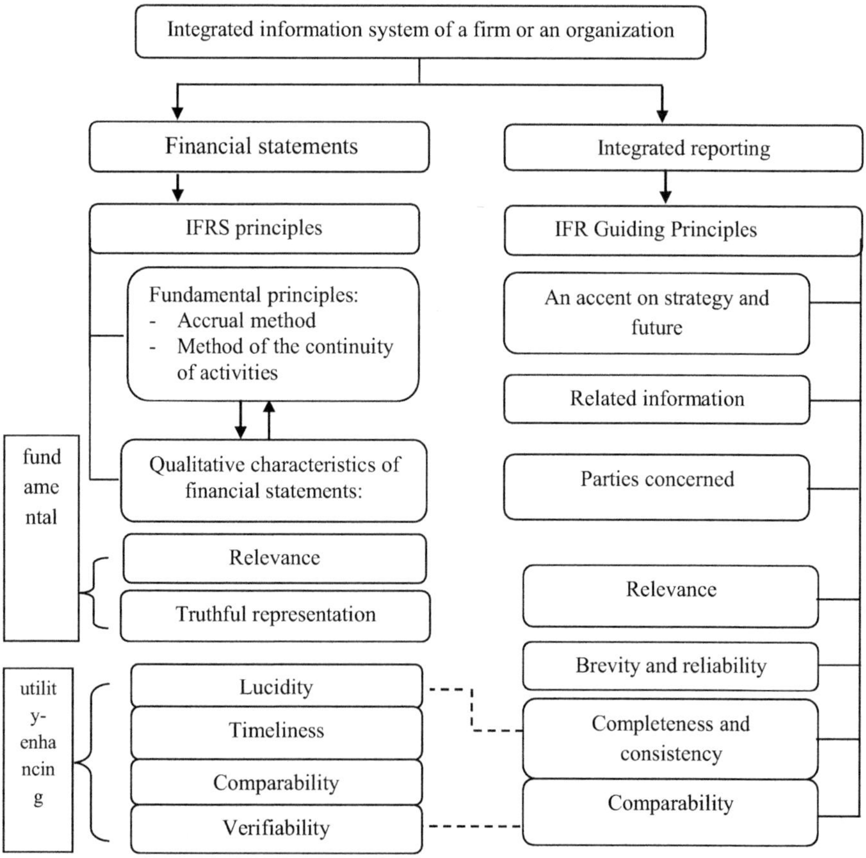

Fig. 2 Fundamental principles of financial and integrated reporting: comparative analysis. (*Source* Developed and compiled by the authors based on [10, 11, 16])

Figure 2 shows the relationship between the qualitative characteristics of financial reporting and IR guidelines. In accounting, the use of the qualitative characteristic "fair presentation" implies that the information must be complete, neutral, and without mistakes. The principle of reliability in the formation of integrated reporting is guiding and means that the information is balanced and there are no significant errors. In MOIE, reliability is often understood as the truthful representation of information. The principle of the absence of significant errors assumes that methods of minimizing accounting and reporting risks were applied and minor errors do not affect the relevant decisions of the users [16, p. 36]. It shows that the reliability of the information in financial and integrated reporting depends on effective internal control and accounting system, internal audit, and independent external audit.

The qualitative characteristic of financial information comparability and the guiding principle of consistency and comparability makes it possible to compare the company's key indicators with other organizations over a certain period. The forms of financial reporting are unified and make it possible to compare only quantitative indicators. The forms of integrated reporting are not regulated and differ because different organizations create different values. However, they can also be compared [16, p. 38].

A comparative analysis of the two reporting models shows that the financial model does not reflect all aspects of business activities and is rooted in historical data. There is no relationship between financial and non-financial indicators.

4 Results

Integrated reporting promotes efficient and productive allocation of capital [13]. The application of a systemic approach makes it possible to highlight the following aspects of integrated reporting:

- The system-element approach consists of disclosing information about eight elements of IR – governance and results, environment, allocation of resources, a new model of business, risks, strategy, etc. [15]. The content elements are interconnected. They can be revealed in any sequence, considering integrated thinking and the application of guidelines and approaches.
- The system-resource approach is oriented on the measurement and disclosure of all types of capital (i.e., financial, industrial, intellectual, human, social, reputational, and natural) and changing its value. This list of capitals is recommended by International Framework for Integrated Reporting (IFR).
- The system-integration approach consists in defining a set of concepts, guiding principles, and content, which ensures the IR's integrity.
- System-communication approach means the need to identify legal, commercial, political, and social external factors, which significantly affect or help the company to create new value [16, p. 40].

The Kyrgyz commercial firms form their financial statements under IFRS rules [1, 2]. The Law of the Kyrgyz Republic "On accounting" was amended on August 9, 2021, to improve reporting practice. These amendments concern the creation of the Public Depository of Financial Statements – a single electronic system for collecting, storing, summarizing, analyzing, and publishing financial statements and audit results [4]. Financial reports of public entities and large businesses are seen in the Kyrgyz Stock Exchange and in the media.

A green economy and sustainable development activities in Kyrgyzstan only start their formation, including a green financial corporation to attract climate finance to the banking sector [24]. The legal base of this work includes the National Development Strategy of the Kyrgyz Republic for 2018–2040 [5], the National Development Program of the Kyrgyz Republic until 2026 [6], and the Green Economy Program in the Kyrgyz Republic for 2019–2023 [3]. The government authorized the Ministry of Economy and Commerce to coordinate this work.

The National Bank of the Kyrgyz Republic is creating a division responsible for ESG financing and the environmental friendliness of the financial system. With the support of the IFC program "Comprehensive Environmental, Social and Governance Standards," the Union of Banks of Kyrgyzstan has developed a roadmap for sustainable financing [7]. Some commercial banks in Kyrgyzstan have created ESG tools. OJSC "Dos-Kredobank" introduced green deposits in the deposit line to launch environmentally friendly technologies and finance eco-projects. CJSC "Bank of Asia" issued gender bonds aimed at developing the business of women entrepreneurs to improve their well-being and livelihoods. Bank OJSC "Aiyl Bank" provides a loan for the development of eco-production.

IFC studied the implementation of sustainable development principles in the Kyrgyz Republic and determined the level of their integration in the financial sector. The conclusion was made that climate change issues and the Paris climate agreement were not mentioned in any bank, website, or annual report. In their annual reports, most banks refer to the issues of CSR, understanding it as charity or voluntary work of employees. None of the banks participate in international initiatives such as CDP UNEP PRI, the global compact, or the equator principles. However, some banks have collaborated with the EBRD or Microfinance companies on green standards or projects. There is no sectoral policy for green industries in the Kyrgyz Republic. Funded projects in the green sectors are mainly limited to energy efficiency projects under EBRD or MFC credit lines. None of the banks has data on the carbon or environmental footprint of the portfolio. The carbon footprint concept is not found in open sources [14].

One way to start developing the green economy is to encourage the Government of Kyrgyzstan to use electric vehicles: its import is exempt from VAT, and the owners are exempt from property tax on the vehicle. In 2023, the country got the opportunity to import duty-free 10,000 electric vehicles. It will help protect the environment and public health. Kyrgyzstan is among the ten countries with the worst air pollution. The concentration of PM2.5 in Kyrgyzstan is well above the limits set by the WHO. This fact also confirms that the work to develop a green economy and implement sustainable development principles has begun in

Kyrgyzstan. It indicates the need to use non-financial reports in the commercial and public sectors. The IIRC notes that IR is intended primarily for use in the private sector by commercial companies of all sizes. However, they can also be used in an adapted form in the public sector and non-profit organizations if necessary [16, p. 7].

Starting IR practice in the private sector and government organizations will force companies to develop and implement ESG strategies to attract ESG investment and funding. This will accelerate the formation of new economic thinking and the awareness of the need to look for new business models. An important driver of such a transition is the transparency and openness of companies, which eliminates corruption, reduces risks, and increases efficiency and profitability. The implementation of the ESG transformation of companies makes it necessary to introduce creative business models that, through optimization and efficiency, allow increasing the value of products, reducing costs, and increasing added value while reducing the price of goods and bringing increased value to all stakeholders [25, pp. 328–329]. These companies become attractive for ESG investment and financing.

There is a direct link between the transition of companies to integrated reporting and ESG principles and their financial and investment opportunities, confirmed by the experience of countries where the ESG transformation began much earlier. According to the report by global accounting firm Moore Global and the Centre for Economics and Business Research, the companies that regard ESG as important demonstrate better growth of profits and investment. Thus, 84% of companies reported that their ability to attract external investment had improved slightly or significantly; more than 86% reported that their brand image had improved [22]. They saw an increase in their income by 10.4% in the USA, 9.3% in Europe, and 9.1% in Australia [23]. Over 2019–2022, their profitability jumped to 11% in the USA, 8.1% in Europe, and 7.4% in Australia. For those who do not rate the importance of ESG, it slipped from revenue growth of 4.9% in the USA and Europe to 2.6% in Australia [23].

The problems eliminated by the transition to integrated reporting and its advantages refer to non-monetary and institutional inflation factors specific to developing and post-Soviet economies. A radical solution to these problems could decrease inflation and inflationary expectations. Thus, a hypothesis about the anti-inflationary effect of integrated reporting has been formulated.

The analysis of macroeconomic indicators over the past two decades based on statistics of Trading economics API has been made. The dynamics, magnitude of inflation rates, and growth rates of the money supply differ significantly, going in different directions. The comparison encompassed GDP dynamics and corruption index, consumer confidence and CPI indices, Kirgizstan CPI Housing Utilities, CPI Transportation, unemployment dynamics, income and wages, productivity, and price dynamics confirmed the presence of institutional and non-monetary factors of inflation in Kyrgyzstan [26]. A conclusion was drawn regarding the potential for an external anti-inflationary effect of the ESG transformation in Kyrgyzstan upon expanding the range of companies implementing integrated reporting.

5 Conclusion

Integrated reporting represents a rapidly evolving concept in comprehensive corporate reporting. It integrates ESG factors and financial performance as a result of integrated thinking. Its advantages include strategic focus, the interests of all stakeholders, and information on increasing all types of capital. Integrated reporting not only reflects the culture of doing business and improvements in company management but also discloses environmental, social, and ethical information and current financial and strategic goals. The principles and contents of IR improve transparency and the institutional characteristics of the company. Hence, companies in developing countries that adopt ESG principles have enjoyed growing revenues, stronger profits, and the greatest benefit in raising funds since 2019.

The widespread adoption of integrated reporting and ESG strategies has the potential to yield an anti-inflationary effect and eliminate institutional factors of inflation by raising the effectiveness of economic institutions and market mechanisms in developing and post-Soviet countries.

References

1. Government of the Kyrgyz Republic (2001) Decree "On international financial reporting standards in the Kyrgyz Republic" (September 28, 2001 No. 593). Bishkek, Kyrgyzstan. Retrieved from http://cbd.minjust.gov.kg/act/view/ky-kg/38280. Accessed 7 Sept 2022
2. Government of the Kyrgyz Republic (2013) Decree "On international financial reporting standards for small and medium enterprises in the Kyrgyz Republic" (October 18, 2013 No. 564). Bishkek, Kyrgyzstan. Retrieved from http://cbd.minjust.gov.kg/act/view/ru-ru/94784. Accessed 7 Sept 2022
3. Government of the Kyrgyz Republic (2019) Decree "On approval of the Green Economy Program in the Kyrgyz Republic for 2019–2023" (November 14, 2019 No. 605). Bishkek, Kyrgyzstan. Retrieved from http://cbd.minjust.gov.kg/act/view/ru-ru/453437. Accessed 7 Sept 2022
4. Kyrgyz Republic (2002) The Law "On accounting" (April 29, 2002 No. 76). Bishkek, Kyrgyzstan. Retrieved from http://cbd.minjust.gov.kg/act/view/ru-ru/1032. Accessed 7 Sept 2022
5. President of the Kyrgyz Republic (2018) Decree "On the National Development Strategy of the Kyrgyz Republic for 2018–2040" (October 31, 2018 UP No. 221). Bishkek, Kyrgyzstan. Retrieved from http://www.stat.kg/ru/ukaz-prezidenta-kyrgyzskoj-respubliki-o-nacionalnoj-strategii-razvitiya-kyrgyzskoj-respubliki-na-2018-2040-gody/. Accessed 7 Sept 2022
6. President of the Kyrgyz Republic (2021) "National Development Program of the "Kyrgyz Republic" (to Decree No. 435 of October 2, 2021). Bishkek, Kyrgyzstan. Retrieved from http://cbd.minjust.gov.kg/act/view/ru-ru/430700. Accessed 7 Sept 2022
7. Abdylda-kyzy M (2023) Green economy in the Kyrgyz Republic: long, expensive, but vital. Financial Portal "Akchabar". 1 January. Retrieved from https://www.akchabar.kg/ru/article/opinion/zelenaya-ekonomika-v-kr-dolgo-dorogo-i-zhiznenno-neobhodimo/. Accessed 2 Jan 2023
8. Derevyankina ES, Yankovskaya DG (2022) Disclosure of ESG factors in the integrated reporting of oil-producing organizations as a basis for making investment decisions. Intell Innov Invest 2:44–56. https://doi.org/10.25198/2077-7175-2022-2-44

9. Global Compact Network Belarus (2022) The new reporting system of the UN Global Compact: what is its peculiarity. 27 January. Retrieved from https://globalcompact.by/news/cop. Accessed 7 Nov 2022
10. IFRS Foundation (2010) Conceptual framework for financial reporting. Retrieved from https://www.ifrs.org/content/dam/ifrs/publications/pdf-standards/english/2021/issued/part-a/conceptual-framework-for-financial-reporting.pdf. Accessed 2 Sept 2022
11. IFRS Foundation (2012) IAS 1 presentation of financial statements. Retrieved from https://www.ifrs.org/content/dam/ifrs/publications/pdf-standards/english/2022/issued/part-a/ias-1-presentation-of-financial-statements.pdf?bypass=on. Accessed 2 Sept 2022
12. IFRS Foundation (2014) IFRS 10 consolidated financial statements. Retrieved from https://www.ifrs.org/content/dam/ifrs/publications/pdf-standards/english/2021/issued/part-a/ifrs-10-consolidated-financial-statements.pdf. Accessed 2 Sept 2022
13. Integrated Reporting Committee (IRC) of South Africa (2014) Preparing an integrated report: a starter's guide. Retrieved from https://docplayer.net/39508673-Preparing-an-integrated-report-a-starter-s-guide.html. Accessed 17 Nov 2022
14. International Finance Corporation (2021) [Report] Sustainable financing in the financial sector of the Kyrgyz Republic. Results of a survey of representatives of the financial sector. IFC, Bishkek. Retrieved from http://ub.kg/wp-content/uploads/2021/09/SF_Report_ru.pdf. Accessed 10 Oct 2022
15. International Integrated Reporting Council (IIRC) (2013) International standard for integrated reporting. Retrieved from https://www.integratedreporting.org/wp-content/uploads/2013/12/13-12-08-THE-INTERNATIONAL-IR-FRAMEWORK-2-1.pdf. Accessed 5 Sept 2022
16. International Integrated Reporting Council (IIRC) (2021) International framework for integrated reporting. Retrieved from https://www.integratedreporting.org/wp-content/uploads/2021/01/InternationalIntegratedReportingFramework.pdf. Accessed 5 Sept 2022
17. International Monetary Fund (2022) IMF annual report 2022. One crisis after another. Retrieved from https://www.imf.org/external/pubs/ft/ar/2022/downloads/imf-annual-report-2022-english.pdf. Accessed 29 Dec 2022
18. Khorin AN, Brovkin AV (2018) Purposes and features of the analysis of the integrated reporting, corporate capital, and complex performance of business process. J Creat Econ 12(4):499–512. https://doi.org/10.18334/ce.12.4.39018
19. Kondrashova OR (2018) Integrated reporting in the management accounting system. Int Account J 21(1):17–29. https://doi.org/10.24891/ia.21.1.17
20. Kondratenko M (2022) ESG principles: what is it and why should companies follow them. RBC Trends. 6 December. Retrieved from https://trends.rbc.ru/trends/green/614b224f9a7947699655a435. Accessed 7 Nov 2022
21. Malinovskaya NV (2021) Problems of the definition of corporate reporting. Account Anal Audit 8(6):31–41. https://doi.org/10.26794/2408-9303-2021-8-6-31-41
22. Moore Global (2020) The $4trillion ESG dividend: bottom line benefits of adopting ESG practices [Executive summary]. Retrieved from https://www.moore-global.com/MediaLibsAndFiles/media/MooreStephens2020/Documents/Moore_ESG_White-Paper_FINAL.pdf. Accessed 14 Feb 2023
23. Murdoch A (2022) Link between ESG and profitability exists: new research. 23 September. Retrieved from https://capitalmonitor.ai/sector/tech/link-between-esg-investment-exists-new-research/. Accessed 14 Feb 2023
24. PJSC Sberbank (2022) Kyrgyzstan: ESG dossier. Retrieved from https://sber.pro/digital/uploads/2022/09/ESG_Kyrgyzstan_A3_482aa3e05b.pdf. Accessed 5 Sept 2022
25. Raju N, Prabhu J (2018) Frugal innovation: how to do more with less (Transl. from English). Olymp-Business, Moscow. (Original work published 2015)
26. Trading Economics (n.d.) Kyrgyzstan – Economic indicators. Retrieved from https://ru.tradingeconomics.com/kyrgyzstan/indicators. Accessed 9 Feb 2023

Part IV
Sociocultural Support for Climate Change and Green Employment

Biotechnological Approaches to Improve the Microclimate and Quality of Life of the Urban Population

Viktor V. Glebov, Dmitry S. Nikitin, Dilyara N. Efremova, Elizaveta V. Anikina, and Olga V. Mareeva

1 Introduction

Contemporary landscapes of large cities exhibit a high degree of saturation with technogenic elements (factories, industrial zones, skyscrapers, vehicles), which to a large extent, negatively affect the visual and physical environment. The ecological condition of many cities in the world is assessed as "unfavorable" and even "dangerous" because toxic substances accumulate in such territories in different environments (air, water, and soil). Consequently, living conditions in such territories pose significant threats to all living organisms, from microorganisms to humans [5].

The processes of urbanization that have engulfed the whole world lead to the quantitative growth of cities, which increasingly distance people from nature, making the urban environment artificial. Everything created in the city for the convenience of people gradually begins to act against them [11].

In urbanized territories, a microclimate is formed, which differs significantly from natural landscapes. For example, urban concrete buildings and paved streets overheat in summer, creating difficult conditions for the population. Additionally,

V. V. Glebov (✉)
V.A. Trapeznikov Institute of Control Sciences of RAS, Moscow, Russia

D. S. Nikitin
Geological Institute of the RAS, Moscow, Russia

D. N. Efremova
Russian State University for the Humanities, Moscow, Russia

E. V. Anikina
RUDN University, Moscow, Russia

O. V. Mareeva
Russian State Agrarian University – Moscow Timiryazev Agricultural Academy, Moscow, Russia

E. G. Popkova (ed.), *Smart Green Innovations in Industry 4.0*, Springer Climate,
https://doi.org/10.1007/978-3-031-45830-9_32

the circulation and humidity of urban air are disrupted. In the natural environment, plant communities (grasses, shrubs, and trees) successfully cope with overheating (due to the evaporation of moisture from the leaves), creating favorable environmental conditions for microclimatic indicators (temperature, humidity, and circulation). The increase in temperatures in the city provides a large fleet of cars and trucks and heat loss from the city's heating mains [11].

Moreover, transport contributes to the accumulation of gas exhaust and the formation of smog, which is dangerous to the health of citizens. Exhaust gases have suspended particles ($P_{2.5}$ P_{10}), which damage the bronchus-pulmonary system when breathing. Smog also reduces insolation, contributing to the growth of depression in the urban population, especially in winter [5, 18].

The development of new territories for urban buildings often leads to deforestation; in the process of urban development (laying roads, sidewalks, dams, sewers, etc.), natural vegetation is destroyed in these areas [18].

The artificial urban environment also creates an increased and high noise level, leading to hearing disorders in citizens, which can cause hearing loss among residents of large cities. It also manifests itself in anxiety, irritability, and a decrease in the psycho-emotional state of individuals. Therefore, the presented problem is extremely relevant in many megacities of the world [13].

In this regard, the research aims to show the possibilities of advanced construction technologies implemented in different countries of the world, making it possible to harmonize the "man-environment" relationship and improve the environment of urban areas and the quality of life of citizens [23].

2 Materials and Method

During the research, the authors apply an integrative approach, which makes it possible to combine different scientific methods (content analysis, statistical analysis of information materials, study of social and historical documents in the field of environmental and architectural concepts of urban development, etc.). The authors summarize the conclusions from scientific publications on the phenomenon presented. The authors actively use abstract and bibliographic databases, including Russian Scientific Citation Index, Google Scholar, Web of Knowledge, Mendeley, ResearchGate, Scopus, and Academia.edu. The search is carried out using terms such as ecopolis, megapolis, green technologies, phytodesign, environmental technologies, and quality of life. During the search, over 100 thousand publications were found that covered this issue.

3 Results

The ideas of harmonious ecological development of cities in Russia originate from the concept of the noosphere developed by V. I. Vernadsky [24], in which there is "a sphere of interaction between society and nature, within which reasonable human activity becomes a determining factor of development" [24].

However, according to Vernadsky's concept, the harmonization of relations between people and nature should be formed from the position of eco-centrism and not the anthropocentrism that still prevails in world society. Despite the warnings made by major scientists in the field of ecology (e.g., Haeckel, Eugene Odum, Monica Turner, George Hutchinson, David Goodall, and others), we observe an increase in crisis processes in the dyad of the "man-environment" relationship [6].

The world is concerned about the impending ecological apocalypse. Many representatives from the field of education, science, and business understand that an important step in changing the environmental situation is the adoption of a nature-centric paradigm, in which there are rational approaches to harmonious relations between people and nature. In the context of this direction, the main thing is environmental approaches in urban development, which is what this research is aimed at.

3.1 Concepts of the Development of Ecological Cities in the Last Century

There is an opinion that the suitability of the natural suburban environment for recreation usually serves as a criterion for the city's success [16]. It was assumed that the successful greening of such cities would be possible under the following conditions:

1. Maintaining close to natural biogeocenoses in a significant part of the city
2. Investing in planting shrubs and trees
3. Reducing the overall population density of the city

However, this idea has not been developed due to the active economic activity of people, the high level of urbanization, and the high density of the urban population. Therefore, a person is concerned about the idea of the possibility of maximum proximity to nature through the creation of elements of the natural environment in the city.

For the first time, the idea of a garden city was proposed by Englishman Ebinzir Howard (1898) in the book "Tomorrow, the peaceful path of social reforms," where the author outlined the concept of the development of the theory and practice of world urban planning. Howard's fantastic ideas about social change impeded reforms in the urban planning legislation of European countries (the UK, France, and Germany), which soon reached the world level. Interest in this problem led to the fact that Howard's work was republished with a different title: "Garden cities at the beginning of the last century of the future in the present" [7].

In Russia, based on these ideas in the last century (1913), the society "Garden City" was created, which was later renamed "garden cities."

Based on this idea, the Russian architect D. A. Lebedev developed town-planning principles that are largely identical to garden cities [10].

The early ideas of a "garden in the city" and a "garden city" developed in the West were followed by the idea of a "city in nature" as a variant of a garden city with dirty production remote from it.

Some results of the integrated greening of urban settlements were achieved in the Soviet Union, although on the scale of individual enterprises and their subordinate urban developments, for example, the scientific center in Pushchino. This project was called Pushchinskaya "Ecopolis Program." The initiator of this movement was biologist B. N. Veprintsev [1].

In the 1960s, a modern garden city, Pushchino, was founded in the Moscow Region, having a rectangular layout and being divided into functional zones [1].

As a result of the project, an ecologically and psychologically optimal urban area was created, which harmoniously combines the needs of a person as a biological and social being and the possibilities of the existence of natural ecosystems in the city [17].

The idea of "ecopolis" was also implemented in the construction project of the settlement of Tchaikovsky – the city of hydraulic builders.

From the drawings of the cities of Pushchino and Tchaikovsky, we see a successful implementation of the greening of urban settlements: the establishment of boundaries between the blocks of green belt developments. A dense ring of green spaces (forest park zones) is being created or restored, where separate islands – settlements (residential, shopping, sports, and other centers) are formed around the perimeter of the green zone. All of them are connected by transport highways.

The continuation of the ideas for creating eco-cities in the Soviet Union is connected with the Kyiv School of Architects, which proposed the concept of "ecological macro zoning of the territory" [14, 17]. They proposed consolidating separate urban areas in different directions; for example, the city could have residential areas, buildings with industrial zones, recreation areas, sports and cultural events, and protected areas [14].

Another interesting environmental project is the "linear cities" project. It was designed by the Spanish architect Arturo Soria. The project was discussed for a long time in the world community (1920–1970) and had different variations.

The main idea of this concept was connected with the idea that the urban population, industrial and administrative institutions, and the entire infrastructure should be located on long transport (communication) corridors. Simultaneously, the main part of the adjacent territories is preserved for natural biosphere processes with a gentle anthropogenic load [3, 14].

3.2 Concepts of Urban Greening in Today's Russia

Presently, the ideas of an ecologically healthy city in Russia are developing on the concept of "minimal anthropogenic influences" on the environment. One of such supporters is professor A. N. Tetior [20]. His idea is that when forming an eco-city, it is necessary to use eco-compatible technological solutions as much as possible (use ecological materials in construction, widely use vertical landscaping, introduce ecological transport, etc.).

It should be noted that despite the development of the principles of environmental friendliness and the ideas of maximum inclusion of elements of the natural environment, the processes of greening cities are curtailed in the direction of sustainable development, which has a deep bias in the economic sphere [15].

In 1996, the Russian Federation officially joined the world community in implementing the concept of sustainable development, which laid down a plan to overcome the global environmental crisis that occurs due to the technogenic evolution of humanity [15, 16].

3.3 Contemporary Eco-biotechnological Solutions in Megacities

In the contemporary idea of urban planning, the smart city concept has been formed, in which designers and builders try to implement new technological solutions in urban areas. According to the plan of the city authorities, this will significantly improve the ecological condition of contemporary cities and the quality of life of the urban population [1, 16, 17].

Vertical landscaping of facades and interiors is being introduced in contemporary cities to improve microclimatic indicators and introduce elements of the natural environment [8].

Japan is an advanced country that widely practices the introduction of natural components into the urban landscape, which helps reduce negative environmental factors. For example, in the Namba Parks residential quarter in Osaka, with the help of a complex of ecological phytotechnologies (shrubby-woody vegetation) and a noise shield, it was possible to separate the residential quarter from the transport zone. According to the results of a survey of residents of this quarter, a favorable effect was obtained due to the formation of an adequate microclimate for residents and a positive green background for the human psyche [2, 12].

An interesting solution for the arrangement of urban roofs of houses is being implemented in many developed countries (the USA, Canada, Taiwan, Japan, etc.). Recently, more than 40 countries of the world have been engaged in phytodesign. According to many experts in urban planning, the phytodesign of roofs of houses in large cities gives a good aesthetic effect, which reduces the anthropogenic impact on the human psyche. The greening of roofs is one of the elements of the formation of the human habitat. Roof landscaping works are underway in Moscow and St. Petersburg. To reduce the anthropogenic load in the city, the city authorities are introducing landscaping elements on the roofs of administrative buildings.

The greening of roofs is of great ecological importance in large cities. However, in addition to the advantages, there are disadvantages of landscaping the roofs of skyscrapers. Below we can list the positive and negative sides of this process, which are considered from different positions.

3.3.1 The Positive Side of Gardening

If we consider landscaping from the city's perspective, then phytodesign improves the attractiveness of urban decoration and the city's prestige and contributes to a greater influx of tourists. Landscaping makes it possible to grow crops. Urban greening is attractive for businesses in terms of investment.

If we consider landscaping from the perspective of building structures, then this significantly decorates the appearance of urban buildings. The positive effect is a reduction in energy for cooling buildings in a hot period and a reduction in the cost of heat losses during the heating of buildings. It is also important to increase the fire safety of city houses. The greening of roofs helps increase the buildings' service life.

From a human perspective, landscaping improves the aesthetic perception of the urban environment. Gardening positively affects a person, reduces morbidity, and improves the psycho-emotional background. In such areas, there is an increase in labor productivity. Landscaping also significantly increases the recreation conditions for people with disabilities.

From the point of view of ecology, there is a decrease in gas pollution and air dustiness of urban air. Landscaping contributes to sound insulation and noise absorption, the reduction of electromagnetic radiation from power lines and cell towers by the ecosystem due to absorption.

From an economic perspective, greening the city provides additional profit from renting land on the roof and increasing the value of the real estate.

3.3.2 The Negative Side

Landscaping increases the cost of apartments with a view of green roofs. When gardening, there is a high cost of professional landscaping of the roof (from 100 to 200 euros per square meter); the costs of maintaining plantings on the roof increase.

Additionally, landscaping leads to a constructive complication of buildings and repair work of city roofs and the need to use special roofing materials.

In addition to these difficulties, there is no possibility of visual control over the condition of the waterproofing layer and control over the wind load.

Various embodiments of ecological phytotechnologies can be seen in different countries; for example, a complex has been built in Germany, where there is a harmonious unity of man with nature, known as the "Forest Spiral" ("Waldspirale") [4, 9, 19].

The cozy house has 1000 windows. They are of different shapes, among which it is impossible to find the same. There are 105 apartments built in the house; the layout and design of each are unique in their own way. The main concept of this amazing house is the idea of the importance of living in harmony with nature [4, 9].

Over the past decades, the environmental situation in China has deteriorated significantly. The emissions of many industrial enterprises poison the country's environment. The Chinese government is developing plans to improve the environmental condition in the country. In this regard, the construction of the world's first forest city (Lanzhou) on an area of 175 hectares is underway in the country's north (Liuzhou Province). This region is one of the most susceptible to smog, so it is quite apparent why this region was chosen to construct a forest city in Lanzhou (China) [21].

The main idea of this project is related to improving the environment since, in many cities in China, the pollution level is repeatedly overestimated due to the high activity of various industrial enterprises. This fact carries a deadly danger to the health and life of the urban population [22].

The construction of an eco-city is part of the project of the Chinese leadership to combat environmental pollution.

It is planned that the eco-city will be completely covered with green spaces, the number of which will be about one million plants from 100 different species, of which 400 thousand will be trees [21].

Such a green urban zone will be able to process about 10 thousand carbon dioxide and more than 50 tons of pollutants (exhaust of transport gases and emissions of thermal power plants, dust, etc.). It is assumed that this green zone will annually supply about 900 tons of oxygen to the atmosphere [4, 9, 15].

Ecologists and climatologists of the country expect that forest plantations in Lanzhou province can solve complex environmental problems: significantly reduce the average air temperature in the city and the region, located in a hot climate zone, create a natural sound-absorbing screen, improve the biodiversity of flora, and create good prerequisites for the quality of urban living.

4 Conclusion

Thus, at the present stage of urban development, an important aspect on the way to greening urbanized territories is the maximum use of natural components (plant-shrub, woody plant communities), which can be achieved with the help of phytotechnologies.

This approach significantly improves the ecological state of the urban environment and makes it possible to solve complex environmental problems (regulation of the microclimate of the urban environment, reducing noise load on the city population, cleaning the air environment, and improving the visual environment of the city).

The widespread adoption of effective eco-phytotechnologies in the construction of new cities and the reconstruction of existing ones can significantly improve the state of the urban environment and significantly improve the quality of life of the urban population.

However, the challenge of greening urbanized territories still persists due to the conditions of economic expediency in urban planning and the problems of the economy in the urban budgets of many countries.

References

1. Bozhukova EE, Kavtaradze DN (1983) Main works on the Ecopolis program (abstracts of publications) 1979–1982. Scientific Center Biol. Research. Academy of Sciences of the Soviet Union, Pushchino, USSR
2. Cilento K (2009) Namba Parks / The Jerde Partnership. ArchDaily. 19 October. Retrieved from https://www.archdaily.com/36987/namba-parks-the-jerde-partnershipm. Accessed 1 Dec 2022
3. Correa AB (1991) Paisaje urbano, Ciudad Lineal y masonería. Villa de Madrid 104:3–22
4. Gibson E (2017) Stefano Boeri designs “vertical forest” city to eat up China. Dezeen. 28 June. Retrieved from https://www.dezeen.com/2017/06/28/liuzhou-forest--city-stefano-boeri-proposes-plant-covered-city-to-eat-up-chinas-smog/. Accessed 1 Dec 2022
5. Glebov VV, Plushikov VG (2021) Urban ecology and monitoring of the urban environment: textbook. RUDN University, Moscow
6. Glebov VV, Rodionova OM, Anikina EV (2020) The use of ecological technologies in the reconstruction of the urban environment. In: Zaitsev PL (ed) Visual images of modern culture: ideals and ideologies (to the 25th anniversary of theological education in Omsk): Collection of scientific articles based on the materials of the VIII All-Russian Scientific and Practical Conference. Dostoevsky Omsk State University, Omsk, pp 111–114
7. Howard E (1902) Garden cities of tomorrow. Sonnenschein Publishing, London, UK
8. Kildisheva SV (2006) Modern methods of urban gardening. Vertical and roof landscaping. In: Ecology of urban areas: international scientific and practical conference. Prima-Press, Moscow, pp 37–38
9. Kira (n.d.) Forest spiral by Hundertwasser; the unique house in Germany. Home Reviews [Blog]
10. Kirichenko EI (2003) Urban planning of Russia, mid–XIX – early XX century. Progress Tradition Publishing, Moscow
11. Koshkina SY, Korchagina OA, Voronkova ES (2013) “Green” construction as a major factor in improving environmental quality and human health. J Prob Contemp Sci Pract. Vernadsky University 3(47):150–158
12. Leksikova VV (2017) Plant as a component of green building. In: Gurin AG, Ignatova GA, Stepanova EI (eds) Ecology and agriculture: proceedings of the international scientific and practical conference. Orel, Russia, pp 83–89
13. Nemcev IA (2014) Green building: eco-settlement in the concept of sustainable development. Urbanistika [Urban Studies] 3:8–25. https://doi.org/10.7256/2310-8673.2014.3.13525
14. Perekladov AA, Plishkin VA (2003) The importance of the ecological approach in the architectural activity of the new millennium. In: Architectural science and education: proceedings of the Moscow Institute of Architecture, vol 1. Stroyizdat, Moscow, pp 238–243
15. Rodionovskaya IS (2018) Modern architecture in the format “green.”. Ecol Urban Territories 2:96–100. https://doi.org/10.24411/1816-1863-2018-12096

16. Safroshkin YV, Farberov VG, Reva BA (1988) Energobiograd is an economically, socially, and ecologically harmonious city of the future based on the energobiological production cycle. Scientific Center Biol. Research. Academy of Sciences of the Soviet Union, Pushchino, USSR
17. Sagalaev AV (2015) Ecological age in architecture. In: Danilov IK (ed) Construction, design, architecture: design decisions of the XIX century: collection of materials of the international scientific e-symposium. International Center for Research Projects, Kirov, pp 96–104
18. Sergienko LI, Podkolzin MM (2010) Green building as an element of sustainable development of Russia. Ecol Urban Areas 1:18–23
19. Tarasova IN, Shalaev DV (2012) Economic assessment of works on gardening of the roof of the organization. Actual Prob Forest Complex 31:209–211
20. Tetior AN (2005) Eco-city: problems, solutions. Tver Publishing House, Tver
21. Travushka Muravushka [Blog] (2017, 13). "Forest city" and vertical forests in China. Retrieved from https://blog-travushka.ru/lesnoj-gorod-i-vertikalnye-lesa-v-kitae.html. Accessed 1 Dec 2022
22. Tsitsilin AN (2006) The main problems of indoor use of plants and their solutions. In: Ecology of urban areas: international scientific and practical conference. Prima-Press, Moscow, pp 82–85
23. Usov, Ya. Yu. (2010). Green areas in bioclimatic buildings. In D. O. Shvidkovsky (Ed.) [et al.]. Science, education, and experimental design at the Moscow Institute of Architecture: theses of scientific-practical conference reports (127–128). Moscow: Architecture-S
24. Vernadsky VI (1988) Philosophical thoughts of a naturalist. Nauka, Moscow

Interstate Climate Change Mitigation Methods as a Global Public Good

Maria A. Kozlova

1 Introduction

Climate change impacts all countries globally. Therefore, a common coordinated approach is required globally to combat climate change effectively. The Kyoto Protocol, an additional document to the 1992 UN Framework Convention on Climate Change, was adopted in 1997. The main goal of the Kyoto Protocol was to reduce greenhouse gas emissions into the Earth's atmosphere to prevent global warming. According to experts, a temperature rise on Earth by more than 2 °C will lead to serious consequences, including lack of water in a number of regions, rising sea levels, and the extinction of several animals and plants species. The Protocol came into effect in 2005; by 2009, 192 countries signed and ratified it (with the US signing but not ratifying the Protocol, and Canada ending its participation in 2012). The signing of the Paris Agreement adopted under the UN Framework Convention on Climate Change was the next step. It entered into force in 2016 and was ratified by 193 countries, including the EU.

The battle against climate change stands as one of the 17 Sustainable Development Goals developed by the UN General Assembly in 2015. Within the framework of the United Nations, the period from 2020 to 2030 has been declared the Decade of Action. During this period, actions must be taken at the global level to achieve the Global Sustainable Development Goals, including the allocation of more funds and the implementation of more effective leadership in this area.

Currently, many scientists argue that combating climate change should be considered as a global public good (while the problems associated with climate change are a global public evil). A global public good is one whose benefits extend across state, social, and generational boundaries [11, p. 22]. M. B. Bättig and T. Bernauer [1], M. Grasso [4], T. W. Hertel [5], M. R. Khan and S. Munira [6], Y. M. Li and

M. A. Kozlova (✉)
MGIMO University, Moscow, Russia

E. G. Popkova (ed.), *Smart Green Innovations in Industry 4.0*, Springer Climate,
https://doi.org/10.1007/978-3-031-45830-9_33

R. Schwarze [10], J. Quiggin [13, 14], A. Sandmo [15], and H. Welsch [22] are among the scientists who have studied the battle against climate change as a global public good.

The following strategies exist for funding global public goods:

1. "Summation" – this strategy makes it possible to coordinate the efforts of a wide number of countries. Simultaneously, it increases the greatest risk of free riders appearing.
2. "Weak link" – the amount of the public good created is measured by the contribution of the least active and successful participant in the process of its creation, while cooperation and assistance to the weak links play a vital role.
3. "Leader" – the creation of public good is the privilege of developed countries.

In the case of combating climate change, the "leader" strategy is the most effective because developed countries have the greatest influence on climate change. Consequently, they should make greater efforts to address this problem and have more developed technological and financial capabilities.

Several documents and the formation of a number of organizations provided the groundwork for international cooperation between developed and developing countries, including the UN Development Programme [19]. For example, the World Bank requested the study Collaborative Solutions 2025 to develop a collaborative platform. The Global Economic Facility (GEF) is contributing funds to cover additional costs associated with making the project environmentally friendly. Its activities are implemented through the UN Development Program, the UN Environment Program, and the World Bank. The Adaptation Fund supports projects and programs that assist developing countries in adapting to the harmful effects of climate change. The Least Developed Countries Expert Group also coordinates work in the field of climate change.

Because Industry 4.0 technologies are especially important in the battle against climate change, we will look in more detail at interstate methods of combating climate change using the most advanced technologies in this chapter. Such authors as A.B. Youssef [23]; K. Fritzche, S. Niehoff, and G. Beier [3]; J. Van den Bergh [21]; T. Stock, M. Obenaus, S. Kunz, and H. Kohl [16]; R. T. Munodawafa and S. K. Johl [12]; K. H. Lee and B. Min [9]; M. Kozlova, A. Gorbacheva, and P. Fedosov [8]; M. Kozlova and T. Dianova [7], as well as UNIDO [20], considered this issue. In accordance with the "leader" strategy of creating a global public good, developed countries create and transfer advanced technologies to developing countries to combat climate change and finance the implementation of these technologies.

Technologies 4.0 have the following effects on climate change:

- Change the intensity of resource use
- Improve the energy efficiency of production
- Contribute to the dematerialization of manufactured products (using fewer materials)
- Contribute to the faster spread of renewable technologies
- Contribute to the transition to a circular economy
- Provide a transition to eco-innovations

Organizations such as the LDC Fund, the Special Climate Change Fund, the Climate Resilience and Adaptation Finance and Technology Transfer Facility (CRAFT), and the LAB provide financial support for the transfer of technologies in the field of climate change.

2 Methodology

An analysis of programs in the field of interstate cooperation in the fight against climate change, including the use of Industry 4.0 technologies, was conducted as part of this chapter.

The author identified the countries with the highest number of climate change projects receiving international funding. Additionally, a correlation analysis was also carried out to assess the relationship between the number of projects funded in various countries and the countries' ranking in combating climate change. The author used correlation analysis to determine if there was a relationship between the number of projects and the country's rating. Conclusions about the outcome were drawn.

3 Results

Let us consider some of the major climate change programs that have received international funding.

The LAB consists of a group of high-level public and private sector leaders who provide expertise, political capital, and financial capital for specific climate change tools, including technology transfer. Since its launch in 2014, the organization has collected more than $3.5 billion in climate change-related investments, more than $1 billion of which has come from private investors. In 2021, the total amount of investments allocated by the LAB amounted to $816 million. The distribution of collected investments by financing projects is presented in Fig. 1.

In this case, the Climate Investor One program accounts for the majority of investments (72.27%). This program promotes increased investment in developing renewable energy projects (hydropower, solar, and wind energy development). Programs related to CRAFT (Climate Resilience and Adaptation Finance & Technology Transfer) take second place (17.64%).

The technology transfer is primarily related to the impact of technology 4.0 on climate change [21]. For example, the Internet of Things (IoT) makes it possible to reduce the volume of materials and energy used. The intelligent network also helps reduce energy consumption. Robotics reduces the volume of materials used and waste. Artificial intelligence allows optimizing the manufacturing process. Big data

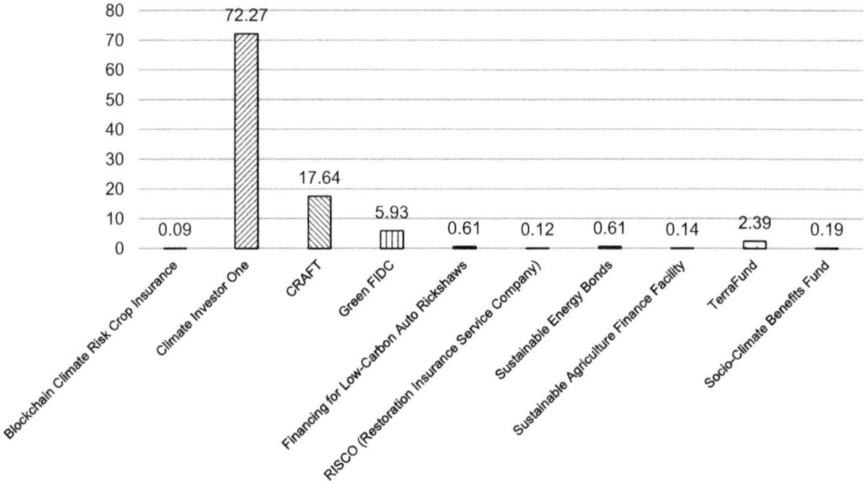

Fig. 1 Percentage allocation of the LAB investments by financing projects in 2021. (*Source* Calculated by the author based on [18])

analysis allows developing eco-innovations and making long-term forecasts. Blockchain enables climate risk innovation. Additive manufacturing and 3D printing promote the development of a circular economy and resource recycling.

CBIT Trust Fund (18 projects), GEF Trust Fund (489 projects), Least Developed Countries Fund (254 projects), Multi Trust Fund (40 projects), and Special Climate Change Fund (63 projects) are among the organizations and funds involved in financing climate change projects.

Let us closely consider the activities of the GEF (Global Environment Facility). Over the last three decades, the organization has provided over $22 billion in grants and blended funding, mobilized another $120 billion to co-finance more than 5000 national and regional projects, and organized 27,000 community-based initiatives through its Program of Small Subsidies.

Figure 2 shows the distribution of GEF programs by funded project types.

As a result, climate change programs account for 67% of the total number of the GEF-funded projects.

Let us highlight the countries with the highest number of projects funded (Table 2) and see how this compares to their position in the Climate Change Performance Index CCPI for 2022 (because the ranking does not include all countries, we will conduct a correlation analysis for the countries for which data is available). The ranking of countries in combating climate change is made up based on the volume of greenhouse gas emissions, the level of development of renewable energy, the level of electricity consumption, the total reserves of primary energy, and the general foreign and domestic state policy in the field of combating climate change. Table 1 lists 24 countries with a total number of national projects equaling 40 or more.

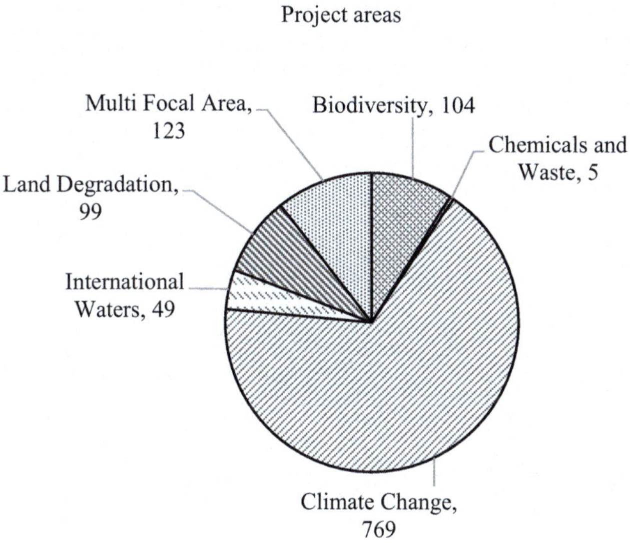

Fig. 2 Industry-wise GEF project financing. (*Source* Calculated by the author based on [15])

Let us test the hypothesis that project funding should be proportional to the country's performance on climate change. Countries that have made significant progress in combating climate change have a higher CCPI score than countries that have made less progress. In this case, let us see if there will be a correlation between the number of projects in the country (national projects separately and the total number of projects) and the CCPI ranking, and if so, what that relationship is. If there were a relationship between the number of projects and a country's ranking, countries with lower rankings would require more international assistance, implying an inverse relationship between these indicators. The results of the correlation analysis are presented in Table 2.

As a result, the data did not confirm a strong relationship between the number of projects and a country's performance on climate change.

This can be explained in several ways:

1. To begin with, the data on the number of projects does not reflect the total amount of financial assistance received by different countries because funding for different projects varies greatly.
2. Because global and regional projects do not provide data on the distribution of funds among individual participating countries, a comparison based solely on participation in global and regional projects does not account for the unique characteristics of individual countries.
3. Because the sizes of the countries differ significantly, the large number of projects funded in China and India demonstrate that these countries are much larger than the small countries listed.

Table 1 Countries with the highest number of GEF-funded projects

No.	Country	Number of national projects	Number of global and regional projects	Total	Position in the CCPI ranking for 2022
1	China	185	19	204	52.66
2	India	100	17	117	69.22
3	Brazil	80	16	96	55.17
4	Indonesia	77	17	94	57.39
5	Vietnam	71	12	83	49.35
6	Colombia	67	13	80	58.11
7	Philippines	66	7	73	58.98
8	South Africa	61	12	73	51.56
9	Russia	59	10	69	35.00
10	Ecuador	57	14	71	–
11	Kenya	52	20	72	–
12	Tanzania	48	19	67	–
13	Morocco	47	16	63	71.64
14–16	Argentina	46	11	57	47.50
14–16	Madagascar	46	13	59	–
14–16	Chile	46	8	54	69.66
17	Thailand	45	7	52	55.28
18	Egypt	43	11	54	59.83
19–21	Cambodia	42	20	62	–
19–21	Pakistan	42	11	53	–
19–21	Costa Rica	42	12	54	–
22–23	Kazakhstan	41	9	50	19.81
22–23	Burkina Faso	41	17	58	–
24	Cameroon	40	10	50	–

Source Compiled by the author based on [2, 17]

Table 2 Correlation analysis of the number of GEF-funded projects and climate change ranking

Indicators	Correlation coefficient value	Tightness of correlation
Number of national projects and climate change rating	0.105188	Weak
Number of regional and global projects and climate change ranking	0.289999	Weak
Total number of projects and climate change rating	0.126682	Weak

Source Calculated by the author based on [2, 17]

4 Conclusion

To sum up, it should be noted that global coordination should play a vital role in climate policy. Therefore, coordination between all countries is critical for the successful implementation of policies to address climate change globally.

Climate change can be viewed as a global public good because it affects all countries and the interests of current and future generations. The "leader" strategy is the most effective approach to financing climate change as a global public good, as developed countries have advanced technologies that can be transferred to developing ones through technology transfer and financing programs related to the introduction of new technologies that contribute to climate change mitigation.

The analysis revealed that there is currently no close relationship between the position of countries in the Climate Change Index and the assistance received by countries through national, regional, and global projects. As a result, in the future, it will be necessary to focus on the fact that assistance is received by those countries where the problem is most severe. It is also appropriate to focus on the "weak link" strategy for financing global public good, as progress in the least successful countries should be prioritized.

Industry 4.0 technologies (e.g., the Internet of Things, smart grids, blockchain, big data analytics, robotics, artificial intelligence, and additive manufacturing) play an important role in assisting the least developed countries. Therefore, within the framework of assistance to developing countries, greater emphasis should be placed on the transfer of cutting-edge technology.

In the condition when countries manage to coordinate their efforts and take measures to tackle existing and potential challenges, climate policy will be most successfully implemented in the leading countries and the world.

References

1. Bättig MB, Bernauer T (2009) National institutions and global public goods: are democracies more cooperative in climate change policy? Int Organ 63(2):281–308
2. Burck J, Uhlich Th, Bals C, Höhne N, Nascimento L, Wong J, … Reuther J (2021) Climate Change Performance Index 2022. Retrieved from https://ccpi.org/download/climate-change-performance-index-2022-2/. Accessed 12 Jan 2023
3. Fritzche K, Niehoff S, Beier G (2018) Industry 4.0 and climate change – Exploring the science-policy gap. Sustainability 10(12):4511. https://doi.org/10.3390/su10124511
4. Grasso M (2004) Climate change: the global public good. Working Papers 75. University of Milano-Bicocca, Milan
5. Hertel TW (2013) Land, environment and climate: contributing to the global public good. WIDER Working Paper No. 2013/107. UNU-WIDER, Helsinki, Finland. Retrieved from https://www.wider.unu.edu/sites/default/files/WP2013-107.pdf. Accessed 29 Dec 2022
6. Khan MR, Munira S (2021) Climate change adaptation as a global public good: implications for financing. Clim Chang 167(3):50. https://doi.org/10.1007/s10584-021-03195-w

7. Kozlova M, Dianova T (2023) Comparative analysis of climate programs in Germany and the USA. In: Popkova EG, Sergi BS (eds) Current problems of the global environmental economy under the conditions of climate change and the perspectives of sustainable development. Springer, Cham
8. Kozlova M, Gorbacheva A, Fedosov P (2021) Impact of digital technologies on the transition to a "green" economy. In: Zavyalova EB, Popkova EG (eds) Industry 4.0: exploring the consequences of climate change. Palgrave Macmillan, Cham, pp 323–335. https://doi.org/10.1007/978-3-030-75405-1_29
9. Lee K-H, Min BS (2015) Green R&D for eco-innovation and its impact on carbon emissions and firm performance. J Clean Prod 108(Part A):534–542. https://doi.org/10.1016/j.jclepro.2015.05.114
10. Li YM, Schwarze R (2013) From global public good to regional economic services: a comparative study on the development of climate change as economic goods in China and the EU. UFZ Discussion Paper No. 12/2013. Helmholtz Centre for Environmental Research (UFZ), Leipzig. Retrieved from https://www.econstor.eu/bitstream/10419/80340/1/757441580.pdf. Accessed 29 Dec 2022
11. Medvedev SA, Tomashov IA (2009) The concept of global public goods. Int Organ Res J 4(2):21–28
12. Munodawafa RT, Johl SK (2019) Big Data analytics capabilities and eco-innovation: a study of energy companies. Sustainability 11(15):4254. https://doi.org/10.3390/su11154254
13. Quiggin J (2010) Agriculture and global climate stabilization: a public good analysis. Agric Econ 41(S1):121–132. https://doi.org/10.1111/j.1574-0862.2010.00494.x
14. Quiggin J (2013) Is it too late to stabilize the global climate? Aust J Agric Resour Econ 57(1):1–14. https://doi.org/10.1111/j.1467-8489.2012.00617.x
15. Sandmo, A. (2015). The public economics of climate change. NHH Dept. of Economics Discussion Paper No 27/2015. Bergen: Norwegian School of Economics
16. Stock T, Obenaus M, Kunz S, Kohl H (2018) Industry 4.0 as enabler for a sustainable development: a qualitative assessment of its ecological and social potential. Process Saf Environ Prot 118:254–267. https://doi.org/10.1016/j.psep.2018.06.026
17. The GEF (2022) Project database. Retrieved from https://www.thegef.org/projects-operations/database. Accessed 12 Jan 2022
18. The LAB (2022) LAB instruments mobilized USD 816 million in 2021. 4 February. Retrieved from https://www.climatefinancelab.org/news/lab-instruments-mobilized-usd-816-million-in-2021/. Accessed 12 Jan 2022
19. UN Environment Programme (2022) Adaptation Gap Report 2022: too little, too slow – Climate adaptation failure puts world at risk. UNEP, Nairobi. Retrieved from https://www.unep.org/resources/adaptation-gap-report-2022. Accessed 29 Dec 2022
20. UNIDO (2017) Accelerating clean energy through Industry 4.0: manufacturing the next revolution. UNIDO, Vienna. Retrieved from https://www.unido.org/sites/default/files/2017-08/REPORT_Accelerating_clean_energy_through_Industry_4.0.Final_0.pdf. Accessed 29 Dec 2022
21. van den Bergh JCJM (2013) Environmental and climate innovation: limitations, policies and prices. Technol Forecast Soc Chang 80(1):11–23. https://doi.org/10.1016/j.techfore.2012.08.004
22. Welsch H (2020) Moral foundations and voluntary public good provision: the case of climate change. Ecol Econ 175:106696. https://doi.org/10.1016/j.ecolecon.2020.106696
23. Youssef AB (2020) How can industry 4.0 contribute to combatting climate change? Rev Econ Ind 169:161–193. https://doi.org/10.4000/rei.8911

Human Resource Management Based on ESG Principles in Entrepreneurship of the Digital Economy to Support Sustainable Development

Inna V. Kushnareva, Olga Yu. Malinina, Anna V. Fedorkova, and Ekaterina S. Alekhina

1 Introduction

In the Decade of Action, the most progressive economic systems increasingly emphasize on entrepreneurial management through ESG principles. This means that businesses demonstrate a high level of corporate responsibility and actively support the Sustainable Development Goals (SDGs): environmental (E), social (S), and economic (G). Management of personnel (HR management) deserves particular attention because it is based on ESG principles that provide comprehensive support for the interests of employees in protecting the environment, realizing their labor potential, and maximizing their income.

In the digital economy, digital practices of HR management are becoming available and increasingly applied in business management practices. The uniqueness of these HR management practices lies in leveraging new opportunities provided by scientific and technological progress. However, the digital economy also presents contradictory social consequences.

On the one hand, digitalization complicates business processes and demands employees to have a higher level of knowledge and skills and show a more creative approach to professional duties. This enhances the value of highly skilled human resources and improves their working conditions, professional income, and career building.

On the other hand, digitalization eliminates the involvement of human resources in economic processes. This creates psychological pressure on workers, depriving them of job security in the long term and forcing them to continue their lifelong

I. V. Kushnareva (✉) · O. Y. Malinina · A. V. Fedorkova
Institute of Service and Entrepreneurship (Branch) of Don State Technical University, Shakhty, Russia

E. S. Alekhina
Don State Technical University, Rostov-on-Don, Russia

E. G. Popkova (ed.), *Smart Green Innovations in Industry 4.0*, Springer Climate,
https://doi.org/10.1007/978-3-031-45830-9_34

learning and skill development. The novelty of this research is that it rethinks digital HR practices from the perspective of their alignment with ESG principles. The research aims to study the international experiences and prospects for improving HR management based on ESG principles in entrepreneurship of the digital economy to support sustainable development.

2 Literature Review

The theoretical basis of this research encompasses two concepts. The first concept is the concept of responsible HR management. According to this concept, HR management based on ESG principles provides a contribution to sustainable development, expressed in the following ways:

- Supporting gender equality (SDG 5) [4, 7]
- Promoting the discovery of human potential (SDG 8) [6, 19]
- Reducing income inequality (SDG 10) [3, 12]
- Creating green jobs (SDG 12) [14, 18]

The second concept revolves around HR management in the digital economy. According to this concept, the following digital practices of HR management stand out:

- Attraction of digital workforce [2, 15]
- Use of digital (smart) technologies to automate HR management, for example, to optimize the system of incentives through intelligent support for personnel decision-making and machine vision as a means of controlling personnel [1, 10, 11, 16]
- Creation of digital (high-tech) jobs [8, 9, 13]

At the intersection of these concepts lies an underexplored knowledge area that investigates how progressive HR practices in entrepreneurship of the digital economy align with ESG principles and support sustainable development. This contribution is currently underdeveloped and undefined, representing a gap in the literature. This research aims to fill the identified gap and answer the following research question (RQ): "How do digital practices of HR management contribute to sustainable development (implementation of the SDGs based on ESG principles)?" To find an answer to the research question posed, this research explores the links between HR management based on ESG principles of the digital economy with the support of sustainable development goals (SDGs) in entrepreneurship, drawing on best international practices.

3 Materials and Methods

To achieve this goal, the research consistently implements two research tasks. The first research task is to identify the benefits of using the special capabilities of the digital economy to manage human resources in entrepreneurship, relying on ESG principles to support sustainable development. For this purpose, the authors use the method of correlation analysis to explore the relationship between the use of special opportunities of the digital economy in HR management with the support of sustainable development (ESG principles and SDGs) in entrepreneurship.

The following indicators, calculated by IMD, serve as indicators of digital practices of HR management [5]:

1. Digital/technological skills (characterizing the attraction of the digital workforce)
2. Use of big data and analytics (reflecting the use of digital (smart) technologies to automate HR management)
3. Digital transformation in companies (indicating the creation of digital (high-tech) jobs)

Sustainability contribution indicators are based on SDG results from UNSTATS [17]. The study relies on 2022 data for the most advanced digital economies – the G7 and BRICS (Table 1).

The second research task is to develop a forward-looking model of human resource management (HRM) based on ESG principles in entrepreneurship to support sustainable development by combining digital and non-digital practices of HRM. The author's conceptual model is based on the results of correlation analysis.

4 Results

4.1 Benefits of Using Special Capabilities of the Digital Economy to Manage Employees in ESG-Based Entrepreneurship to Support Sustainable Development

To address the first research problem concerning the determination of the benefits of utilizing the special features of the digital economy for HR management in entrepreneurship based on ESG principles to support sustainable development, the authors processed statistics from Table 1 using the method of correlation analysis. The identified relationship of the use of special opportunities of the digital economy in HR management with the support of sustainable development (ESG principles and SDGs) in entrepreneurship is depicted in Fig. 1.

The results shown in Fig. 1 indicate that the most promising digital practice of HR management is the engagement of digital human resources, which by 3.98% correlates with SDG 5, by 2.21% correlates with SDG 8, by 5.60% correlates with SDG 10, and by 2.12% correlates with SDG 12. The use of digital (smart)

Table 1 HRM* statistics based on ESG principles in entrepreneurship of the digital economy in G7 and BRICS countries in 2022 (points)

		Digital practices of HRM*			Contribution to sustainable development			
		Digital/Techno-logical skills	Use of big data and analytics	Digital transformation in companies				
Category of countries	Country	Attraction of digital workforce	Use of digital (smart) technology to automate HRM*	Creation of digital (high-tech) jobs	Supporting gender equality (SDG 5)	Promoting human potential (SDG 8)	Contribution to reducing income inequality (SDG 10)	Creating green jobs (SDG 12)
G7 countries	UK	7.13	5.95	6.29	83.6	79.2	68.4	65.7
	Germany	5.80	4.47	4.69	80.5	86.9	89.1	59.4
	Italy	6.02	4.64	4.92	73.4	74.4	76.9	75.9
	Canada	7.85	6.62	6.96	80.0	81.8	85.9	61.2
	USA	8.02	7.20	7.03	76.6	79.1	52.0	67.8
	France	7.01	4.75	5.25	87.4	81.2	87.5	63.9
	Japan	4.02	3.22	4.42	61.6	86.8	80.5	67.3
BRICS countries	Brazil	5.05	4.13	5.03	69.2	65.3	13.5	84.6
	Russia	6.13	5.17	4.56	66.6	76.9	76.9	84.7
	India	7.69	6.27	7.18	33.9	70.1	38.5	96.2
	China	7.89	6.58	6.98	77.1	72.7	34.5	90.6
	South Africa	5.48	5.44	5.19	83.5	60.9	0.0	88.7

Source Compiled by the authors based on IMD [5] and UNSTATS [17] materials
Note *HRM – human resource (personnel) management

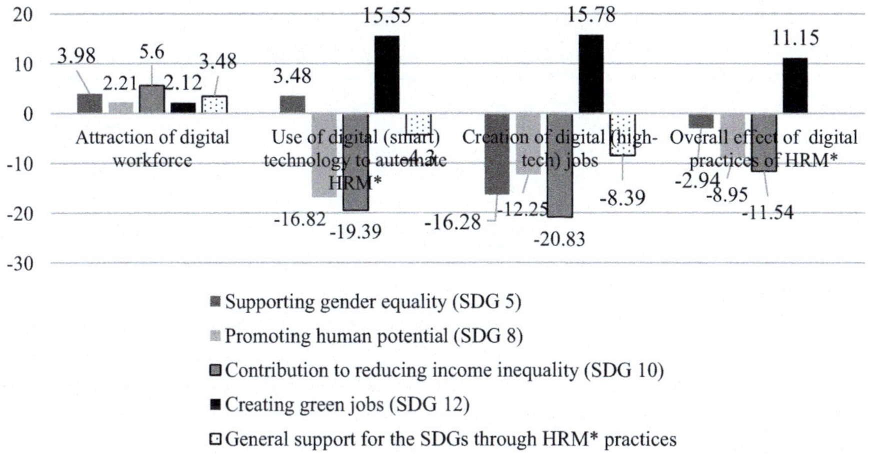

Fig. 1 Contribution of special opportunities of the digital economy to HR management in entrepreneurship based on ESG principles to support sustainable development (correlation), %. *Note* *HRM – human resource (personnel) management. (*Source* Calculated and compiled by the authors)

technologies to automate HR management supports only SDG 5 (correlation of 3.48%) and SDG 12 (correlation of 15.55%).

The creation of digital (high-tech) jobs supports only the implementation of SDG 12 (correlation of 15.78%). Overall, digital practices of HR management most fully and consistently support green jobs (implementation of SDG 12), with an average correlation of 11.15%.

4.2 *Promising Model of HRM Based on ESG Principles in Entrepreneurship to Support Sustainable Development with the Combination of Digital and Non-Digital HRM Practices*

Based on the results of correlation analysis, the author developed a model of HR management based on ESG principles in entrepreneurship to support sustainable development (Fig. 2).

In Fig. 2, the thickened arrows show the decreasing priority and effectiveness of supporting the SDGs through alternative HRM practices. The involvement of digital personnel in the author's model provides systemic support for sustainable development (ESG principles and a set of four considered SDGs) from entrepreneurship.

In their totality, digital practices of HR management are applied to support the SDGs in the following order of decreasing priority and effectiveness:

1. Creation of green jobs (SDG 12)
2. Support of gender equality (SDG 5)

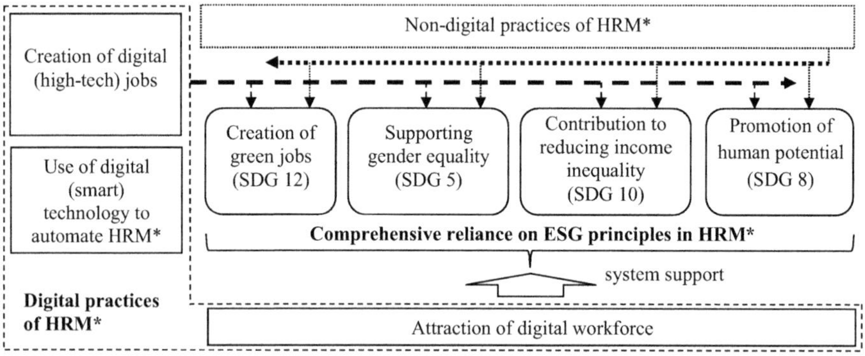

Fig. 2 Prospective model of human resource management (HRM) based on ESG principles in entrepreneurship to support sustainable development with a combination of digital and non-digital practices of HRM. *Note* *HRM – human resource (personnel) management. (*Source* Developed by the authors)

3. Contribution to reducing income inequality (SDG 10)
4. Promotion of human potential (SDG 8)

Conversely, non-digital practices of HR management fill gaps in digital practice. They are applied to support the SDGs in the following decreasing order of priority: (1) SDG 8, (2) SDG 10, (3) SDG 5, and (4) SDG 12. This provides a comprehensive reliance on ESG principles in HR management and the most comprehensive support for sustainable development in entrepreneurship.

5 Discussion

This research contributes to the literature by filling a knowledge gap at the intersection of the concept of responsible HR management and the concept of HR management in the digital economy. The obtained results clearly identified and specified the contribution of progressive practices of HR management in entrepreneurship of the digital economy contribute to the implementation of ESG principles and support of sustainable development (Table 2).

According to Table 2, in support of Cruz and Gameiro [2] and Stockinger et al. [15], the research proved the most promising and systemic support of the SDGs through the involvement of digital personnel. In contrast to Al-Kharabsheh et al. [1], Ramachandran et al. [10], Rodgers et al. [11], and Theres and Strohmeier [16], the research substantiated that the use of digital (smart) technologies to automate HR management makes a limited contribution to sustainable development, supporting only SDG 5 and SDG 12. In contrast to Popkova [8], Popkova and Sergi [9], and Sergi and Popkova [13], the research revealed that the creation of digital (high-tech) jobs weakly contributes to sustainable development; it supports only SDG 12.

Table 2 Author's interpretation of the contribution of digital practices of HR management to support sustainable development (ESG principles and SDGs) compared with the existing literature

Digital practices of HRM*		Literature noting the promise of the practice of their benefits	Refined interpretation of the practices and their benefits for sustainable development
Digital practices HRM*	Attraction of digital workforce	Evans [4] and Martin [7]	The most promising practice; systemically supports the SDGs
	Use of digital (smart) technology to automate HRM*	Langer and König [6] and Zhao et al. [19]	Limited contribution (supports only SDG 5 and SDG 12)
	Creation of digital (high-tech) jobs	Doerrenberg et al. [3] and Schröpf [12]	Weak contribution to sustainable development (supports only SDG 12)
Contribution to sustainable development (support for the SDGs)	Supporting gender equality (SDG 5)	Song et al. [14] and Veerasamy et al. [18]	Benefits are modest
	Promoting human potential (SDG 8)	Cruz and Gameiro [2] and Stockinger et al. [15]	Weak advantages are extracted
	Contribution to reducing income inequality (SDG 10)	Al-Kharabsheh et al. [1], Ramachandran et al. [10], Rodgers et al. [11], and Theres and Strohmeier [16]	Weak advantages are extracted
	Creating green jobs (SDG 12)	Popkova [8], Popkova and Sergi, [9], and Sergi and Popkova [13]	The advantages are the most significant

Source Developed by the authors
Note *HRM – human resource (personnel) management

In support of Song et al. [14] and Veerasamy et al. [18], the research proved that the benefits of implementing digital practices of HR management to create green jobs (SDG 12) are significant and most pronounced among all considered SDGs. In contrast to Evans [4] and Martin [7], the research found that the area of support for gender equality (SDG 5) provides the benefits of moderate implementation of digital practices of HR management. In contrast to Langer and König [6] and Zhao et al. [19], the research proves that weak benefits are derived from digital practices of HR management to help unlock human potential (SDG 8). In contrast to Doerrenberg et al. [3] and Schröpf [12], it is established that the contribution of digital practices of HR management to reducing income inequality (SDG 10) is negligible.

6 Conclusion

Thus, a study of the international experience of the most advanced digital economies (using the G7 and BRICS countries as an example) revealed that digital practices of HR management make a fragmented contribution to sustainable development, supporting the realization of certain SDGs. The greatest support from digital practices of HR management is provided for green jobs (SDG 12).

The main conclusion of the research is that ESG-based HR management requires a combination of digital and non-digital HR practices that fill each other's gaps. Prospects for improving HR management based on ESG principles in entrepreneurship of the digital economy to support sustainable development are related to a flexible combination of digital and non-digital practices of HR management.

For this purpose, the authors developed a model of HR management based on ESG principles in entrepreneurship to support sustainable development in a combination of digital and non-digital practices of HR management. The first advantage of the model is its high efficiency, achieved through a clear focus of HR management practices on specific sustainable development results.

The second advantage is the combination of digital and non-digital practices of HR management, considering their strengths and weaknesses to support ESG principles. In this way, alternative practices fill each other's gaps and reinforce each other, allowing for the fullest and most systemic support of the SDGs in HR management, as well as consistent adherence to ESG principles in this management process.

References

1. Al-Kharabsheh SA, Attiany MS, Alshawabkeh ROK, Hamadneh S, Alshurideh MT (2023) The impact of digital HRM on employee performance through employee motivation. Int J Data Netw Sci 7(1):275–282. https://doi.org/10.5267/j.ijdns.2022.10.006
2. Cruz SA, Gameiro A (2023) Digital work platform: understanding platforms, workers, clients in a service relation. Front Sociol 7:1075808. https://doi.org/10.3389/fsoc.2022.1075808
3. Doerrenberg P, Duncan D, Löffler M (2023) Asymmetric labor-supply responses to wage changes: experimental evidence from an online labor market: asymmetric labor-supply responses to wage changes. Labour Econ 81:102305. https://doi.org/10.1016/j.labeco.2022.102305
4. Evans C (2012) Recruitment initiatives aimed at increasing the gender diversity within ITEC employment: not so gender neutral? Equal Divers Incl 31(8):741–752. https://doi.org/10.1108/02610151211277608
5. IMD (2023) World competitiveness online database. Retrieved from https://worldcompetitiveness.imd.org/. Accessed 5 Feb 2023
6. Langer M, König CJ (2023) Introducing a multi-stakeholder perspective on opacity, transparency and strategies to reduce opacity in algorithm-based human resource management. Hum Resour Manag Rev 33(1):100881. https://doi.org/10.1016/j.hrmr.2021.100881
7. Martin H (2022) Stagnating female employment rates in France after several decades of growth. Popul Soc 11(606). https://doi.org/10.3917/popsoc.606.0001

8. Popkova E (2021) The social management of human capital: basic principles and methodological approaches. Int J Sociol Soc Policy 41(1–2):24–36. https://doi.org/10.1108/IJSSP-03-2020-0062
9. Popkova EG, Sergi BS (2022) High-tech economic growth from the standpoint of the theory of economic time: modelling and reducing space–time inequality. In: Inshakova EI, Inshakova AO (eds) New technology for inclusive and sustainable growth. Springer, Singapore, pp 15–22. https://doi.org/10.1007/978-981-16-9804-0_2
10. Ramachandran R, Babu V, Murugesan VP (2023) The role of blockchain technology in the process of decision-making in human resource management: a review and future research agenda. Bus Process Manag J 29(1):116–139. https://doi.org/10.1108/BPMJ-07-2022-0351
11. Rodgers W, Murray JM, Stefanidis A, Degbey WY, Tarba SY (2023) An artificial intelligence algorithmic approach to ethical decision-making in human resource management processes. Hum Resour Manag Rev 33(1):100925. https://doi.org/10.1016/j.hrmr.2022.100925
12. Schröpf B (2023) The dynamics of wage dispersion between firms: the role of firm entry and exit. J Labour Market Res 57(1):1. https://doi.org/10.1186/s12651-022-00326-3
13. Sergi BS, Popkova EG (2022) Towards a 'wide' role for venture capital in OECD countries' industry 4.0. Heliyon 8(1):e08700. https://doi.org/10.1016/j.heliyon.2021.e08700
14. Song D, Bai Y, Wu H, Wang X (2023) How does the perceived green human resource management impact employee's green innovative behavior? —From the perspective of theory of planned behavior. Front Psychol 13:1106494. https://doi.org/10.3389/fpsyg.2022.1106494
15. Stockinger C, Polanski-Schräder L, Subtil I (2023) The effect of information level of digital worker guidance systems on assembly performance, user experience and strain. Appl Ergon 106:103896. https://doi.org/10.1016/j.apergo.2022.103896
16. Theres C, Strohmeier S (2023) Met the expectations? A meta-analysis of the performance consequences of digital HRM. Int J Hum Resour Manag. https://doi.org/10.1080/09585192.2022.2161324
17. UNSTATS (2023) The sustainable development goals report 2022. Retrieved from https://unstats.un.org/sdgs/report/2022/. Accessed 5 Feb 2023
18. Veerasamy U, Joseph MS, Parayitam S (2023) Green human resource management and employee green behavior: participation and involvement, and training and development as moderators. South Asian J Human Resour Manag. https://doi.org/10.1177/23220937221144361
19. Zhao S, Liu M, Xi M, Zhu CJ, Liu H (2023) The role of leadership in human resource management: perspectives and evidence from China. Asia Pac Bus Rev 29(1):1–10. https://doi.org/10.1080/13602381.2023.2146901

Improving Entrepreneurial Efficiency Through the Implementation of ESG Principles in Human Resource Management

Marina V. Bugaeva, Natalia G. Tregulova, Sergey L. Vasenev, and Artem V. Lukomets

1 Introduction

In the Decade of Action, the market environment is unstable because of the increased cyclicality of the economy. This fact complicates the business environment and critically exacerbates competition for market positions. With limited room for maneuvering (e.g., a minimum margin of financial strength), businesses are forced to build their activities to maximize efficiency. With a high level of competition, the inability of enterprises to influence the price, and low solvent demand, the growth of entrepreneurial efficiency cannot be achieved by preferential growth of the result. Therefore, this growth comes at the expense of cost reduction.

The specificity of the Decade of Action is that the Sustainable Development Goals (SDGs) are deeply integrated into the management of entrepreneurship. Business initiatives to protect the environment are widely supported in sustainable communities; they are also heavily regulated by strict environmental laws. Therefore, a reduction in green investments is not always affordable and justified. Similarly, initiatives of enterprises related to innovation and technological modernization cannot be canceled in a highly digital competitive environment.

In this regard, human resource (HR) management is the most vulnerable area of business management in the urgent need to reduce costs. In the COVID-19 pandemic and global crisis, savings on human resources were achieved by transferring workers to remote employment. However, in the context of the worsening international sanctions crisis, many companies cannot apply this measure of saving costs

M. V. Bugaeva (✉) · N. G. Tregulova · S. L. Vasenev
Institute of Service and Entrepreneurship (Branch) of Don State Technical University, Shakhty, Russia

A. V. Lukomets
Federal Scientific Center "V.S. Pustovoit All-Russian Research Institute of Oil Crops", Krasnodar, Russia

E. G. Popkova (ed.), *Smart Green Innovations in Industry 4.0*, Springer Climate,
https://doi.org/10.1007/978-3-031-45830-9_35

on HR management. Therefore, there is a risk of massive staff reductions and, consequently, a large-scale increase in unemployment.

This fact determines the relevance of scientific studies on the issues of improving the efficiency of entrepreneurial activity while maintaining the support of the SDGs in HR management. The novelty of this research is that it proposes an innovative approach to HR management supported by the SDGs – relying on ESG principles. The research aims to investigate the prospects for improving the efficiency of entrepreneurial activity by introducing ESG principles in HR management.

2 Literature Review

The fundamental basis of the research is the theory of HR management. In accordance with this theory, the support of the SDGs in HR management implies a responsible attitude of the enterprise toward its employees. Two approaches to HR management supported by the SDGs have emerged nowadays.

The first approach, which is the most widespread in international practice, involves the isolated implementation of corporate social responsibility in HR management. This approach involves creating knowledge-intensive jobs to support SDG 4, as well as increasing employment and career opportunities for women (including those with advanced degrees) to support SDG 5.

He and Kim [3], Ramos-González et al. [10], and Vu [12] note the non-commercial and commercial benefits of human resources management on the principles of corporate social responsibility. The non-commercial benefits include strengthening the company's reputation as a responsible employer and attracting and retaining the best personnel due to their high loyalty to the company. Commercial advantages are expressed in the growth of the market capitalization of enterprises.

Due to the above advantages, Latan et al. [4], Liao et al. [5], Lu et al. [6], Omidi and Dal Zotto [7], and Popkova [8, 9] indicate that HR management based on the principles of corporate social responsibility is the preferred approach to HR management to improve the efficiency of entrepreneurial activity.

The second approach has emerged in recent years and is not yet widespread. However, it is becoming increasingly popular among enterprises worldwide. This approach is related to HR management based on ESG principles. In this case, corporate social responsibility is carried out not in isolation but together with other forms of enterprise responsibility to employees.

Green jobs are created (environmental performance is improved: E) in support of SDG 12, research talent is unlocked (S) in support of SDG 9, and an enabling environment and incentives for productivity growth are created (G) in support of SDG 8.

Adeneye and Kammoun [1], Chouaibi and Zouari [2], and Velte [11] note that the benefits of HR management on ESG principles are similar to the benefits of HR management based on the principles of corporate social responsibility. HR management based on the principles of corporate social responsibility is preferable due to the high cost of ESG management from the perspective of business efficiency.

The literature review shows that the theoretical issues of HR management in support of the SDGs have been studied in detail, which allows us to characterize the degree of penetration of the research problem as high. Simultaneously, the contribution of alternative approaches to HR management to the efficiency of entrepreneurial activity is insufficiently studied from an empirical point of view and remains uncertain.

This gap in the literature triggers the research question of how to manage human resources to improve business performance. The working hypothesis of the research is that the introduction of ESG principles in HR management contributes to improving the efficiency of entrepreneurial activity. To fill a gap in the literature, this research models and compares the contribution of HR management based on the principles of corporate social responsibility and HR management on ESG principles to business performance.

3 Materials and Methods

The research addresses two main tasks. First, it determines the contribution of alternative approaches to HR management to the efficiency of entrepreneurial activity. The authors studied the influence of the factors of corporate social responsibility and HR management on ESG principles in the efficiency of entrepreneurial activity.

Second, the authors used the method of regression analysis. The working hypothesis is considered proven if the contribution of HR management factors on ESG principles in the efficiency of entrepreneurial activity is greater than the contribution of corporate social responsibility factors in it. The information-empirical base of the research is the official statistics WIPO [13] for 2022 (Table 1).

The reliability of the regression model is evaluated using Fisher's F-test and Student's t-test. The second task is to determine the prospects for improving the efficiency of entrepreneurial activity by introducing ESG principles in HR management based on the example of Russia in the Decade of Action. For this purpose, based on the resulting regression model, the authors defined a Pareto-optimal combination of factors of human resource management, which achieves the highest (100 points) efficiency of entrepreneurial activity in Russia.

Table 1 Statistics of business performance and human resource management in the sample countries in 2022, points 1–100

	Efficiency of entrepreneurial activity	Factors of corporate social responsibility		Factors of HR management on ESG principles		
	Market capitalization, % GDP	Knowledge-intensive employment, %	Females employed with advanced degrees, %	Environmental performance	Research talent, % in businesses	Labor productivity growth, %
	Market capitalization of companies	Knowledge-intensive employment	Employment of women	Green jobs	Discovering the research talents of employees	Increase in labor productivity
Country	MCL	Csr_1	Csr_2	E	S	G
South Africa	100	31.2	33.0	37.2	22.4	60.2
Switzerland	89.1	79.3	68.9	65.9	58.9	53.9
Singapore	70.1	93.9	92.8	50.9	63.6	70.9
USA	62.5	72.5	92.6	51.1	88.3	59.1
Canada	51.4	67.6	64.3	50.0	71.2	49.8
United Kingdom	47.2	78.7	79.8	77.7	50.9	48.9
Japan	44.8	38.5	75.8	57.2	91.3	39.4
Malaysia	43.9	42.3	48.4	35.0	19.1	54.4
Netherlands	41.2	81.7	73.4	62.6	86.1	46.2
Russia	16.0	71.1	87.6	37.5	56.7	63.0

Source Compiled by the authors based on WIPO materials [13]

4 Results

4.1 Contribution of Alternative Approaches to HR Management to Entrepreneurial Efficiency

In the first research task, the authors conduct a regression analysis of the data from Table 1 to determine the contribution of alternative approaches to human resource management in the efficiency of entrepreneurial activity. As a result, the following regression model is obtained:

$$\begin{aligned} \mathrm{MCL} = & -192.9259 - 0.8478^{*}\,\mathrm{Csr}_1 - 1.7880^{*}\,\mathrm{Csr}_2 + 2.3296^{*}\,\mathrm{E} \\ & + 0.9784^{*}\,\mathrm{S} + 4.6078^{*}\,\mathrm{G} \end{aligned} \tag{1}$$

Model (1) demonstrates that the factors of corporate social responsibility do not positively contribute to the efficiency of entrepreneurial activity: the regression coefficient for knowledge-intensive employment (Csr_1) was −0.8478; for female employment (Csr_2), it was −1.7880. Simultaneously, all factors of HR management on ESG principles significantly contribute to improving the efficiency of entrepreneurial activity.

Thus, if the number of green jobs increases by 1 point, the market capitalization of enterprises increases by 2.3296 points. A one-point increase in the disclosure of employees' research talents increases the market capitalization of businesses by 0.9784 points. When labor productivity increases by 1 point, the market capitalization of enterprises increases by 4.6078 points. The results of the regression analysis are detailed in Table 1.

According to Table 2, 95.43% of business performance is due to HR management supported by the SDGs. The significance of F is 0.0319; consequently, model (1) corresponds to a significance level of 0.05. The reliability of model (1) is confirmed by a successful Fisher's F-test. Only the factor variables E, S, and G passed the Student's t-test, for which the t-statistic was 4.5146, 3.4527, and 5.3239, respectively, while the tabular t was 2.26.

Thus, based on the results obtained, the working hypothesis is considered proven. The contribution of HR management based on ESG factors to the effectiveness of entrepreneurial activity is greater than the contribution of corporate social responsibility factors to it.

Table 2 Results of regression analysis

Regression statistics						
Multiple R	0.9543					
R^2	0.9106					
Normalized R^2	0.7988					
Standard error	11.0451					
Observation	10					
Variance analysis						
	df	SS	MS	Ffact	Ftab	Significance of F
Regression	5	4970.3366	994.0673	8.1484	6.2561	0.0319
Residual	4	487.9794	121.9949	F-test passed		
Total	9	5458.316				
Regression coefficients						
	Coefficients	Standard error	t-statistics	*P*-value	Lower 95%	Upper 95%
Y-intersection	−192.9259	55.0675	−3.5034	0.0248	−345.8179	−40.0339
Csr_1	−0.8478	0.3782	−2.2419	0.0884	−1.8978	0.2022
Csr_2	−1.7880	0.4061	−4.4026	0.0117	−2.9156	−0.6604
E	**2.3296**	0.5160	4.5146	0.0107	0.8969	3.7623
S	**0.9784**	0.2834	3.4527	0.0260	0.1916	1.7652
G	**4.6078**	0.8655	5.3239	0.0060	2.2048	7.0108

Source Calculated and compiled by the authors

4.2 *Perspective on Improving Entrepreneurial Efficiency Through the Implementation of ESG Principles in HR Management: The Case of Russia*

In the second research task, based on the obtained regression model, the authors defined a pareto-optimal combination of factors of HR management, in which the maximum high (100 points) efficiency of business activities in Russia is achieved (Fig. 1). The purpose of determining this combination of factors is to determine the prospects for improving the efficiency of entrepreneurial activity through the introduction of ESG principles in HR management on the example of Russia in the Decade of Action.

As shown in Fig. 1, the market capitalization of Russian companies increases to the maximum possible 100 points (by 525% compared with 2022). Simultaneously, the following points are observed:

- Increasing the number of green jobs to 40.15 (up 7.06% from 2022)
- Increasing the level of disclosure of research talents of employees to 59.24 points (by 4.48% compared to 2022)
- Increasing productivity to 77.77 points (23.44% more than in 2022)

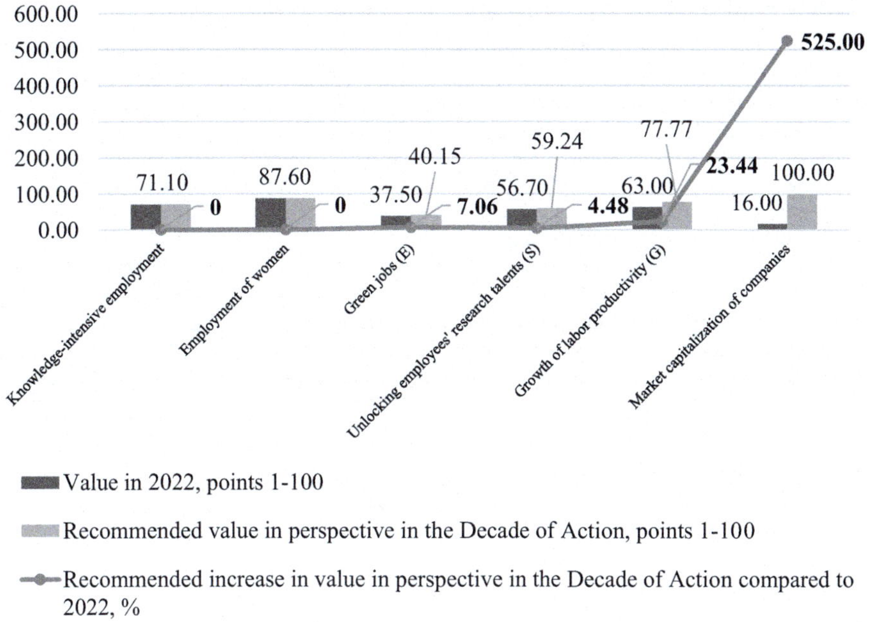

Fig. 1 The prospect of improving the efficiency of entrepreneurial activity through introducing ESG principles in HR management on the example of Russia in the Decade of Action. (*Source* Calculated and compiled by the authors)

Simultaneously, knowledge-intensive employment and female employment remain unchanged at the 2022 level of 71.10 points and 87.60 points, respectively.

5 Discussion

The contribution of the research to the literature lies in the development of HR management theory by clarifying the contribution of corporate social responsibility and HR management based on ESG principles to the effectiveness of entrepreneurship. The research results are shown in Table 3 and compared with the existing literature.

In contrast to He and Kim [3], Ramos-González et al. [10], and Vu [12], the research proves that HR management based on the principles of corporate social responsibility does not contribute to the growth of the market capitalization of the enterprise. Therefore, it provides mainly non-commercial benefits.

In contrast to Adeneye and Kammoun [1], Chouaibi and Zouari [2], and Velte [11], the research substantiates that the benefits of HR management based on ESG principles are not illogical but far exceed the benefits of HR management based on the principles of corporate social responsibility.

Table 3 Results of the research compared with the existing literature

Comparison area	Existing literature		A new interpretation derived from the research
	Existing interpretation	Source	
The benefits of HR management based on the principles of corporate social responsibility	Non-commercial and commercial – expressed in the growth of the market capitalization of enterprises	He and Kim [3], Ramos-González et al. [10], and Vu [12]	Mainly non-commercial: no contribution to the growth of the company's market capitalization has been identified
The benefits of HR management based on ESG principles	Similar to the benefits of HR management based on the principles of corporate social responsibility	Adeneye and Kammoun [1], Chouaibi and Zouari [2], and Velte [11]	Significantly exceed the benefits of HR management based on the principles of corporate social responsibility
Approach to HR management, the preferred approach to improving the efficiency of entrepreneurial activity	HR management based on the principles of corporate social responsibility	Latan et al. [4], Liao et al. [5], Lu et al. [6], Omidi and Dal Zotto [7], and Popkova [8, 9]	HR management based on ESG principles

Source Compiled by the authors

In contrast to Latan et al. [4], Liao et al. [5], Lu et al. [6], Omidi and Dal Zotto [7], and Popkova [8, 9], the research proves that HR management based on ESG principles is the preferable approach to increase the efficiency of entrepreneurial activity.

6 Conclusion

In conclusion, it should be noted that the hypothesis is confirmed by the research results, which proved that introducing ESG principles in HR management help improve the efficiency of business activities. The theoretical significance of the research results is related to the fact that they have made it possible to rethink the essence of corporate social responsibility from the perspective of business efficiency. The research clarified the cause-and-effect relationship of personnel refinement and substantiated the preference for relying on ESG principles to grow entrepreneurial efficiency.

The practical significance of the results lies in the fact that the perspective and the author's recommendations will improve the effectiveness of business through the introduction of ESG principles in HR management on the example of Russia in the Decade of Action. Based on the compiled econometric model, it is similarly possible to determine the prospects and development of applied solutions to improve the efficiency of business activities in other countries based on HR management on ESG principles. The social significance of the research is that it reveals the prospects of strengthening support for SDG 8, SDG 9, and SDG 12 in HR management in entrepreneurship.

References

1. Adeneye Y, Kammoun I (2022) Real earnings management and capital structure: does environmental, social and governance (ESG) performance matter? Cogent Bus Manag 9(1):2130134. https://doi.org/10.1080/23311975.2022.2130134
2. Chouaibi Y, Zouari G (2022) The effect of corporate social responsibility practices on real earnings management: evidence from a European ESG data. Int J Discl Gov 19(1):11–30. https://doi.org/10.1057/s41310-021-00125-1
3. He J, Kim H (2021) The effect of socially responsible HRM on organizational citizenship behavior for the environment: a proactive motivation model. Sustainability 13(14):7958. https://doi.org/10.3390/su13147958
4. Latan H, Chiappetta Jabbour CJ, de Sousa L, Jabbour AB, Ali M, Pereira V (2022) Career satisfaction in the public sector: implications for a more sustainable and socially responsible human resource management. Hum Resour Manag J 32(4):844–863. https://doi.org/10.1111/1748-8583.12469
5. Liao Z, Cheng J, Chen Q (2022) Socially responsible human resource management and employee ethical voice: roles of employee ethical self-efficacy and organizational identification. Corp Soc Responsib Environ Manag 29(4):820–829. https://doi.org/10.1002/csr.2236
6. Lu H, Xu W, Cai S, Yang F, Chen Q (2022) Does top management team responsible leadership help employees go green? The role of green human resource management and environmental felt-responsibility. Corp Soc Responsib Environ Manag 29(4):843–859. https://doi.org/10.1002/csr.2239
7. Omidi A, Dal Zotto C (2022) Socially responsible human resource management: a systematic literature review and research agenda. Sustainability 14(4):2116. https://doi.org/10.3390/su14042116
8. Popkova E (2021a) The social management of human capital: basic principles and methodological approaches. Int J Sociol Soc Policy 41(1–2):24–36. https://doi.org/10.1108/IJSSP-03-2020-0062
9. Popkova EG (2021b) The scientific and methodological approach to classification of economic systems according to the criterion of pleasure. Int J Trade Global Markets 14(4–5):450–458. https://doi.org/10.1504/IJTGM.2021.116728
10. Ramos-González MM, Rubio-Andrés M, Sastre-Castillo MÁ (2022) Effects of socially responsible human resource management (SR-HRM) on innovation and reputation in entrepreneurial SMEs. Int Entrep Manag J 18(3):1205–1233. https://doi.org/10.1007/s11365-020-00720-8
11. Velte P (2016) Sustainable management compensation and ESG performance – The German case. Probl Perspect Manag 14(4):17–24. https://doi.org/10.21511/ppm.14(4).2016.02
12. Vu T-V (2022) Perceived socially responsible HRM, employee organizational identification, and job performance: the moderating effect of perceived organizational response to a global crisis. Heliyon 8(11):e11563. https://doi.org/10.1016/j.heliyon.2022.e11563
13. WIPO (2023) Global Innovation Index 2022, 15th Edition. What is the future of innovation-driven growth? Retrieved from https://www.wipo.int/publications/en/details.jsp?id=4622. Accessed 3 Feb 2023

Social Management in the Innovation Space: Empirical Experience of Research

Maria P. Pavlova, Nikita Y. Gulyaev, Olga D. Parshina, Tatiana N. Ivanova, and Mikhail N. Ivanov

1 Introduction

The intensity of innovation processes within a territory significantly depends on the functioning of the socio-economic system when the provision of the balance in the system is supported, and innovation is introduced in all areas of society while maintaining the priority of existing trends and a comfortable state of society.

The objective of social management within the context of innovative development of the territory is to ensure the ability of society to self-renewal and its progressive development through the impulse of innovation.

The conceptual framework for modeling the social management of the innovation space of the territory is based on the following principles:

- Elements of the social environment (family microenvironment – the interaction and relationship of family members, the intellectual level of the family; interpersonal interaction with samples of intellectual behavior as a relay of domestic and international intellectual culture; information and communication enrichment that helps increase the amount of information needed for the creation of innovation) affect the process of formation and development of the individual as a subject of innovation space.
- The creative activity of potential innovators is a completely uncontrollable activity on the part of the socio-economic system.
- The model of innovative space of the territory includes a set of individual innovative spaces (locums) of the person as a source of creative activity formed through socio-economic interactions in the society.

M. P. Pavlova · N. Y. Gulyaev · O. D. Parshina · T. N. Ivanova (✉) · M. N. Ivanov
Togliatti State University, Tolyatti, Russia
e-mail: IvanovaT2005@tltsu.ru

E. G. Popkova (ed.), *Smart Green Innovations in Industry 4.0*, Springer Climate,
https://doi.org/10.1007/978-3-031-45830-9_36

- The integration of these locums determines the level of human potential of the territory, the ability to accumulate and reproduce knowledge, the accretion of a person, and the self-development of a person. Human potential acts as a generator of innovative transformations in the territory.
- The presence of people with sufficient academic and functional knowledge and practical readiness, capable of creating, reproducing, and transmitting knowledge, ensures the progressive development of society through motivated continuous processes of intellectual excellence.

To design the system of social management of the innovation space of the territory, a comprehensive empirical study of socio-economic processes impacting the monitoring and innovation processes of the territory is necessary. This will reflect the regularities and features of the development of the participants in the managerial activity.

1.1 Educational Environment of the Innovation Space

The empirical study for modeling the system of social management of innovation space should encompass two key aspects:

- Assessment and dynamics of human potential of the territory as a determinant of innovation space.
- The state of material support for innovative transformations in the territory.

The empirical study of the human potential of the territory is appropriate in the context of the three main categories using the statistical socio-economic information provided by the Federal State Statistics Service of the Russian Federation (Rosstat) for the territories.

The study in the first category, "Analysis of the number and composition of the population by level of education," is based on the hypothesis that people with higher education as having sufficient knowledge are more inclined to creative activity.

People involved in the innovative economy should possess certain knowledge and skills, independently generate new ideas, and ensure practical implementation of these ideas. Through an enriched educational environment, the higher education system makes it possible to develop certain competencies and characteristics necessary for implementing activities directed at the innovative development of the territory. The established productive inseparable interrelation of higher education and science provides innovative personalities with certain fundamental and applied knowledge and skills as a source of new ideas and innovations.

The statistical sample includes the composition of the population by level of education, both employed and unemployed in the region's economic activity. The potential of the unemployed also participates in the synthesis and formation of the human potential of the territory.

Particular attention should be paid to analyzing data on students in higher education institutions. This analysis will make it possible to predict the level of education in the studied area at any point in time. It should be noted that the higher education system includes Bachelor's, Specialist, and Master's programs. The analysis results can present two types of reflection of the development in the innovation territory.

The first type is when there is a tendency to prioritize secondary vocational education. The negative dynamics of the number of students enrolled in Bachelor's, Specialist, and Master's programs will be additional evidence of this type of development. The choice of applicants in obtaining a working profession and specialties of secondary vocational education is largely due to the goal of further achieving stability in life, characterized by a greater guarantee of high wages with minimal time spent on the educational process.

The second type is observed when higher professional education is a priority in the territory among applicants. This indicates a stable positive trend in understanding the need for education to realize oneself as an active participant in innovative processes. Consequently, the active formation of interpersonal relationships and interactions between the learner and the teacher as carriers of intellectual behavior and intellectual culture is fixed on the territory.

Additional research on the age group of the able-bodied population will make it possible to define a group of people and the prospects of its development. These people have a sufficient level of ability to operate their accumulated knowledge, possess active cognition, and show a high level of use and adaptation of new information. They can be considered as potential innovators of the territory because their economic and creative productivity is the highest. In foreign studies, it is the age of 47. It should be noted that the median age of the unemployed and working population in the territory (in the boundaries up to 47 years) can be regarded as a period of building up the amount of knowledge and information through their labor and intellectual activity to implement them and further transform them into an innovative product. In this case, the actor is characterized by the presence of certain qualifications, knowledge, and skills.

1.2 Reproduction Processes in the Innovation Space

The research data in the second category, "Diagnosis of reproduction processes of human potential," will enable us to trace the perspective of the intensity of reproduction processes and assess their dynamics at the present time, affecting the translation of innovation culture in the territory.

Apparently, these reproduction processes are constant and natural. They depend on the socio-economic conditions of the development of the territory []. Natural reproduction processes of labor resources are characterized by the coefficients of natural population growth and fertility and mortality rates, establishing the wave demographic processes of the territory. The study of the working-age population by gender and age will provide information on the ratio of men and women in the territory and will indirectly present the dynamics of the number of women of fertile (reproductive) age, which will predetermine the projected birth rate processes in the coming years. This analysis is complemented by studies of the total fertility rate (i.e., the number of children per woman), which illustrate the process of simple reproduction through the prism of the complete replacement of the generation of parents by generations of children. These data will make it possible to determine the dynamics of reproduction processes in the innovation territory. It will also make it possible to analyze the potential human capital, which will be involved in the economy of the territory in 10–15 years, and determine the projected number of carriers of human capital, which can act as translators of innovation culture in the territory [6].

The speed of the population's reproductive processes characterizes either rapid processes of forming a young society capable of social cataclysms (revolutions, revolts) or, on the contrary, slow reproductive processes that lead to the formation of an old society with prevailing stagnation processes. This is reflected in the study of the dynamics of demographic processes on the age structure of the territory's population. It is apparent that the increase in population and the increase in young people of working age in the territory definitely shows the progressive dynamics of social development in the innovative territory; the decline in these indicators illustrates stagnant processes.

It is necessary to note the paradoxical impact of innovation on society. On the one hand, innovation promotes societal changes as elements, structures, connections, and interactions move from one state to another. Then, the transformation of society's values entails crisis phenomena. Society needs stability by its nature; any innovative activity is a potential threat to changes in the existing order [3].

On the contrary, in his scientific concept of anomie, R. Merton argues that the crisis and dysfunctionality of the social system encourage the growth of individuals-innovators, who tend to use the means leading to success in the shortest way to solve their life tasks and achieve existing goals. Consequently, the innovative activity of individuals (innovators) does not seek to change the structure of society; the person-innovator only selects the means to take a higher position in this structure [5]. Thus, innovative activity is viewed as risky and stressful.

Studies of demographic processes help determine the type of reproductive process of the population (regressive or progressive), which directly impact the preparation of a new generation to carry out socially necessary functions for the formation of an innovative socio-economic system within the studied territory.

The formation of human resources in the territory can be influenced not only by natural demographic processes but also by migration. Migration processes of the territory correlate with demographic processes, increasing the working-age population and participating in the processes of fertility and mortality. The study of the dynamics of the rate of migration growth will make it possible to trace the artificial increase in population through migration.

2 Methodology of Empirical Research

The third category of the study, "Assessment of the quality of human potential," is characterized by the physical, moral, spiritual, and socio-professional development of human resources in the territory in relation to the innovation activity of the territory.

The study is based on the assertion proved by domestic scientists N. N. Golub, A. G. Shkurat, T. I. Zaslavskaya, and V. A. Yadov. According to these scholars, when meeting basic needs, a person is aware of the importance of creating and implementing innovations in all areas of life [5, 8].

Basic human needs are systematized in individual regular sequences, where cycles of their satisfaction are independently defined. The natural process of satisfying individual basic human needs is infinite. Thus, it provides the states in which an individual's sense of inner comfort and balance will be achieved [2].

In his works, V. S. Klemes defines the state of an individual, in which the motivation for a new higher-level need will be formed through the concept of "measure of need satisfaction" as a value that captures the level of satisfaction of needs [4]. The conducted research proves that when the measure of satisfaction of one need increases, the motivation for a new need begins to emerge, defining a gradual and slow process of emergence of a new need.

Determining the moment of the emergence of the need for innovation is quite a controversial issue. On the one hand, this need arises when physiological needs and a sense of physical and financial security are met. On the other hand, it is not a need but rather a part of human activity [7].

2.1 Innovative Behavior in Society

Analysis of human potential in the context of innovation needs will allow us to determine the level of readiness of the population to create and implement new solutions, as well as the speed of assimilation of innovations in all areas of life. The important socially significant process of assimilation integrates the areas of science and creativity with real everyday socio-economic processes in the dynamic development of society, reflecting the social effect of introducing innovation. The end result of this process is considered innovation, which, having passed into the category of tradition, norms, and stereotypes, becomes the basis for creating the next innovation, which allows society to make a qualitative leap in its progressive development.

Statistical indicators of innovation activity of the territory (the number of innovation-active enterprises, the amount of funding for innovation activity, data on the introduction of the use of intellectual property, the scientific potential of the territory involved in innovative processes, and the number of personnel involved in R&D in the territory) will determine the calculated level of innovation activity of socio-economic processes of the territory, illustrating a complete picture of perspective or regressive development of innovation activity in the territory [1].

It should be noted that unstable financing of innovative activity negatively affects the volume of sold innovative goods, works, and services. However, with positive trends in the use of intellectual property objects in socio-economic processes, this situation can be a signal that society is at the stage of assimilation of innovations. Simultaneously, if the volume of realization of innovative goods, works, and services increases in the territory with a decrease in the financing of innovation activities, it may indicate that the stage of assimilation of innovations in the territory is completed and now they are the direct attributes of the population's life.

Additional research of statistical data on the funding of all types of scientific work (fundamental and applied development by field of science) in the fields of natural, technical, medical, and agricultural sciences will make it possible to predict a positive correlation of these data with the dynamics of the gross domestic product of the territory (since the funding of works from the state budget is carried out in the interests of the development of the territory, considering its resource potential) for the construction of a comfortable socio-cultural space. The level of comfort is confirmed by the increase in the integrating indicator – life expectancy at birth for the territory.

The progressive dynamics of the use of intellectual property objects (IPO) in the territory indicates the formation of a sufficient scientific and research base and a high level of literacy of inventors, perhaps even in conditions of the declining popularity of higher education among applicants and the median age of the working population. Data on the scientific potential of the territory is presented by the ranking of scientific figures (Doctor of Sciences and Candidate of Sciences) who are directly engaged in research innovation activities. Insignificant fluctuations in the values of the statistical sample of the number of personnel employed in R&D and its qualitative characteristic (the number of researchers with scientific degrees) allow us to conclude about the formation and the stable development of the scientific backbone of society in the territory, which can generate and implement new ideas and solutions.

3 Results

Material provision of the territory provides qualitative and quantitative characteristics of socio-economic processes and makes it possible to evaluate and control resource flows to achieve a synergistic effect of the territory of socio-economic processes [8].

The scale and complexity of the organizational and functional infrastructure of the territory largely determine the study of the resource base of the territory in terms of an integrated approach. We propose to evaluate the material support of innovation activity of the territory according to the algorithm with the allocation of natural resources (includes the totality of natural resources and infrastructure necessary for their use, as well as measures for their reproduction and conservation) and industrial-economic support (includes industrial subjects of economic activity, the functioning and activity of which is provided by the presence of supporting infrastructure and communication ties in the territory).

3.1 Resource Provision for the Development of the Territory Under the Innovation Scenario

The development of the territory through extensive economic development, the organization and activities of research and production bases, and technical and technological support of socio-economic processes make it possible to develop competitive advantages, which provide a high level of market and social activity of the subjects of the territory. During the formation of the knowledge economy, when new types of potentials, innovative resources, information, knowledge, and advanced technologies are prioritized, resource provision of socio-economic processes recedes into the background. However, without material support, providing economic and political independence of socio-economic processes, a qualitative leap in the progressive development of society and the territory can be stretched over time.

It should be noted that by supplementing the empirical research with data on the material provision of the territory, there is an opportunity to assess the prospects for the development of innovation space in modern realities.

4 Conclusion

The obtained results determine the rational type of social management of the innovation space of the territory, ensuring the mobilization of the potential of society in the socio-cultural environment, considering the mechanisms of self-regulation and homeostasis. Informatization of the management process, namely, obtaining reliable and timely information, makes it possible to determine the dynamics of the development of the system of social management of the innovation space of the territory and the choice of management tools, ensuring the achievement of the goals of the entire system at each stage of its activity.

References

1. Arkhipova MU, Lebedev AV (2012) Innovations and population well-being: relationship, trend, perspectives. Economics, Statistics, and Informatics. UMO Bulletin 6:91–95

2. Golub NN, Shkurat AG (2021) The impact of socio-economic factors on the standard of living of the population. In: Rasskazov FD (ed) Modern studies in the humanities and natural sciences: a collection of scientific articles (Part VII). Pero, Moscow, pp 25–29
3. Karpova YA (2004) Introduction to the sociology of innovation: textbook. Piter, St. Petersburg
4. Klemes VS (2017) Analysis of virtual relationships as a source of satisfaction of basic human needs. Bulletin of Amur State University. Ser: Hum 78:94–98
5. Merton RK (1966) Social structure and anomie (trans: Samarskaya EA, Gretzky MN, from French). In: Nikiforov BS (eds) Sociology of crime: modern bourgeois theories. Progress, Moscow, pp 299–313
6. Pak TI (2016) Social and economic factors contributing to demographic indicators of life expectancy in Russia and the G20 countries. Sci Notes Young Res 4–5:80–86. Retrieved from http://elib.fa.ru/art2016/bv4174.pdf/download/bv4174.pdf. Accessed 12 Nov 2022
7. Ushakova DV (ed) (2011) Creativity: from biological foundations to social and cultural phenomena. Institute of Psychology of RAS, Moscow
8. Zaslavskaya TI, Yadov VA (2008) Social transformations in Russia in the epoch of global changes. Sociol J 4:8–22

Digital Transformation of the Labor Market in an Environmentally Oriented Perspective

Anastasia V. Danskaya

1 Introduction

The acceleration of the scientific and technological process in the conditions of the Fourth Industrial Revolution stimulates the strengthening of the state policy orientation in the direction of new technology studies: artificial intelligence, digitalization, robotics, etc. Contemporary processes form a highly flexible world architecture of new types of organizational structures and management technologies at the macro- and micro levels, requiring the application of sustainable development principles to the contemporary world system to balance its elements.

An emerging trend in the world is the promotion of job creation activities based on the new conditions of a green economy. The main feature of a green economy policy (according to UNEP) is the formation of a market oriented to the environmental agenda and the process of creating green jobs. The jobs of an originally different digital-based formation contribute to the reduction of negative emissions into the atmosphere and have a healthier effect on the environment.

Digitalization and green are the two main sectors of the new economy that provide the foundations for the Fourth Green Industrial Revolution (according to the IMF). The digital economy forms a balance between the economic and social aspects of life (education, health, and resource management), reducing the added value of providing services and making the benefits of creating new companies (workplaces) available. A green economy forms a sustainable development of the economic and environmental aspects of society by reducing environmental damage and introducing new technologies for renewable energy (a low-carbon economy) and the production of environmentally friendly materials.

This chapter aims to study the trends toward the formation of a green employment market in the conditions of the Fourth Industrial Revolution.

A. V. Danskaya (✉)
MGIMO University, Moscow, Russia

E. G. Popkova (ed.), *Smart Green Innovations in Industry 4.0*, Springer Climate,
https://doi.org/10.1007/978-3-031-45830-9_37

2 Materials and Method

The specific features and perspectives of the climate impacts on employment market trends are studied in the works of Abrakova [3], Fisher and Bubola [6], Glazyev [7], Konina [13], Konina and Sapir [14], Konina et al. [15], Popkova and Zavyalova [20], Schuh and Scheuer [21], Syvorotkin [24], Tinbergen et al. [25], Veredyuk [26], Walter and Lee [27], Zavyalova and Krotova [29], and Zavyalova et al. [30].

Fourth Industrial Revolution – the environmental agenda market is studied in the works of Corfe [4], Keynes [12], Kudinova et al. [16], Liu et al. [17], Mitrović et al. [18], and Schwab [22]. Despite the large number of publications on adjacent topics, economic science has not sufficiently studied and solved the issues of the formation of green employment structure through digital technologies.

The study on the ozone layer of the earth, conducted in the mid-1970s [19] found that high levels of certain chemicals negatively affect the planet's atmosphere. Depletion of the ozone layer leads to increased ultraviolet radiation, which is detrimental to public health: skin diseases (skin cancer and burns) and eye diseases (cataracts and decreased sharpness of vision). It also greatly affects the flora and fauna of the planet: disruption of plant photosynthesis, increased crop losses (fires), and a reduction in plankton in the ocean [23].

Montreal Protocol (1987) was supposed to be a new plan for clearing the atmosphere by implementing measures to reduce emissions of ozone-depleting substances (ODS). The results were a fourfold reduction in ODS emissions after 1988, compared to a projected reduction of 10 times by 2050. However, it is worth noting that this protocol to the 1985 Vienna Convention for the Protection of the Ozone Layer, in its latest version (Montreal Adjustment 2010–2014), is of an advisory nature. The United Nations Environment Program (UNEP) and its staff provide no assurance on the accuracy of indicators or strategies, policies, and methodologies for calculating emissions of ozone-depleting potential (ODP) hydrochlorofluorocarbons (HCFCs), nor have they identified specific control methods [8].

The World Bank experts have identified the most serious risks on a global scale, which will affect the state of the world community for the next decade (the most considerable global risks can be found at the link listed under "Data availability") [28]. The most considerable threats to the world community are unnatural and extreme weather conditions and inefficiency of measures taken to curb climate deviations and return them to normal values. The world's diversity of flora and fauna (IUCN Red List species), lack of household livelihoods, financial sector (debt) crisis, and geo-economic regional conflicts were also noted.

These changes entail the emergence of a new business environment, which is reflected in the socio-economic processes of society through new adapted (sustainable) forms of implementation, public policy, and business strategy [5, 6, 10].

Sustainable development is formed based on legal, economic, and institutional conditions that ensure green employment. The legal framework [1, 2] for a green economy involves the creation of institutions to regulate the transition to the digital and clean energy era. Perhaps the most important institutional eco-initiative at the

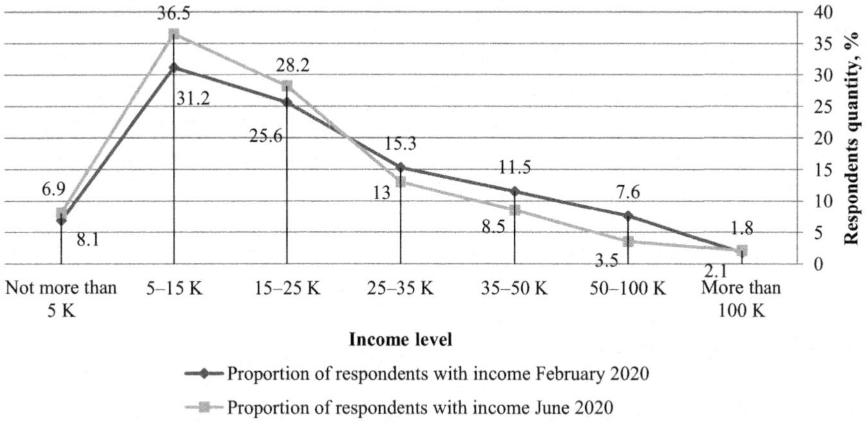

Fig. 1 Population income level 2020. (*Source* Compiled by the authors based on [9])

international level is the Green Jobs Initiative, adopted in 2007 as part of an agreement between the United Nations Environment Program (UNEP), the International Labor Organization (ILO) [11], the International Trade Union Confederation (ITUC), and the International Organization of Employers (IEO). The first report, "Green Jobs: Towards Decent Work in a Sustainable, Low-Carbon World," explored the challenges of transitioning to green jobs and emphasized the need to strengthen employment and income generation programs as a tool for poverty eradication (Fig. 1).

Under COVID-19, health and income issues have become more acute. As of April 2020, the ILO's projection model showed a reduction in hours worked (4.5% globally) and a 60% reduction in income for 1.6 billion workers. No change in earnings (according to a survey by IC Rosgosstrakh Life) was reported by 50.3% of the respondents. However, statistics show that approximately 72.8% of respondents lived on incomes of no more than 25,000 rubles per person and their number increased by 9.1% from February to June. In 2020, the federal minimum wage in Russia was 12,130 rubles. The results show that shrinking income levels exacerbate the need to control social inequality between people and businesses, which is a key issue in the concept of a sustainable and green economy.

Sustainable development (UNEP) refers to the creation of resilient jobs, a concept that includes the total number of jobs that address poverty within the concept of a sustainable world development. This type of jobs changes the nature of ecological and economic chains by increasing the mobility of people and the turnover of capital through tourism and, more generally, the service sector. The green employment includes the protection and conservation of biodiversity, the reduction of the consumption of energy, materials, and water resources, and the minimization of greenhouse gas emissions and pollution of any kind into the environment.

3 Results

Climate change is becoming a major agenda on a global scale. The climate (according to the global risk map economic forum "Davos") negatively affect the most sensitive resources (human resources) and industries (agriculture, fisheries, and tourism) in the medium and long term.

A new trend in the contemporary world is the implementation of measures to promote the development of employment based on the new conditions of the sustainable use on the basis of high-tech technologies to stabilize vulnerability. The main feature of a green economy policy (according to UNEP) is the formation of a market oriented to the environmental agenda and the creation of green jobs. The new type (jobs) will contribute to the reduction of negative emissions into the atmosphere and bring the effect of "recovery" of the environment.

Document of a legal and consultancy framework verify a new criterion for the green digital workplace for highly competent workers: a fair wage level, safe working conditions, opportunities for career advancement, and respect for workers' rights.

Regarding the legislative framework, public policy on environmental protection is not developed in the control and monitoring of cyclical actions due to the innovativeness of the high variability and turbulence of contemporary living conditions. However, there are already several treaties and programs (UN) that define a framework of digital transformation in the field of green employment and are already implemented by countries at the national level, although these measures for the development of green infrastructure are not yet sufficient.

The formation of a green economy labor market will involve measures to implement the basic principles of environmental management and reduce environmental pollution, which will contribute to the sustainable development of the economy.

The government needs to structure the following activities:

1. Planning and financing green employment processes
2. Formation of public policies
3. Formation of an education system aimed at shaping the competences of the economy of the future
4. Providing an economic rationale for the greening of production for business
5. Preparation and formation of the legislative basis of the green labor market

4 Conclusion

The processes of shaping the new digital economic conditions of the future, with a focus on environmental security, are completely changing the architecture of the global economy and leading to transformational changes in the economy of employment. Due to the transition to a low-carbon economy, changing the structure of economic entities, the state policy of the countries will take a course on the principles of sustainable development, which will ensure economic and environmental security. In the context of creating new digital eco-jobs, income growth will provide a decent standard of living and make it possible to preserve and restore the natural capital of the planet.

Thus, the development of the market of green employment in a digital economy is directly related to structural changes in the labor market and the emergence of new types and ways of employment for workers, leading to the necessity of the identification of strategic priorities and the implementation of different measures to a newly structured labor market. It is necessary to reorient the system of education to ecological orientation to implement training programs for the future economy in conditions of increasing the value of environmental initiatives.

Green jobs positively affect the economy. Socio-economic measures provide favorable conditions in the labor market, keep the environment safe, and allow for environmentally friendly living conditions (reduction of hydrocarbon emissions, increase in clean water).Data AvailabilityData on digitalization, economic growth, and serious global risks that support the study's findings are available in https://figshare.com/ with an identifier https://doi.org/10.6084/m9.figshare.21997673.v1

References

1. Government of the Russian Federation (2017) Passport of the national program "Digital Economy of the Russian Federation" (Approved by order dated July 28, 2017 No. 1632-r), Moscow. Retrieved from http://static.government.ru/media/files/urKHm0gTPPnzJlaKw3M5cNLo6gczMkPF.pdf. Accessed 17 Oct 2022
2. Government of the Russian Federation (2020) Passport of the Program of fundamental scientific research in the Russian Federation for the long-term period (2021–2030) (Approved by order dated December 31, 2020 No. 3684-r), Moscow. Retrieved from http://static.government.ru/media/files/skzO0DEvyFOIBtXobzPA3zTyC71cRAOi.pdf. Accessed 17 Oct 2022
3. Abrakova TA (2020) Modern Russian historiography on some reasons of Soviet society crisis in 1970's–1980's of XX century. Mod Sci Thought 4:89–94
4. Corfe S (2020) 4IR and the Environment: how the Fourth Industrial Revolution can curb air pollution and decarbonise the economy [Report]. Social Market Foundation, London. Retrieved from https://www.smf.co.uk/wp-content/uploads/2020/01/4IR-and-the-Environment-Report-Jan-2020-1.pdf. Accessed 17 Oct 2022
5. Deloitte (2021) International study on trends in personnel management. Socially responsible business in the new reality. Retrieved from https://www2.deloitte.com/kz/ru/pages/human-capital/articles/human-capital-trends-2021.html?ysclid=ldka4iyfp3370633230. Accessed 15 Oct 2022
6. Fisher M, Bubola E (2020) As coronavirus deepens inequality, inequality worsens its spread. The New York Times, 15 Mar 2020. Retrieved from https://www.nytimes.com/2020/03/15/world/europe/coronavirus-inequality.html. Accessed 21 Oct 2022
7. Glazyev SY (2016) National economy structures in the global economic development. Econ Math Methods 52(2):3–29
8. Herweijer C, Combes B, Jackson B, Johnson L, McCargow R, Bhardwaj S (2017) Enabling a sustainable Fourth Industrial Revolution: how G20 countries can create the conditions for emerging technologies to benefit people and the planet, 22 May 2017. Retrieved from

https://www.g20-insights.org/wp-content/uploads/2017/05/G20-Insights-PwC-Enabling-a-Sustainable-4IR.pdf. Accessed 17 Oct 2022
9. Insurance Company "Roosgosstrakh Life" (2020) During the pandemic, 10% of Russians have lost their sources of income, but there are more who earn over 100,000 rubles, 7 July 2020. Retrieved from https://rgsl.ru/news/za-vremya-pandemii-10-rossiyan-poteryali-istochniki-dokhoda-no-stalo-bolshe-tekh-kto-zarabatyvaet-sv/. Accessed 20 Oct 2022
10. International Labour Organization (2013) Report V: sustainable development, decent work and green jobs. In: International Labor Conference, 102nd Session. ILO, Geneva. Retrieved from https://www.ilo.org/wcmsp5/groups/public/%2D%2D-ed_norm/%2D%2D-relconf/documents/meetingdocument/wcms_207370.pdf. Accessed 21 Oct 2022
11. International Labour Organization (2020) Climate action for jobs initiative. New climate action for jobs board calls for a sustainable recovery from the COVID-19 crisis, 9 Sept 2020. Retrieved from https://www.ilo.org/global/about-the-ilo/newsroom/news/WCMS_754932/lang%2D%2Den/index.htm. Accessed 2 Nov 2022
12. Keynes JM (1931) Economic possibilities for our grandchildren. In: Essays in persuasion. Palgrave Macmillan, London, pp 321–332. https://doi.org/10.1007/978-1-349-59072-8_25
13. Konina NY (2018) Issues of modern geoeconomics. Int Trends 16(2):186–196. https://doi.org/10.17994/it.2018.16.2.53.11
14. Konina NY, Sapir EV (2021) Geo-economic aspects of the "Green economy" in Industry 4.0. In: Zavyalova EB, Popkova EG (eds) Industry 4.0: exploring the consequences of climate change. Palgrave Macmillan, Cham, pp 337–352. https://doi.org/10.1007/978-3-030-75405-1_30
15. Konina N, Vladimirova I, Stonehouse G (2021) The evolution of conglomerates in the digital era: adapting strategies to the requirements of the fourth industrial revolution. In: Konina N (ed) Digital strategies in a global market. Palgrave Macmillan, Cham, pp 145–155. https://doi.org/10.1007/978-3-030-58267-8_11
16. Kudinova GE, Rosenberg GS, Yurina VS (2012) Towards a green economy: pathways to sustainable development and poverty. Nairobi, Kenya; Geneva, Switzerland; Moscow, Russia: UNEP, 2011. Princ Ecol 4(4):41–48
17. Liu J, Liu L, Qian Y, Song S (2022) The effect of artificial intelligence on carbon intensity: evidence from China's industrial sector. Socio Econ Plan Sci 83:101002. https://doi.org/10.1016/j.seps.2020.101002
18. Mitrović Z, Bjelica D, Borozan T (2022) Are future project managers ready for a digitalized world? In: Mihić M, Jednak S, Savić G (eds) Sustainable business management and digital transformation: challenges and opportunities in the post-COVID era. Springer, Cham, pp 525–543. https://doi.org/10.1007/978-3-031-18645-5_33
19. OECD (2011) Towards green growth. A summary for policy makers. Retrieved from https://www.oecd.org/greengrowth/48012345.pdf. Accessed 21 Oct 2022
20. Popkova EG, Zavyalova EB (eds) (2021) New institutions for socio-economic development: the change of paradigm from rationality and stability to responsibility and dynamism. De Gruyter, Berlin/Boston. https://doi.org/10.1515/9783110699869
21. Schuh G, Scheuer T (2022) Increasing labor productivity in intelligent manufacturing enterprises: an approach to identify relevant capability bundles. In: Proceedings of PICMET 2022: Portland international conference on management of engineering and technology "Technology management and leadership in digital transformation – looking ahead to post-COVID era", Portland, pp 1–12. https://doi.org/10.23919/PICMET53225.2022.9882721
22. Schwab K (2016) The fourth industrial revolution (trans: from English). Eksmo, Moscow. (Original work published 2016)
23. State Council of the Russian Federation (2016) Report on ecological development of the Russian Federation for future generations, Moscow. Retrieved from http://ecoline.ru/wp-content/uploads/report-on-the-ecological-development-of-the-russian-federation-in-the-interests-of-future-generations-2016.pdf?ysclid=ldk9xbkzji425457910. Accessed 21 Oct 2022
24. Syvorotkin VL (2014) Environmental threats of the Montreal Protocol. Space Time 4(18):211–221

25. Tinbergen J, Dolman A, Ettinger JV (1976) Reshaping the international order: a report to the Club of Rome. Dutton, New York
26. Veredyuk OB (2011) The impact of climate change on employment. St. Petersburg Univ J Econ Stud 4(5):22–29
27. Walter S, Lee JD (2022) How susceptible are skills to obsolescence? A task-based perspective of human capital depreciation. Foresight STI Gov 16(2):32–41. https://doi.org/10.17323/2500-2597.2022.2.32.41
28. World Economic Forum (2011) Global risks 2011: an initiative of the risk response network, 6th edn. WEF, Geneva. Retrieved from https://www3.weforum.org/docs/WEF_Global_Risks_Report_2011.pdf. Accessed 23 Oct 2022
29. Zavyalova EB, Krotova TG (2021) Methods for achieving SDG 1, poverty eradication. In: Popkova EG, Sergi BS (eds) Modern global economic system: evolutional development vs. revolutionary leap. Springer, Cham, pp 1987–2001. https://doi.org/10.1007/978-3-030-69415-9_218
30. Zavyalova EB, Lee JS, Ostrovskaya E (2020) Corporate social responsibility. In: Dementieva A, Zavyalova E (eds) Corporate governance in Russia: quo vadis? De Gruyter, Berlin/Boston, pp 173–187. https://doi.org/10.1515/9783110695816-016

Green Human Resources to Support Climate-Responsible Entrepreneurship in Digital Economy Markets Through the Integration of the Markets of Education and Labor

Aleksei V. Bogoviz, Elena L. Kharitonova, Vyacheslav D. Komissarov, and Vitaly V. Sych

1 Introduction

Human resources management is one of the leading sub-systems of entrepreneurship. It forms preconditions and ensures the achievement of the goals of economic subjects, determines the level of economic activity and innovativeness, and is the key factor of development. This sub-system is an important internal element of entrepreneurial structures. It is significantly influenced by the external environment and reacts to its dynamics. Therefore, climate change and the related concepts of sustainable development, climate responsibility, and eco-friendly behavior form new approaches to labor relations in the entrepreneurial environment.

Considering humans as a component of human resources or personnel, this research relies on the three following statements:

1. Human labor activity is the reason for climate change.
2. Climate change influences the conditions of labor activity.
3. Climate-responsible labor activity of humans is a tool for tackling climate change.

The benchmarks on the direction of labor efforts, determined by the third statement, define the necessity for forming green competencies with employees in different sectors of the economy. This sets special requirements for education and labor markets, which requires the coordination and integration of the mechanisms of education, employment, and development. In this context, the environment, which includes climate change and deterioration of the environment, makes individuals and organizations adopt a green policy [12].

A. V. Bogoviz (✉)
Independent Researcher, Moscow, Russia

E. L. Kharitonova · V. D. Komissarov · V. V. Sych
Platov South-Russian State Polytechnic University, Novocherkassk, Russia

E. G. Popkova (ed.), *Smart Green Innovations in Industry 4.0*, Springer Climate,
https://doi.org/10.1007/978-3-031-45830-9_38

Development of green skills is treated as a mandatory condition for climate-responsible entrepreneurship and growth, which are often considered from the position of "transition", that is, quick qualitative change in value and production and economic processes. To alleviate the consequences of the impact of the transition to green growth on labor markets, it is necessary to overcome the potential lack of skills and obstacles to the mobility of the workforce [15]. Thus, training, retraining, and mobility between sectors are the basis for the development of green human resources, which, under the conditions of the digital economy, is extended to the similar staff shortage and requires the development of digital skills with employees.

2 Materials and Methods

Aspects that are considered in this chapter belong to topical scientific problems, which are actively researched and interpreted. Over the last 20 years, these problems were studied in detail from the position of ecology, the concept of sustainable development, green information technologies, education for sustainable development, tackling climate change, digitalization, and digital competencies, digital transformation for sustainable development, green transition, etc. Therefore, the methodology of this research focuses on the concept of the green economy and covers the spheres of digital innovations, sustainable development and ecology, climate change, education, and employment.

The concept of the green economy is considered not only from the position of reduction of emissions and sustainability of natural resources use but also from the position of its influence on population employment. The labor market and educational sphere are positioned as the key factors of ensuring the economy's need for green human resources; processes connected with climate change are positioned as the initial condition for change of profession or receipt of environmental competencies. The main results of this work are demonstrated in the form of a scheme, which characterizes the interconnection between climate change, processes in labor and educational markets, government initiatives and the change in value and behavioral priorities.

The key aspects of the problem of this research were studied in works devoted to the concept of sustainable development and reporting [6, 10] and green transition [9]; directions for the development of environmental management of personnel management [12]; problems of ecologization of the labor market [4, 13], its influence on the economy [5, 8, 15] and climate change [7]; policy of active digitalization of economy and society [1] and digital sustainability of entrepreneurship [2]; as well as the role of educational establishments in the processes connected with sustainable development and climate change [3] and the interconnection between education and environmental employment [11] and digitalization [14].

The purpose of this chapter is to find the key features and regularities that reflect the influence of green human resources on climate-responsible entrepreneurship in the markets of the digital economy from the position of integration of the effects of education and the labor market.

3 Results

Global discussion on economy development in the twenty-first century is closely connected with the value of sustainable development and the fight against climate change. It is aimed at the increase in environmental responsibility of companies within the concepts of the green economy, green growth, or green development. They are based on the understanding that under the current conditions, modernization and development of the economy are impossible with the use of previous approaches. Conservation of natural resources and a responsible attitude toward the natural environment is an important condition for the survival of mankind. Thus, the green economy is an objective factor of strategic planning, within which economic interests should conform to the needs of the natural environment [5].

The concept of green human resources in the environment of climate-responsible entrepreneurship in the digital market is an element of the general process of green transformation of economic systems. Its main goal is to ensure sustainable conditions for the decarbonization of production, increase in the level of acceptance of the values of sustainable development, etc. Green transformation is considered from the position of changes in capital infrastructure and labor market, which are required to achieve zero emissions, including due to redistribution of employees between professions and sectors [4].

Redistribution of the labor market is performed in several directions. First, it is aimed at the creation of jobs in the sphere of renewable energy and waste management. In these spheres, more jobs are created per unit of investment compared to the sphere of traditional energy based on fossil fuels [9]. Another direction of ecologization of employment is implemented through the adaptation of labor conditions, modernization of technologies, increase in the level of monitoring and tracking of emissions, correction of the value chain, etc. [11]. In both cases, personnel training is the key problem, a "bottleneck", which often restrains the green transition or partial modernization of the subjects of entrepreneurship in the direction of climate responsibility.

One of the factors in tackling climate change is digital economy development. According to this, the role of green human resources in entrepreneurship in digital markets is particularly important. Thus, when creating preconditions for the future, European institutions, focus on human and sustainable and digital development. They strive toward joining efforts around governments, business, infrastructure, and individual and group skills to achieve a high level of quality of life, comprehensive use of digital skills, and harmonious interaction between human and natural environment. In this context, The Digital Decade policy program focuses, among other goals, on the formation of the necessary digital skills with at least 80% of the population, achievement of the basic level of digital intensity with 90% of SMEs, and

100% digitalization of the main government services and medical registration of citizens. The important focus here is made on support for stable interaction between humans, technologies and ecosystems and the use of technology to fight climate change and environmental challenges [1].

The problem of the development of environmental skills with employees has many aspects. On the one hand, it is closely connected with climate change, which drives employees from the environment that transforms because of the change in the temperature or water regime or because of the natural disaster. Another aspect of the problem is connected with the striving toward "green transition", within which processes of reduction of greenhouse emissions, which is often accompanied by a decrease in employment in professions that deal damage to ecology and influence climate, take place. The third aspect of the problem is connected with a stable view of climate and environmental responsibility in business and governance. According to this, representatives of business and government encourage employees toward the conscious transition to employment that is less aggressive toward climate and ecosystems. Each aspect is based on different mechanisms of change in employment and has different motivational nature. However, a common feature for each of them is the need for the development of labor skills according to changes that have already taken place or that are about to happen. In this context, the formation of skills based on climate or environmental responsibility with employees conforms to the common trend of economic systems' development and better fits the future or desirable state of the labor market.

It is expected that competent retraining of employees based on the development of their climate-responsible skills will allow ensuring net growth of 18 million jobs in the energy sector alone. The main demands that are set to labor markets from the position of the green transition are related to the high level of mobility and flexibility, the possibility to satisfy the economy's needs for a skilled workforce, high functionality of the labor market due to the connection with complex educational programs and the ability to create inclusive systems of social protection.

On the whole, the processes of climate change and training of green personnel are closely interconnected (Fig. 1). Climate change, as an element of understanding of negative phenomena, is the initial process, which leads to one of the three events aimed at support of climate-responsible activity through the training of responsible personnel. The first event is related to climate change and changes in the conditions of doing business in particularly sensitive sectors and regions. A consequence of such actions is learning new specialities, which include environmental and digital competencies. Another event is a form of reaction of global, government, and regional institutions to climate change and adopted international agreements. It starts the processes of centralized transformation of the labor market, which is ensured by the use of programs of re-training of employees. The third event is the environmental behavior of employees, entrepreneurs, or managers. It may be based on own values or come from strategic interests. In any case, it implies support for the self-development of employees and stimulation of managers, employees, or organizations for climate-responsible activities through the acquisition of the appropriate skills.

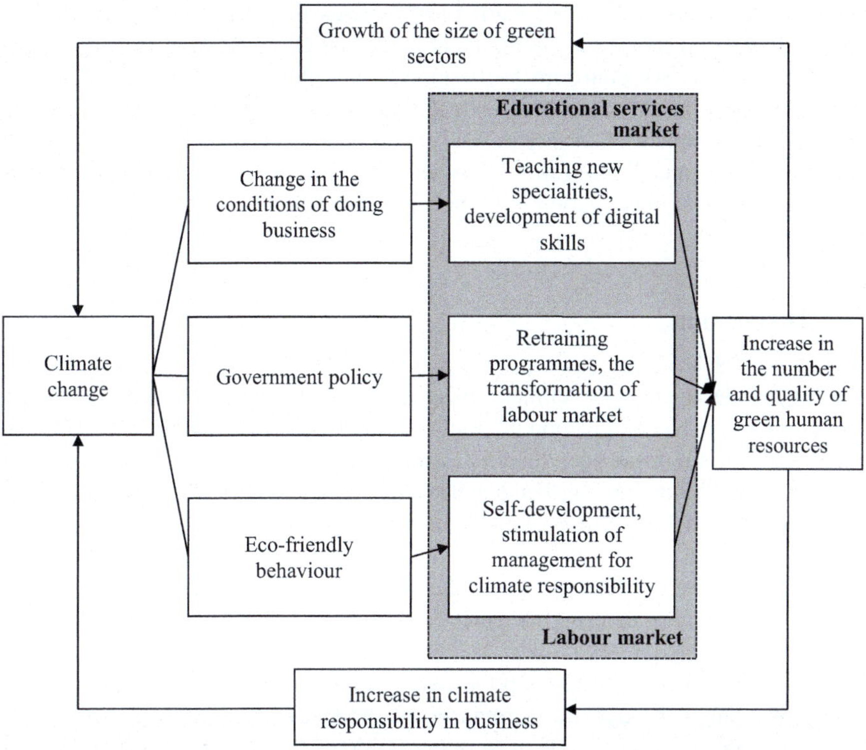

Fig. 1 Influence of the labor and educational markets on the training of green human resources and the fight against climate change. (*Source*: Developed by the authors based on Refs. [5, 8])

Training, retraining, and self-development of employees from the position of formation of green human resources are elements of one system, which includes the labor market and educational sphere, which interact with each other. The labor market sets a requirement of the set of competencies and level of qualification, and educational establishments develop the appropriate programs and conduct training. On the other hand, the educational sphere is a multiplier of qualitative changes, popularizing the values of sustainability. The result of such symbiosis is an increase in the number and quality of green human resources, which satisfy the personnel hunger of new eco-friendly sectors or raise the level of climate responsibility in traditional spheres. Such actions facilitate decarbonization and ecologization of the economy, opposing climate change and raising the level of adaptability to changes.

An important feature of training of green human resources for the digital economy is the acute shortage of skilled personnel in green and digital environments. Dynamic development of both sectors needs a rational decision on the integration of their training courses and programs. As a result, training of ICT specialists with important environmental competencies or teaching employees of eco-oriented professions digital skills will substantially raise the level of their readiness to the current challenges of the labor market and increase the corresponding capabilities of the digital and ecological environment in the context of tackling climate change.

Such an approach conforms to the modern interpretation of the role of educational establishments, according to which they have a key responsibility in the sustainable development of society [6] and are an interested party in the promotion of the UN Agenda and digitalization of society [14] and agents of changes for sustainable development in the period of global climate action [3].

4 Discussion

In this chapter, the factor of climate change is considered one of the main conditions for an increase in ecologization of employees and their involvement with the green economy. Within this aspect, there is a range of provisions that pose a large scientific interest and require further detailed consideration. The main aspects of this problem are based on a different understanding of the role of low-skilled and skilled labor. According to this view, low-skilled labor is most often an object of the largest number of created or released jobs. The sphere of the use of skilled personnel is characterized by much higher rates of creation of jobs compared to their liquidation.

Therefore, there is an assumption that skilled personnel acquire the status of green personnel through the formation of additional competencies without the change in profile or profession most often. While in the case of low-skilled personnel, the acquisition of environmental competencies comes with dismissal or change in the profile of employment. That is why this group of employees is characterized by frequent movement between sectors. These aspects require additional research and must be taken into account during the evaluation of the scales of the change in the labor market in the short term.

The problems of assessment of the scale of costs of retraining of personnel, development and implementation of the mechanisms of their training, adaptation, and self-realization under the conditions of the green economy are equally important. Here the most acute issue is the correct identification of the requirements of the labor market regarding environmental competencies and their maximum implementation in the effective methodologies of the educational process. The scientific resolution of these problems will allow creating effective mechanisms, which will provide the market of the digital economy with green personnel that possess a large potential for comprehensive implementation of the climate responsibility concept in the market of digital entrepreneurship.

5 Conclusions

Climate change is one of the key strategic factors in the economy. More and more entrepreneurial structures when planning their activity, take into account the indicators of climate-responsible behavior and implement the sustainability elements in their work. Despite the topicality of such an approach, efforts on tackling climate change still bring limited effect. The reason for this lies in personnel hunger, which is inherent to the green economy.

On the whole, climate change influences the labor market in different ways. Direct influence is seen in the change of climate conditions and dismissal of many employees because of the change in the conditions of doing business: drought, floods, natural disasters, etc. The indirect influence of climate change is seen in global or government institutions or the popularization of eco-friendly behavior. As a result of each of the above actions, the most optimal – from the position of the labor market and strategy – is the formation or increase in employees' environmental competencies with their further involvement in the sphere of the green economy. From the position of digital markets, it is expedient to combine these processes with the development of digital competencies. The basis for the achievement of these goals is the integration of the mechanisms of the educational and labor markets, which will lead to the formation of the need for green personnel and the mechanisms of its satisfaction through the expansion and improvement of educational programs and methodologies.

References

1. EC (2022) Europe's Digital Decade: digital targets for 2030. European Comission. https://commission.europa.eu/strategy-and-policy/priorities-2019-2024/europe-fit-digital-age/europes-digital-decade-digital-targets-2030_en. Accessed 30 Jan 2023
2. George G, Merrill RK, Schillebeeckx SJD (2021) Digital sustainability and entrepreneurship: how digital innovations are helping tackle climate change and sustainable development. Entrep Theory Pract 45(5):999–1027. https://doi.org/10.1177/1042258719899425
3. Giesenbauer B, Mueller-Christ G (2020) University 4.0: promoting the transformation of higher education institutions toward sustainable development. Sustainability 12:3371. https://doi.org/10.3390/su12083371
4. IMF (2022) A greener labor market: employment, policies, and economic transformation. Chapter 3. In: World economic outlook: war sets back the global recovery. International Monetary Fund. Research Department. https://www.elibrary.imf.org/display/book/9781616359423/CH003.xml?tabs=abstract?cid=ca-com-compd-elib_rotator. Accessed 31 Jan 2023
5. Jacob K, Quitzow R, Bär H (2015) Green jobs: impacts of a green economy on employment. Environmental Policy Research Centre. Freie Universität Berlin. https://energypedia.info/images/f/fc/Green_Jobs_-_Impacts_of_a_Green_Economy_on_Employment.pdf. Accessed 29 Jan 2023
6. Kräusche K, Pilz S (2017) Integrated sustainability reporting at HNE Eberswalde – a practice report. Int J Sustain High Educ 19(2):291–312. https://doi.org/10.1108/IJSHE-07-2016-0145
7. Martinez-Fernandez C, Hinojosa C, Miranda G (2010) Greening jobs and skills. Labour market implications of addressing climate change. OECD Local Economic and Employment

Development (LEED) Working Paper Series. https://www.oecd.org/cfe/leed/45484420.pdf. Accessed 29 Jan 2023
8. OECD (2017) Employment implications of green growth: linking jobs, growth, and green policies. OECD Report for the G7 Environment Ministers. https://www.oecd.org/environment/Employment-Implications-of-Green-Growth-OECD-Report-G7-Environment-Ministers.pdf. Accessed 29 Jan 2023
9. OECD (2020) Making the green recovery work for jobs, income and growth. OECD Policy Responses to Coronavirus (COVID-19). https://www.oecd.org/coronavirus/policy-responses/making-the-green-recovery-work-for-jobs-income-and-growth-a505f3e7/#snotes-d4e99. Accessed 30 Jan 2023
10. OECD (2022) Environment at a glance indicators. OECD Publishing, Paris. https://doi.org/10.1787/ac4b8b89-en
11. Raine C (2022) Sustainable learning: the green jobs market landscape. Encoura. All Topics, Program Spotlight. https://encoura.org/sustainable-learning-the-green-jobs-market-landscape/. Accessed 31 Jan 2023
12. Sharma C, Sakhuja S, Nijjer S (2022) Recent trends of green human resource management: text mining and network analysis. Environ Sci Pollut Res 29:84916–84935. https://doi.org/10.1007/s11356-022-21471-9
13. The Adecco Group (2021) Skills for the green economy. https://www.adeccogroup.com/future-of-work/latest-research/skills-for-the-green-economy/www-prd-amz930-com.azureedge.net/-/media/project/adeccogroup/pdf-files/tag-skills-for-the-green-economy-paper.pdf/. Accessed 29 Jan 2023
14. Trevisan LV, Eustachio JHPP, Dias BG, Filho WL, Pedrozo EA (2023) Digital transformation towards sustainability in higher education: state-of-the-art and future research insights. Environ Dev Sustain:1–22. https://doi.org/10.1007/s10668-022-02874-7
15. World Bank (2012) Human capital: implications of green growth policies for labor markets and job creation. In: Inclusive green growth. The pathway to sustainable development. World Bank, Washington, DC, pp 91–104. https://documents1.worldbank.org/curated/en/368361468313515918/pdf/691250PUB0Publ067902B09780821395516.pdf. Accessed 1 Feb 2023

The Modern Experience in Achieving Cultural Inclusiveness for the Development of Carbon Landfills on the Basis of Universities in the Eurasian Economic Union

Margarita A. Meretukova, Nadezda V. Gamulinskaya, Marina L. Nechaeva, and Irina B. Khakonova

1 Introduction

The problem of climate change is very topical. It threatens the existence of sustainable ecosystems, which are suitable for living, recreation, and economic use. An increase in the rates of global warming leads to more frequent natural disasters, with the growth of their destructive force. An increase in risks and threats to mankind, especially for future generations, predetermined the joining of the civilization's efforts around the idea of tackling climate change, which involves primarily the reduction of greenhouse emissions and their removal from the atmosphere. According to the agreements achieved, each country has a task of reducing greenhouse emissions and promoting the idea and values of eco-friendly behavior.

Countries of the Eurasian Economic Union (Armenia, Belarus, Kazakhstan, Kyrgyzstan, and the Russian Federation) differ in their contributions to the global volume of greenhouse emissions and by the extent of their obligations regarding decarbonization of the economy. Russia's role is the largest, with CO_2 accounting for more than 85% of all greenhouse emissions produced by the Russian economy, primarily in the energy sector.

Despite the identified priorities and increasing threats related to climate change, the rates of reduction of emissions at the global scale are still insufficient. Consequently, certain countries are developing the mechanisms to financially incentivize carbon neutrality. The tools of such stimulation include the possibility of the introduction of a carbon tax on products aimed at international trade. Given the

M. A. Meretukova (✉) · I. B. Khakonova
The Adyghe State University, Maykop, Russia

N. V. Gamulinskaya · M. L. Nechaeva
Vyatka State University, Kirov, Russia
e-mail: nv_gamulinsky@vyatsu.ru

E. G. Popkova (ed.), *Smart Green Innovations in Industry 4.0*, Springer Climate,
https://doi.org/10.1007/978-3-031-45830-9_39

resource character of the economies of EAEU countries, they are taking active measures to avoid this.

One of the most active tools for reducing greenhouse emissions is the creation of carbon landfills. A specific feature here is the high level of their inclusiveness, which is based on the ability of universities (main operators of landfills) to create an environment of equality, accessibility, and cooperation among various social groups.

2 Materials and Methods

The methodology employed in this chapter is realized by combining the sustainable development concept and the provisions of functioning of global institutes of tackling climate change, which are set onto the processes of organizational interaction, sociology, and behavioral economics. The issues of the technological character of the functioning of carbon landfills and governance in the EAEU with an emphasis on the regulatory and institutional support for the processes related to tackling climate change are studied. Most of the research tools in this chapter are applied at the empirical level, they include the methods of analysis synthesis, network analysis, and graphical method.

This chapter explores various aspects of the problem of formation and use of the policy of cultural inclusiveness in the context of the development of carbon landfills on the basis of universities in countries of the EAEU. The outlined aspects cover the identification of the essence and role of inclusiveness in modern social [10], the issues of the policy of inclusiveness in the context of civil society formation [1], coordination of the processes of social interaction and tackling climate change [8], and transformation of views of inclusiveness [12]. From the position of ensuring the institutional ability to fight climate change, we study the conditions of the functioning of the system of the fight against climate change [6, 7, 13–16], theoretical and practical aspects of development and functioning of carbon [9], experience and specifics of the work of carbon landfills [2–5], and prospects for their development [11].

The primary objective of this study is to describe the experience and potential of implementing the policy of cultural inclusiveness within specific tools for tackling climate change – in this case, carbon landfills based at universities in EAEU countries – and to emphasize the key role of educational institutions in these processes.

3 Results

In the modern world, new ethical norms become more popular; they are expressed through the notions of equality, tolerance, and diversity. This predetermines the topicality of processes that are connected with the provision of equal rights of access to opportunities and the maximum involvement of citizens or organizations in important social processes. This approach is implemented most fully through the tools of inclusiveness. Inclusiveness is a form of inclusion of someone or something in the social process. The word "inclusion" is derived from the Latin "Includo," which means "to include" [10].

Inclusiveness was traditionally treated as the integration of handicapped people, people with special needs or characteristics from the position of religion, sex or other physiological or psychological features in social groups and processes. However, the generation of millennials, which soon will be the majority in the environment of economically active people, perceives this notion differently. They value inclusiveness not as an abstract ideal, which makes everyone feel good, but as important tools that ensure competitiveness and growth of the business. This generation cares more about cognitive diversity or diversity in thoughts, ideas, and philosophies, which allows solving business problems through a culture of cooperation. For the millennials, inclusiveness is unifying people with different values and features, the formation of teams in which everyone has a right to speak, and the use of different points of view to make the influence on business stronger [12].

In this treatment, from the position of the problem of climate change, inclusiveness may be viewed as a culture that ensures cooperation, accessibility, and equality. According to this, the values of tackling climate change or adaptation to them should be received well by different social and economic groups of the society, and the tools and mechanisms for solving climate problems should be open and available for a wide circle of participants.

An example of the culture of inclusiveness within the studied context is cooperation between the bodies of civil society and government, which includes different forms of communication, exchange of thoughts, criticism, and involvement in processes. Therefore, work meetings with representatives of public authorities and participation in discussions of decision projects are an important tool of the work, which allows ensuring control over public authorities and expanding the vision and potential, as well as increasing the society's reaction to certain targeted actions of the public authorities [1]. This approach conforms to the European Climate Pact [6].

The cornerstone of actions aimed at combating climate change is the so-called carbon policy, which is implemented at different levels and in different countries worldwide. Its key provisions derive from SDG 13 [14] and are implemented through the UN Framework Convention on Climate Change [13] and the Kyoto Protocol [15], which sets international mandatory and differentiated goals for the reduction of emissions for six greenhouse emissions, and through the Paris Agreement [16], which elevates global responsibility for climate threats and establishes nationally determined contributions (NDC) for each country.

The achievement of the goals of reduction of greenhouse emissions varies by country. As of early 2021, EAEU countries demonstrated positive dynamics in this regard, successfully reducing CO_2-equivalent emissions compared to 1990 (Fig. 1).

One of the key factors in the above dynamics is the reduction of industrial production compared to 1990. However, stabilization of the economy and restored growth demand additional measures aimed not only at the reduction of emissions but also at their depositing.

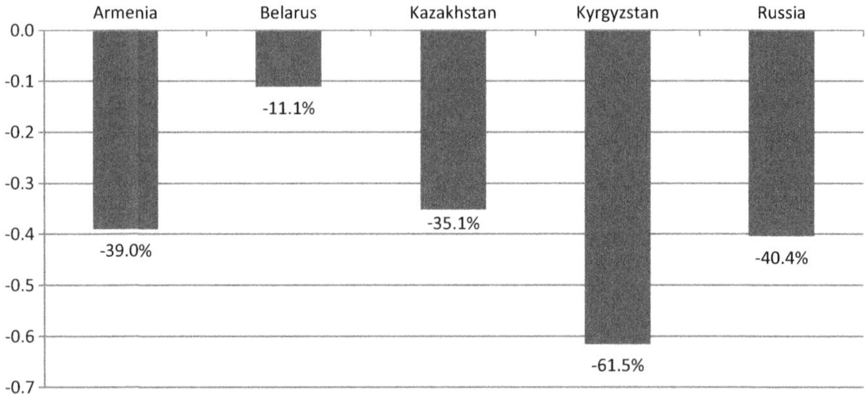

Fig. 1 Reduction of greenhouse emissions in countries of the EAEU in 2020 compared to 1990, %, CO_2-equivalent. (*Source*: Created by the authors based on Ref. [7])

Tools that allow removing excessive CO_2 from the atmosphere include carbon landfills. The specific feature of these landfills is that they allow using biological and biotechnological properties of territories and ecosystems for sequestering greenhouse emissions. They involve the development and adaptation of technologies on field, forest, and other measuring of emission and absorption of greenhouse emissions by available and modified ecosystems, study the possibility to use remote sensing technologies, and deal with the development and adaptation of mathematical models on emission and absorption of greenhouse emissions in different conditions [11].

The functioning of carbon landfills involves the use of soil's ability to secure carbon, which is much higher than the similar ability of the air. Thus, studying the technologies of effective use of soils allows extrapolation of the mechanisms of carbon capture from the atmosphere by plants and improves capabilities in the reduction of the greenhouse effect [9].

Among EAEU countries, carbon landfills are fully operational only in the Russian Federation and Kazakhstan. A project of carbon landfill has been developed in Belarus, while such projects are still at the stage of discussion in Armenia and Kyrgyzstan. As of early 2023, there are 15 carbon landfills in Russia, with a total area of 39,000 km^2 [3]. The area of the first carbon landfill in Kazakhstan is 2 hectares [2].

A distinctive feature of carbon landfills is the high level of their inclusiveness. Thus, for example, the carbon landfill "FOR&ST CARBON" (Voronezh Region, Russia) has an area of 181.3 hectares and covers forests and arable land. The main operator of the landfill is Voronezh State University of Forestry and Technologies named after G.F. Morozov. Apart from the operator, industrial partner SIBUR Holding PJSC, technological partner V.V. Dokuchaev Voronezh Federal Agrarian Scientific Centre are involved in the activity of the landfill, as well as the participants: St. Petersburg Research Institute of Forestry, All-Russian Research Institute for Silviculture and Mechanization of Forestry and the Institute of Problems of Ecology and Evolution named after O.M. Severtsov of the Russian Academy of Sciences [4].

Within this project, several functions, which form its inclusiveness, are performed: research work, educational activity, and international integration of the carbon landfill. Apart from its direct tasks, research activity influences the managerial decisions of regional and central authorities. The obtained scientific results also serve as a conceptual basis for certain sectors of economic activity, including forestry, agriculture, waste management, etc.

The role of universities in the achievement of cultural inclusion is very high. This is connected primarily with the implementation of educational programs, within which knowledge and values of climate sustainability are disseminated in society. Involving postgraduates in this process improves the prospects of further inclusion through in-depth research of scientific problems of climate change. A large potential for inclusiveness belongs to schools and workshops, implemented by universities, which involve a wide circle of students, scholars, public figures, and representatives of interested parties in the process of decarbonization. The direction of international integration allows for the achievement of synergy through active integration with other scientific establishments of the world.

The rest of the carbon landfills function by a similar principle and implement, to a certain extent, cultural inclusiveness through involving more people from different spheres in the processes of decarbonization. Due to this, the values of environmental behavior and the mechanism of its implementation become more popular and are successfully implemented.

4 Discussion

Despite its large dissemination in social processes, cognitive diversity is still implemented in a limited way in the systems of corporate management. In most cases, its use in a company's activities is declared only formally. However, the hierarchy and bureaucracy of management, which are traditional elements at most companies, hinder the flexibility and combination of ideas and approaches. The existing studies [12] show that millennials, as opposed to the older generation, are rather liberal to alternative views and approaches. However, there is a lack of studies that would provide a comprehensive assessment of the perception of cultural inclusiveness by millennials compared to previous generations of economic agents. As for scientific works on this problem in the context of the development of carbon landfills on the basis of universities of EAEU countries, they are still at the initial stage.

Thus, the implementation of the policy of inclusiveness at carbon landfills is an underexplored topic. In this context, potential opportunities for scientific cognition are present in many spheres. Attention should be paid to the issue of achievement of maximum inclusiveness, which would allow involving public organizations, representatives of local communities, activists, mass media, etc., in the process of carbon landfills development. Apart from this, the use of the diversity of the process's participants in the projects' interests also requires attention, given the possibilities of social diffusion and communication networks.

These important scientific tasks could be solved by developing models that would coordinate the results of the activity of carbon landfills with the level of their implementation of the cultural inclusiveness policy. In this case, it is expedient to go beyond the limits of carbon landfills and extend the research results to other spheres of social life, including social behavior, purchasing power and other activity, which influences – directly or indirectly – ecology and climate processes.

5 Conclusions

The issues of climate change form an important direction of scientific research due to the large number of threats created by them. These issues are global, but have different influences depending on the country and region of the world. The climate policy, which has been developed at the global, international, and national levels, requires the involvement of the largest possible number of participants for the achievement of a positive effect in tackling climate change. The basis for such involvement is the policy of cultural inclusiveness, which creates preconditions for equal and accessible cooperation of all representatives of society in projects aimed at tackling climate change.

The realization of the cultural inclusiveness policy could take place through different social institutes, including public authorities and public organizations. However, the largest potential for the success of inclusiveness measures rests with universities. Ensuring scientific, research, educational, and international activity, universities can involve a wide range of individuals and organizations of different profiles and spheres in the resolution of climate problems. An example of the implementation of such projects is carbon landfills. Their activities are aimed at the research for the most effective means for the absorption of carbon by natural or modified ecosystems.

References

1. Andrusevich A, Andrusevich N, Kozak Z, Romanko S (2020) Klimaticheskaia politika i grazhdanskoe obshchestvo: Budushchee stran Vostochnogo partnerstva v kontekste Evropeiskogo zelenogo kursa [Climate policy and civil society: the future of the eastern partnership countries in the context of the European green deal]. Analiticheskii dokument. Predstavitelstvo Fonda Konrada Adenauera v Ukraine [Analytical document. Representation of the Konrad Adenauer Foundation in Ukraine], 60 p
2. CP RF (2022) V Kazakhstane otkrylsia pervyi v Tsentralnoi Azii karbonovyi poligon [The first carbon landfill in Central Asia opened in Kazakhstan]. Karbonovye poligony Rossiiskoi

Federatsii [Carbon polygons of the Russian Federation]. https://carbon-polygons.ru/news/v--kazaxstane-otkryilsya-pervyij-v-czentralnoj-azii-karbonovyij-poligon. Accessed 2 Feb 2023
3. CP RF (2023a) Karbonovye poligony Rossiiskoi Federatsii [Carbon polygons of the Russian Federation]. https://carbon-polygons.ru/. Accessed 3 Feb 2023
4. CP RF (2023b) Karbonovyi poligon «FOR&ST CARBON» [Carbon polygon "FOR&ST CARBON"]. https://carbon-polygons.ru/polygons/forest-carbon. Accessed 2 Feb 2023
5. CSR (2022) Karbonovye poligony Rossii: nastoiashchee i budushchee [Carbon polygons in Russia: present and future]. Tsentr strategicheskikh razrabotok [Center for Strategic Research]. https://in.minenergo.gov.ru/upload/iblock/b57/b57b83b358ea6bcf9495d219a3677423.pdf. Accessed 1 Feb 2023
6. EC (2023) European Climate Pact. European Commission. https://climate.ec.europa.eu/eu-action/european-green-deal/european-climate-pact_en. Accessed 1 Feb 2023
7. EEC (2022) Okruzhaiushchaia sreda. Statistika Evraziiskogo ekonomicheskogo soiuza. Evraziiskaia ekonomicheskaia komissiia [Environment. Statistics of the Eurasian Economic Union; Eurasian Economic Commission], Moscow, 58 p
8. FBS (2020) Borba s izmeneniem klimata i sotsialnaia spravedlivost [Climate action and social justice]. Sbornik argumentov dlia razvitiia spravedlivogo i ekologichnogo obshchestva [Collection of arguments for the development of a just and environmentally friendly society]. Friedrich Ebert Stiftung. https://library.fes.de/pdf-files/bueros/kiew/17374.pdf. Accessed 1 Feb 2023
9. Kerimov IA, Makhmudova LS, Elzhaev AS (2022) Karbonovyi poligon Chechenskoi Respubliki: nachalo issledovanii [Carbon polygon of the Chechen Republic: the beginning of research]. In: Kerimova IA, Shirokova VA (eds) Sovremennye problemy geologii, geofiziki i geoekologii Severnogo Kavkaza [Modern problems of geology, geophysics and geoecology of the North Caucasus], vol XII. IIET RAN, Moscow, pp 651–658
10. Osipian A (2022) Inkliuzivnost – prostymi slovami: v biznese, obrazovanii i obshchestve [Inclusion – in simple words: in business, education and society]. RBK. https://trends.rbc.ru/trends/social/6260fb9d9a794768ff6beb0f. Accessed 2 Feb 2023
11. Saveleva RI, Gazizova OV (2021) Karbonovye poligony – nizkouglerodnoe budushchee ekonomiki Rossii [Carbon landfills are the low-carbon future of the Russian economy]. In: Ekonomicheskoe razvitie v XXI veke: tendentsii, vyzovy i perspektivy: sbornik nauchnykh trudov IX Mezhdunarodnoi nauchno-prakticheskoi konferentsii «Gorizonty Rossii» [Economic development in the 21st century: trends, challenges and prospects: a collection of scientific papers of the IX International Scientific and Practical Conference "Horizons of Russia"]. April 23, 2021. Plekhanov Russian University of Economics, Moscow, pp 175–179
12. Smith C, Turner S (2015) The radical transformation of diversity and inclusion. The millennial influence for inclusion. Deloitte University Leadership Center for Inclusion. https://launchbox365.com/wp-content/uploads/2017/03/us-inclus-millennial-influence-120215.pdf. Accessed 1 Feb 2023
13. UN (1992) United Nations Framework Convention on Climate Change. United Nations. https://unfccc.int/sites/default/files/convention_text_with_annexes_english_for_posting.pdf. Accessed 3 Feb 2023
14. UN (2023) Goals. 13. Take urgent action to combat climate change and its impacts. United Nations. Department of Economic and Social Affairs. Sustainable Development. https://sdgs.un.org/goals/goal13. Accessed 3 Feb 2023
15. UNFCCC (1997) Kyoto Protocol to the United Nations Framework Convention on Climate Change. United Nations Framework Convention on Climate Change. https://unfccc.int/sites/default/files/resource/docs/cop3/l07a01.pdf. Accessed 3 Feb 2023
16. UNFCCC (2015) The Paris Agreement. United Nations. https://unfccc.int/sites/default/files/english_paris_agreement.pdf. Accessed 2 Feb 2023

Regularities and Features of Overcoming Cultural Differences and Ensuring the Inclusiveness of Higher Education for the Development of Climate-Smart Green Innovations in Industry 4.0 in the Eurasian Economic Union

Svetlana A. Tikhonovskova, Anastasia A. Sozinova, Elena V. Karanina, and Marina G. Shadzhe

1 Introduction

The development of climate-smart green innovations in Industry 4.0 is connected with the financial capabilities of companies, government investments in this sphere and the partnership of higher education and subjects of industry in their development, approbation, and implementation. Inclusive conditions and the focus on overcoming cultural differences in universities are important preconditions for further participation of the latter in R&D and design, including in the sphere of the creation of innovative climate-oriented products (services). Inclusiveness implies the access of students, researchers, and academic staff to participation in scientific and practical and educational activities regardless of possible limitations of social contacts, caused by pandemics and other events, and problems with mobility. As for cultural differences, students, lecturers, and researchers may be foreigners and have differences in culture, language, behavior, and communication. The population of the member countries of the EAEU, which includes Kazakhstan, Russia, Kyrgyzstan, Armenia, and Belarus, is acquainted with the culture and language specifics of each other. Thus, students and academic staff of each of the above countries can effectively adapt to the work in universities on the territory of the EAEU. Participation of universities of the above countries in the creation of various climate-smart green

S. A. Tikhonovskova
Platov South-Russian State Polytechnic University (NPI), Novocherkassk, Russia

A. A. Sozinova (✉) · E. V. Karanina
Vyatka State University, Kirov, Russia
e-mail: aa_sozinova@vyatsu.ru; karanina@vyatsu.ru

M. G. Shadzhe
The Adyghe State University, Maykop, Russia

E. G. Popkova (ed.), *Smart Green Innovations in Industry 4.0*, Springer Climate,
https://doi.org/10.1007/978-3-031-45830-9_40

technologies depends on the level of adaptability of these organizations and their personnel in the sphere of inclusiveness and overcoming cultural differences.

This research aims at determining the regularities and specific features of overcoming cultural differences and supporting the inclusiveness of the systems of higher education for the creation of climate-smart green innovations in Industry 4.0 in EAEU countries. The tasks set for reaching this aim include the determination of the achievements in the sphere of climate protection as a result of the implementation of climate-smart green innovations in Industry 4.0 and describing regularities and features of dealing with cultural differences and achieving inclusiveness in the universities of the EAEU countries for the development of the considered innovations.

2 Materials and Methods

To solve the tasks set in this research, we analyzed materials on the formation of the higher education systems, including the works devoted to supporting its inclusiveness and eliminating cultural differences, namely [6, 8, 9, 18, 19]. Tomyuk [18] is devoted to the analysis of the stages of development of modern universities in view of information, technological, socio-economic, and organizational transformations. The author demonstrated the role of universities' participation in these transformations, as the basis for the growth of their competitiveness in the market of scientific and educational services. Khasanov and Anufrieva [6] dwell on the participation of green universities in the achievement of environmental SDGs and the creation and implementation of climate-oriented initiatives at the level of the government, region, and territories. Nosonov [8] elaborates on the specifics of the innovative development of Russian universities and determines the problems and perspectives in this sphere [9] describes the features of green construction in Armenia with the use of innovative materials and technologies, which stimulate climate protection and dwell on the directions of the national policy in energy efficiency. The author of [19] identifies the characteristics and features of the development of modern science in Armenia and Turkey and distinguishes the problems of formation and commercialization of R&D.

As for the methods applied in this research, the method of classification is used to identify data characterizing the level of the selected indicators of the higher education system and indicators of ecologization, which were reached due to the development of climate-smart green innovations in Industry 4.0 in countries of the EAEU; the comparative analysis allows comparing the indicators of the systems of higher education, ecologization, and development of innovations; the systems approach enables us to find the main regularities and features of the designated characteristics of the selected countries' higher education.

3 Results

By 2023, five countries of the EAEU achieved certain results in the unification of standards of the higher education system, though, at the current stage, the activity of national universities has its specifics and different effectiveness (Table 1).

The Russian system of higher education has the best indicators of efficiency in the EAEU (Table 1). There are scientific and educational consortia at the level of Russia and each country of the EAEU, e.g., Russian-Kyrgyz Consortium of Technical Universities (11 Kyrgyz universities and 40 Russian universities) [17]. There exist also bilateral scientific & technical and educational partnerships at the level of the universities of Belarus, Armenia, Kazakhstan, and Kyrgyzstan. There are plans to create a Belarus-Armenia scientific and educational center [2]. On the whole, the main directions of partnership in the EAEU are connected with the central role of the Russian Federation and the creation of scientific and educational centers in each EAEU country. Russian universities have their branches in Belarus (three universities) [12]; Armenia (three universities); Kyrgyzstan (three universities) [7]; Kazakhstan (six universities) [11].

We should also mention the Eurasian Network University, which includes universities from five countries [3]. The concept of the Eurasian Network University involves training of personnel in the scientific and educational sphere with the use of the network form, which includes remote programs, which facilitate inclusiveness and are more universal. This will help deal with cultural differences. The goals of this project are the formation of personnel reserves in the labor market of the EAEU countries, an increase in the level of R&D for raising competitiveness and export of digital technologies (including climate-smart innovative products), provision of higher education services in the EAEU, constant growth of knowledge and skills and personal growth of students [14].

The features and regularities of the interaction between Russian universities and universities of the other four countries of the EAEU are as follows:

Table 1 Level of the development of higher education systems of the EAEU members

		Values				
	Indicators	Armenia	Belarus	Kazakhstan	Kyrgyz Republic	Russian Federation
1	The Times Higher Education Universities Ranking: Average score of top 3 universities, worst 0–100 best (2022)	0	16.45	16.45	0	52
2	Expenditure on research and development, % of GDP (2018)	0.19	0.6	0.12	0.1	0.98
3	University-industry R&D collaboration, score (2022)	35.7	No data	29.6	24.4	46.5

Source: Formed by the authors based on Refs. [16, 20]

- Traditional use of the Russian language in these countries, which stimulates the integration of educational services and partnership in R&D.
- The long presence of Russian educational services in the markets of higher education of four countries of the EAEU.

Let us consider the state of climate characteristics, connected with the achievement of the UN SDGs (Table 2). Results in this sphere are linked to the implementation of climate-smart green innovations in Industry 4.0, which is conducted with the help of partnerships at the level of universities and industry.

We see certain positive values on municipal solid waste in Armenia, Kyrgyzstan, and Kazakhstan. As for CO_2 emissions, the lowest level is observed with Armenia and Kyrgyzstan, which was achieved due to the use of green innovations.

As for the indicator "mean area that is protected...," the most threatening situation is with Kazakhstan [15, 16]. In recent years, Kazakhstan started reforms for the biodiversity protection (creation of a special body responsible for programs in this sphere) and achieved an increase in investments for this purpose (420 million tenge in 2017, 3.6 billion tenge in 2018 and 4.1 billion tenge in 2019) [5]. We should note the contribution of climate-responsible entrepreneurship to the protection of land biodiversity in Kazakhstan. There are projects of management of hunting grounds with the help of open digital service platforms, which allowed this sphere to leave the shadow economy and opened prospects for regulated use of biodiversity [5]. The transition to e-commerce in this sphere involves accounting for the use of hunting grounds' resources, which is an element of climate-smart green innovations. Since the unions of entrepreneurs in this sector are aimed at the growth of international tourist flows, there are also projects of service promotion with the use of digital apps and online advertising.

Implementation of climate-smart green innovations in Industry 4.0 in countries of the EAEU takes place based on partnerships between universities and

Table 2 Level of climate characteristics of implementing the UN SDGs in countries of the EAEU

		Values				
	Indicators	Armenia	Belarus	Kazakhstan	Kyrgyz Republic	Russian Federation
1	CO_2 emissions from fossil fuel combustion and cement production, tCO_2/capita (2020). SDG 13. Long-term goal – 0.	1.99	6.08	15.52	1.76	10.81
2	Municipal solid waste, kg/capita/day. SDG 12. Long-term goal – 0.1.	0.46 (2014)	1.24 (2015)	0.76 (2012)	0.51 (2015)	1.14 (2012)
3	Mean area that is protected in terrestrial sites important to biodiversity, %. SDG 15. Long-term goal – 100.	22.57	47.09	13.05	23.6	25.12

Source: Compiled by the authors based on Ref. [16]

entrepreneurship. As shown in Table 1, the best results in this sphere are observed in the Russian Federation and Armenia.

The Association of Green Universities of Russia was created in 2017 with the following purposes: development of innovative green practices based on the tools of Industry 4.0; positioning of climate-smart management of organizations with the use of green innovations in the sphere of energy efficiency for the reduction of the negative impact on the environment due to the technologies of circular use of waste [4, 6]. According to reference [4], climate-smart green innovations in Industry 4.0, which are created by the Association of Green Universities of Russia include digital programs for control of energy and water consumption; technologies of waste processing without environmental pollution and others.

The main problem with the generation of knowledge and solutions in R&D by Russian universities is the organizational, technological, and material issues of commercialization of innovative technologies. Despite the effective implementation of inclusiveness and overcoming of cultural differences, there are still problems with commercialization. Since the process of commercialization of innovative technologies (including climate-smart green innovations) takes a lot of time and involves bureaucratic problems, certain innovations may lose their importance and may not be used at all [8]. Another problem in the sphere of partnership at the university-business level is the technological and technical base of universities, which, because of the insufficient level of upgrades and investments, does not allow for the desired generation of new knowledge in innovative green technologies [8].

The positive feature of the implementation of inclusive participation of universities in the creation of climate-smart green innovations is their cooperation with large businesses in projects aimed at the achievement of certain economic and environmental indicators. An example of such an effective partnership is the cooperation between Rusagro (one of Russia's largest manufacturers of agrarian products) and Michurinsk State Agrarian University in modeling of crop rotation for various categories of fields considering natural changes and changes in seeds (involving researchers and students on the constant basis in the remote digital regime and using AI); remote dispatch of organization of fieldwork (including control over spending water and the use of fertilizers and pesticides) [1, 13]. This partnership is very effective. It is based on the network interaction, with partners' participation in the remote regime and the goal of commercialization of R&D for the needs of the commercial activity of Rusagro.

In Armenia, we see the organization of partnerships at the level of the entrepreneurial sector and technological universities in the sphere of creation of climate-smart green innovations in Industry 4.0 in the following directions: construction (design, control over construction and exploitation of "passive" buildings (built with the use of special eco-friendly materials, including thermal bridges and thermal insulation)) [9, 10]. Armenian universities also have the practice of development and realization of climate-smart green innovations aimed at the achievement of energy efficiency. Like Russian universities, Armenian universities also face the problem of commercialization of R&D, which is partly due to the absence of a strategy for the commercialization of R&D [19].

4 Discussion

We distinguished patterns in overcoming cultural differences and supporting the inclusiveness of universities for the development of climate-smart green innovations in Industry 4.0 in the countries of the EAEU, which are connected with historical and territorial aspects of the interaction of the scientific and educational sector and countries on the whole. This sphere has its advantages and problems.

Economic, material and technical & technological problems are peculiar to the commercialization of R&D in Russia and Armenia and the other three countries of the EAEU. This is proved by the indicator "University-industry R&D collaboration" (Table 1). Because of the absence of comprehensive strategies for the commercialization of R&D in these countries, many prominent ideas and solutions, which could have raised the productivity of business activity in the main sectors of the economy and would have facilitated the achievement of the key SDGs in the sphere of climate protection, are not fully developed.

An important positive feature in ensuring the inclusiveness of universities' participation in the creation of innovations is their collaboration with entrepreneurship in local and regional projects (management of automatized systems in the remote form, monitoring of the management of processes and creation and use of digital applications).

5 Conclusion

To sum up, we discovered the features of overcoming cultural differences and ensuring inclusiveness of higher education for the development of climate-smart green innovations in Industry 4.0 in the EAEU. We dwelt on the prospects for joining the activities of universities of five EAEU countries for the achievement of the UN SDGs. Despite low barriers in the sphere of culture and implementation of inclusiveness in higher education, the universities of Kyrgyzstan and Armenia have not achieved high results.

At present, partnership at the university-business level in each country of the EAEU is very important. However, more effort should be invested in ecologization within the interaction of all countries of the EAEU. Since the material, technical, technological, and economic bases of universities in Kyrgyzstan, Armenia, Kazakhstan, and Belarus are lower compared to the Russian Federation, universities of these four countries cannot compete with the Russian universities Federation in R&D. Further cooperation of universities of the EAEU countries could be based on measures from national governments in legislative support for commercialization of R&D and drawing external investments in the development of the scientific and technological base.

References

1. Analytical Center for the Government of the Russian Federation (2020) Voluntary national overview of the course of implementation of the 2030 Agenda for Sustainable Development. https://sustainabledevelopment.un.org/content/documents/26420VNR_2020_Russia_Report_Russian.pdf. Accessed 31 Jan 2023
2. Bsu.by (2022) Creation of the Belarus-Armenia scientific and educational centre was discussed in BSU. https://bsu.by/news/sozdanie-belorussko-armyanskogo-nauchno-obrazovatelnogo-tsentra-obsudili-v-bgu-d/. Accessed 31 Jan 2023
3. EEC.EAEUNIO (2022) The Eurasian Network University accepted four universities from Kazakhstan and Russia. https://eec.eaeunion.org/news/v-evraziyskiy-setevoy-universitet-vklyucheno-eshche-chetyre-vuza-iz-kazakhstana-i-rossii. Accessed 31 Jan 2023
4. Greenuniversity (2022) The Association of Green Universities of Russia. https://зеленыевузы.рф. Accessed 31 Jan 2023
5. Kansonar (2021) Kazakhstan ranked 113th out of 166 countries in the ranking of countries by preservation of land ecosystems in 2020. https://almaty-kansonar.kz/v-rejtinge-stran-po-sohraneniju-jekosistem-sushi-kazahstan-v-2020-godu-zanjal-lish-113-e-mesto-iz-166/. Accessed 31 Jan 2023
6. Khasanov V, Anufrieva EI (2019) Green universities in the context of safety. Culture and ecology as the basis for sustainable development of Russia. In: Green bridge through generations. Part I: Materials of the international forum (Ekaterinburg, 12–15 April 2019). UrFU, Ekaterinburg, pp 323–328. https://elar.urfu.ru/handle/10995/80151. Accessed 31 Jan 2023
7. Minobrnauki (2022) Russia-Kyrgyzstan cooperation: inter-university cooperation, Kyrgyz-Russian Slavic University, research of high mountain Tian Shan. https://minobrnauki.gov.ru/press-center/news/?ELEMENT_ID=46952. Accessed 31 Jan 2023
8. Nosonov AM (2020) Strategy for realisation of the innovative potential of Russian universities. Modern Problems of Science and Education, No. 5. https://science-education.ru/ru/article/view?id=30234. Accessed 31 Jan 2023
9. Papoyan AS, Changhong Z (2021) The research on green building development in Armenia. Young Res Don 3(30):11–15. https://cyberleninka.ru/article/n/the-research-on-green-building-developmen-in-armenia. Accessed 31 Jan 2023
10. RA Government (2014) Armenia development strategy for 2014–2025. https://policy.asiapacificenergy.org/sites/default/files/Development%20Strategy%20of%20the%20Republic%20of%20Armenia%20for%202014-2025_ENG.pdf. Accessed 31 Jan 2023
11. Rfembassy.kz (2022) Branches of Russian universities. http://www.rfembassy.kz/tm/russian_mission_in_kazakhstan/filialy_rossiiskih_vuzov/. Accessed 31 Jan 2023
12. RSR (2021) Russian education in the Republic of Belarus. https://rus-bel.rsr-online.ru/. Accessed 31 Jan 2023
13. Rusagrogroup (2022) Annual and quarterly reporting of Rusagro. https://www.rusagrogroup.ru/ru/investoram/otchety-i-publikacii/. Accessed 31 Jan 2023
14. Scientificrussia (2022) Memorandum on the creation of the Eurasian Network University signed on 25 May. https://scientificrussia.ru/articles/25-maa-podpisan-memorandum-o-sozdanii-evrazijskogo-setevogo-universiteta. Accessed 31 Jan 2023
15. Sdgindex (2021) Sustainable Development Report 2021. The Decade of Action for the Sustainable Development Goals. https://www.sdgindex.org/reports/sustainable-development-report-2021/. Accessed 31 Jan 2023
16. Sdgindex (2022) Sustainable Development Report 2022. From Crisis to Sustainable Development, the SDGs as Roadmap to 2030 and Beyond. https://www.sdgindex.org/reports/sustainable-development-report-2022/. Accessed 31 Jan 2023
17. Studyabroad (2022) Russian-Kyrgyz consortium of technical universities. http://studyabroad.mpei.ru/eng/RKKTU/Pages/default.aspx. Accessed 31 Jan 2023

18. Tomyuk ON (2022) Strategic tracks for the development of a modern university in the conditions of transformation. In: Russian regions in the focus of changes: collection of reports in two volumes (18–20 November 2021, Ekaterinburg), vol II. UrFu, Ekaterinburg, pp 217–225
19. Vahanyan GA (2022) Role and place of science of Turkey and Armenia (comparative analysis). Large Eurasia: Development, Security, Cooperation, pp 1014–1019. https://cyberleninka.ru/article/n/rol-i-mesto-nauki-turtsii-i-armenii-sravnitelnyy-analiz. Accessed 31 Jan 2023
20. WIPO (2022) Global Innovation Index 2022. What is the future of innovation-driven growth? https://www.wipo.int/global_innovation_index/en/2022/. Accessed 31 Jan 2023

Student Tourism as a Mechanism for Overcoming Cultural Differences and Ensuring the Inclusiveness of Higher Education to Achieve SDGs in Universities

Anastasia A. Sozinova, Anna A. Grabar, Olga N. Soboleva, and Natalia V. Gritsuk

1 Introduction

Modern students belong to Generation Z, who are called digital natives [1] because of their focus on the digitalization of all spheres of life, a high level of gadget proficiency, preferences in communication, and work in a digital environment. This category of the world's population can find a common language with each other through the use of digital technologies for quick communication (electronic translation functions), and they also have a good command of English, which is most common during international scientific and student conferences, and training programs in various specialties.

Overcoming problems related to cultural differences in universities at the present stage is realized through student exchanges between universities, which also act as a form of tourism. These exchanges can influence the creation of joint sustainable development projects (start-ups) in various areas, namely, in the field of new technologies in renewable energy, in the field of greening universities, and in creating an energy-efficient smart environment. Also, universities aimed at attracting foreign tourists are striving to create an academic environment that would allow them to maintain standard norms and rules that contribute to achieving inclusive conditions and leveling cultural differences.

In view of the need to achieve the SDGs of universities, it is relevant to assess the existing practices of the impact of student tourism on overcoming cultural differences and ensuring inclusive conditions for higher education.

The aim of this article is to identify the features of the formation of this type of tourism as a mechanism for combating cultural differences and achieving inclusiveness in the higher education system for the purposes of the SDGs of

A. A. Sozinova (✉) · A. A. Grabar · O. N. Soboleva · N. V. Gritsuk
Vyatka State University, Kirov, Russia
e-mail: aa_sozinova@vyatsu.ru; grabar@vyatsu.ru; soboleva@vyatsu.ru; nv_gritsuk@vyatsu.ru

E. G. Popkova (ed.), *Smart Green Innovations in Industry 4.0*, Springer Climate,
https://doi.org/10.1007/978-3-031-45830-9_41

modern universities. The implementation of the formulated article is connected with the need to deal with the following tasks: assessment of the ranking of higher educational institutions of the world for the implementation of the SDGs; establishing the existence of a connection between the impact of student tourism on overcoming the manifestations of cultural differences and achievement of inclusiveness.

2 Materials and Methods

The issues of the influence of student tourism on the development of inclusiveness and overcoming cultural differences in order to achieve the UN SDGs of universities were studied by the authors at the level of individual countries or in the context of theoretical foundations. Namely, the article [1] presents a study of the characteristics of the behavior of generation Z and X compared the use of mobile applications by these categories of the population. The study [6] allows us to determine the features of the cultural adaptation of Iranian students in Canadian universities, reveals the problems and advantages in this area. The article [7] is aimed at establishing the conditions and opportunities for academic and social implementation of foreign students in Canada.

In this article, the methodological tools necessary for solving research tasks were used. Based on the classification method, a rating assessment of the achievements of world universities in the field of impact on the SDGs is presented. The integrated method made it possible to identify the specifics of the impact through the sphere of student tourism on the inclusiveness of higher education and overcoming cultural differences in universities for the implementation of the SDGs.

3 Results

Universities of the world, guided by the need to maintain prestige, act as pioneers in various fields and directions, providing a public demonstration of the need for change, implementation of concrete steps, and the improvement processes. The implementation of specific SDGs of the university affects the state of sustainable development of both the institution of higher education itself and the achievement of these goals at the level of territories, regions, and the state. The active work of the university in the field of inclusiveness management, in overcoming cultural differences, which ensures the implementation of projects by students related to the achievement of the SDGs, can be improved through the new experience of the latter, acquired on tourist trips related to cultural, educational, environmental, and social spheres. An assessment of the ranking of world universities in terms of the implementation of the SDGs will make it possible to determine the list of leaders in this area (Table 1). These data will later become the basis for characterizing student tourism flows at the level of the countries in which these universities are located.

Table 1 Ranking of universities by the impact on the SDGs for 2022

University	Overall ranking of impact on the SDGs	Total impact on SDGs (score)	Priority impact on SDGs (target)	Results of the impact on the SDGs
Arizona State University (USA)	2	98.5	11	94.6
			14	97.9
			15	97.7
			17	93.3
Chulalongkorn University (Thailand)	16	95.2	3	89
			9	98.1
			15	90.4
			17	97.9
Hokkaido University (Japan)	10	96.2	2	91.4
			9	95.7
			15	90
			17	97.7
King Abdulaziz University (Saudi Arabia)	4	97.5	4	89.9
			9	99.8
			10	93.6
			17	93.8
Kyungpook National University (South Korea)	13	95.8	2	84.7
			9	98.5
			11	92
			17	94.9
Newcastle University (United Kingdom)	8	96.5	9	96.9
			11	91.5
			12	90.6
			17	96.1
University of Technology Sydney (Australia)	15	95.6	6	89.7
			7	79
			12	92.9
			17	96.5
University of British Columbia (Canada)	13	95.8	9	100
			11	88.2
			12	88.3
			17	96.6
University of Alberta (Canada)	11	96	2	86.3
			11	91
			15	95.4
			17	96.1
University of Manchester (United Kingdom)	9	96.4	11	93
			12	91.1
			14	92.9
			17	95.5

(continued)

Table 1 (continued)

University	Overall ranking of impact on the SDGs	Total impact on SDGs (score)	Priority impact on SDGs (target)	Results of the impact on the SDGs
University of Victoria (Canada)	12	95.9	12	89
			13	89.1
			15	95.8
			17	92.11
Queen's University (Canada)	7	96.6	1	90.9
			9	97.6
			11	93.1
			17	93.1
Universiti Sains Malaysia (Malaysia)	4	97.5	5	78.1
			7	85.2
			16	96.4
			17	100
University of Auckland (New Zealand)	6	96.7	2	88
			9	98.3
			15	95.5
			17	94.6
Western University (Canada)	3	97.8	1	93.5
			2	89.8
			9	97.5
			17	95.4
Western Sydney University (Australia)	1	99.1	5	80.3
			6	90
			12	93.3
			17	99

Source: Compiled by the author based on Ref. [8]

Table 1 lists the top 16 universities in the world that have had an impact on the SDGs.

There are five Canadian universities among the 16 leading sustainable development universities. It was determined that each of these universities had a positive impact on SDG 17, related to strengthening the means of implementing the global partnership for sustainability (Table 1). In this case, we mean promoting the development of student tourism (inflows) due to: the admission of immigrants from other countries aimed at realizing their own academic goals and able to be effective in joint student projects related to the SDGs; grants organized by these universities for talented foreign students; organization of participation of own students in international scientific-student conferences, symposiums, practical research projects in which students take part.

Let us consider in more detail the directions of the formation of student tourism, organized by the Canadian higher education system.

We can note such an element as the attraction of immigrants to study in higher educational institutions in Canada, which is a priority for the state. Namely, among the reasons that explain the participation of the Canadian government in creating conditions for the successful adaptation of these citizens in the state, one can name the employment of these students in the social sphere, the field of digital technologies, trade, in which there is a demand for personnel with high efficiency, who speak English and French (170,000 jobs are supported in these infrastructures) [10]; growth of GDP from the livelihoods of foreign students in Canada; increasing the academic achievements of universities (creating and implementing new innovative projects, including in the field of SDGs). To support foreign students in Canada, both those who immigrated and pay for tuition and living at their own expense and for those who were able to enroll in the framework of the grants won at the government level, the following is carried out: attracting the private sector to employ these students; amending federal legislation regarding the time limit that foreign students can spend outside the university, which allows them, through inclusiveness in the higher education system, to work to pay for tuition and living in the country. In 2014, the government of Canada set a goal to increase the number of foreign students to the level of 450,000 citizens. In 2021 this figure was more than 845,000 people, and by the end of September 2022, it was more than 917,000 citizens. The analysis shows that the annual contribution of international students to the national GDP is more than 22,000 million Canadian dollars [10].

A study of empirical materials [6] shows that Canadian universities use the following measures to prevent the oppression of gender, national rights, and the emergence of disagreements on cultural grounds, namely: there are sets of rules that are put forward to students, teachers, researchers regarding equality in communication, study, joint projects; research innovative projects (including in the context of the environmental component of the SDGs) are created without a focus on nationality (they can be initiated by foreign students, participation in such projects is not limited to foreigners). It is determined that these sets of rules are called the principles of equity, diversity, and inclusion (EDI) [7]. Their adoption is a collective effort to identify, combat the problems of exclusion, and marginalization of national minorities in the universities of the state. Despite the existing problems in the field of integration into the academic environment, foreign students have quite attractive conditions for studying and living in Canada, and the best of them receive country citizenship.

Among the main problems that cannot be completely overcome at Canadian universities are insufficient efforts of the university community to understand the need to apply more specific approaches to the adaptation of foreign students from Asia and Africa due to their difficulties in adapting to monocultural rules; the lack of initiatives by university management, teachers regarding the creation of international research groups that implement certain projects [7]. Despite the existence of opportunities for the implementation of student projects, including in the field of SDGs of universities, groups of students in terms of "local-foreigners" are involved in their implementation in practice. Although, foreign projects are also successfully implemented at universities. This, in our opinion, indicates the existence of conditions for competition rather than cooperation and partnership.

All five surveyed universities have their own SDG programs (plans). For example, Queen's University (Canada) has a Climate Action Plan, all the planned measures of which are implemented at the level of a group of students, and teachers belonging to the so-called green campus. This group includes students from different countries. Among the main climate measures achieved through their activities at the university, one can note: declaring the need to preserve the climate in the university; environmentally oriented waste management, recycling with the help of technologies created at the university; collection and processing of food waste in the university canteen; refusal to use bottled drinking water; implementation of energy efficiency measures; advocacy for responsible investment in achieving the SDGs [5].

Universities in Australia, like those in Canada, lead the global SDG rankings, which is largely due to the successful initiatives of the student community, represented by both local and international students. Before the Covid-19 pandemic, there were 580,000 international students in the state (the main subjects of student tourism, contributing to the achievement of the SDGs of the Australian higher education system). The impact of restrictive measures affected the fact that by the end of 2021, this figure amounted to 250,000 people, and only in October 2022 it grew to 370,000 citizens. About 72,000 international students are still not using valid student visas [2]. The problem of a high reduction in the flow of student tourism to Australia is largely due to the fact that the main countries from which foreign students came were India and China, where there are still several restrictions on movement and border crossing.

It was revealed that the attraction of foreign students is directly related to the impact of Australian universities on the SDGs. Namely, the study of one foreign student at prestigious universities in Australia is about 50,000 Australian dollars per year, and the income of universities from studying one local student in these institutions is approximately 15,600 Australian dollars per year [2]. In less prestigious ones, this figure is lower, but the proportion between the cost of studying for local and foreign students remains the same. The analysis shows that most universities in Australia direct a significant part of their income from attracting foreign students to finance R&D and student researches related to the achievement of the SDGs. Therefore, the growth of student tourism directly contributes to the sustainable development of the state's universities.

Among the measures that contribute to attracting foreign students to Australian universities, we note the opportunity to stay in the country after graduation (the right to work for foreign students during the period of study and after graduation is fixed by law); removal of restrictions on the hours that an international student must spend within universities during the educational process [2]. We can note that providing the opportunity to independently manage study time, which is also typical for Canada, is an inclusive opportunity for the state's higher education system.

Western Sydney University (Australia) ranks first in the world in terms of impact on the SDGs, with special achievements noted in the context of the goals of gender equality, clean drinking water, responsible consumption, and global partnership. It should be noted the University's Action Plan for achieving environmental sustainability, according to the goals of which, the participants of the unit that oversees their implementation organize: research on problems in the climate sphere; development and commercialization of technologies to ensure greening in many sectors of the economy; managing sustainable energy, university waste, agriculture for dining needs [12].

New Zealand, like Australia, shows competition in attracting student tourism flows from other countries. This is due to the geographical proximity and the similarity of curricula. Also, the legislation of New Zealand does not provide for such loyal conditions for employment during and after training, which is presented in Australia. Before the pandemic, about 120,000 foreign students studied at New Zealand universities, at the end of July 2022 this figure was about 14,600 people [3]. In New Zealand universities, as in Australia, there is a significant difference between the cost training of local and foreign students, due to the profits from their involvement, various projects are implemented in the field of impact on the SDGs. The introduction of restrictive measures related to the pandemic affected a significant reduction in income from educational services of state universities, which led to a decrease in national GDP. That is, the income of national universities from attracting international students was NZ$1.2 billion in 2019, NZ$0.963 billion in 2020, and NZ$0.594 billion in 2021 [3].

Among the measures to ensure the inclusiveness of higher education in the state, one can note: the legally enshrined right to work during the study and after graduation (for 3 years for citizens who have become graduate students). That is, inclusiveness implies the possibility of providing one's own education and life in the country through employment. At this stage, in order to increase the flow of student tourism, the norms for the presence of students on campus are reviewed, and the possibility of reducing the mandatory hours of stay at the university is considered. If this requirement is removed, New Zealand universities will be more competitive with Australian universities.

It is worth noting that New Zealand universities are implementing both projects in the field of impact on the SDGs within their own space, and to increase environmental and economic sustainability. Specifically, we can note the successful startups of the University of Auckland in the field of energy efficiency, which was implemented at the level of modern companies; wireless power transmission [11].

4 Discussion

In the course of the analysis, the main directions by which universities ensure the leveling of cultural differences and the achievement of the inclusiveness of the higher education system in order to achieve the goals of sustainable development are identified.

Based on the experience of assessing the sustainable development of universities in Canada, it has been determined that this process is implemented due to the participation in this process of the government of the state, which, like institutions of higher education, is interested in the growth of foreign student tourism and the permanent life of foreign students within the country. It is shown that both the government and universities not only realize their needs for academic, sustainable development, and employment in areas that require mobile personnel who speak English and French but also support the aspirations of foreign students in the implementation of their academic, career, and social goals. An important feature has been identified that helps foreign students adapt to the cultural and academic environment in Canada, namely, a support system has been created that does not allow students, teachers, or researchers to discriminate on national or gender grounds (a system of fines, punishments, deductions, and dismissals). This measure is restrictive in terms of the impact on the local scientific, student, and teaching environment. It should also be noted the use of a stimulus measure by the federal government, which implies the expansion of the inclusive possibility of remote learning for foreign students outside university campuses (at the moment, the restriction on study time outside institutions has been lifted).

The effective implementation of incentive measures to introduce an inclusive system of higher education makes it possible to achieve high efficiency in ensuring the growth of student tourism flows from other countries. The high interest of universities in attracting foreign students in combination with their motivation determines the attractiveness of countries for student tourism. Australia, which has a high standard of living, and higher education focused on sustainability, seeks to form an active, employable, mobile society, including through the initiatives and aspirations of international students. In view of this, to attract the students, inclusive conditions are created that stimulate them to academic, labor, and social activities in the country.

Incentive measures in the field of ensuring inclusive conditions for higher education must be constantly analyzed, reviewed, and compared with similar measures that are provided at the level of other countries. The experience of New Zealand has shown that the inability of the state to quickly respond to changes in the educational services market of a neighboring state (Australia), to the growth of problems faced by foreign students, affected the fact that student tourism flows after the pandemic could not recover as quickly as it happened in other countries. In this case, one can support the opinion [9] regarding the need to abandon standardized approaches that involve some criteria and rules that are unchanged for a long time and move toward flexible methods.

The study [4] notes the positive impact of cultural differences on the effectiveness of business projects, this aspect is also important in the creation of university projects related to the achievement of the SDGs. Therefore, in our opinion, overcoming cultural differences in the academic environment at the current stage can acquire new formats, and be considered as an adaptation of this environment to the existing cultural characteristics of foreign students.

5 Conclusion

It should be noted that similar features of the formation of student tourism in relation to the inclusiveness of higher education and overcoming cultural to implement the SDGs of the universities of the world have been identified. Namely, universities in Canada, Australia, and New Zealand, which are highly resilient, demonstrate the use of incentive measures related to ensuring the inclusiveness of higher education, which are reinforced by legislatively regulated support from states. Separately, it was revealed that Canada at the level of the academic environment applies restrictive measures (sets of rules and regulations) regarding the prevention of discriminatory manifestations in the field of cultural differences. This measure is focused on the cultural, social, and academic integration of foreign students, the growth of their influx from other countries, the implementation of innovative projects through non-standard approaches in the field of SDGs, carried out by talented students in the context of interaction between different cultures and traditions.

We have identified the need to implement the SDGs of the universities of the world not within the framework of a focus on overcoming cultural differences, but in the context of the mutual enrichment of cultures and the impact of this process on the emergence of new innovative and technological solutions that will allow us to solve previously unrealizable tasks of sustainable development.

References

1. Agárdi I, Alt M (2022) Do digital natives use mobile payment differently than digital immigrants? A comparative study between generation X and Z. Electron Commer Res:1–28. https://doi.org/10.1007/s10660-022-09537-9
2. Hurley P (2022) International students are returning to Australia, but they are mostly going to more prestigious universities. https://theconversation.com/international-students-are-returning-to-australia-but-they-are-mostly-going-to-more-prestigious-universities-193391. Accessed 7 Feb 2023
3. ICEF (2022) How are international student numbers recovering so far in New Zealand? https://monitor.icef.com/2022/11/how-are-international-student-numbers-recovering-so-far-in-new-zealand/. Accessed 7 Feb 2023
4. Kozhakhmetova A, Zhidebekkyzy A, Turginbayeva A, Akhmetova Z (2019) Modelling of project success factors: a cross-cultural comparison. Econ Sociol 12(2):219–234. https://www.economics-sociology.eu/?673,en_modelling-of-project-success-factors-a-cross-cultural-comparison. Accessed 7 Feb 2023
5. Queen's University (2022) Green campus. https://www.queensu.ca/about/greencampus. Accessed 7 Feb 2023
6. Razavipour R (2022) Gendered experiences of human dignity within Canadian universities: the case of Iranian international students. J Comp Int High Educ 14(5s):38–45
7. Tavares V (2021) Feeling excluded: international students experience equity, diversity and inclusion. Int J Incl Educ. https://doi.org/10.1080/13603116.2021.2008536
8. Times Higher Education (2022) Impact rankings 2022. https://www.timeshighereducation.com/impactrankings. Accessed 7 Feb 2023

9. Turginbayeva A, Nurseitova G, Zhakupbekova G, Doszhanov K, Konysbay A (2020) Ability to use flexible project management in the hotel business. E3S Web Conf 159:04009. https://doi.org/10.1051/e3sconf/202015904009
10. Tustin S (2022) International students are a key part of Canada's future. Rentals for Newcomers. https://www.rentalsfornewcomers.com/blog/international-students-are-canadas-future. Accessed 7 Feb 2023
11. University of Auckland (2022) Sustainability. https://www.auckland.ac.nz/en/about-us/about-the-university/the-university/sustainability-and-environment.html. Accessed 7 Feb 2023
12. Western Sydney University (2022) Environmental sustainability. https://www.westernsydney.edu.au/environmental_sustainability/home. Accessed 7 Feb 2023

Edtech and Its Contribution to Overcoming Cultural Differences and Ensuring the Inclusiveness of Higher Education to Create Climate-Smart Green Innovations Based on Universities

Olesya P. Kazachenok, Alexander V. Bespyatykh, Meerimai Z. Karbekova, and Ilham M. Saipidinov

1 Introduction

The growth of the world economy and world population and improvements in the quality of life are the results of targeted economic policies based on materialism. Natural resources have become the key factor in the economic progress of human civilization. However, their reserves are being depleted, and active interference with natural ecosystems leads to dangerous changes. The key issue of the twenty-first century is how 9–10 billion people can coexist on this planet with limited space, resources, and climate [9].

Future tendencies are formed now. Thus, an important direction for the resolution of the given problem is the search for tools that will allow reducing risks or ensuring their reduction to acceptable levels. Under such conditions, education and science play a key role. Universities occupy the central role as centers of education, innovation, and research [6]. There are roles and means through which higher educational establishments can actively influence the resolution of current problems, including climate change. Back in 1992, it was announced within the UN Framework

O. P. Kazachenok (✉)
Volgograd State University, Volgograd, Russia

A. V. Bespyatykh
Vyatka State University, Kirov, Russia
e-mail: bespyatyh@vyatsu.ru

M. Z. Karbekova
Jalal-Abad College of the Jalal-Abad State University named after B. Osmonov, Jalal-Abad, Kyrgyzstan

I. M. Saipidinov
Osh State University, Osh, Kyrgyzstan

E. G. Popkova (ed.), *Smart Green Innovations in Industry 4.0*, Springer Climate,
https://doi.org/10.1007/978-3-031-45830-9_42

Convention on Climate Change that “education is an important element for an adequate global response to climate change” [15].

In this context, universities must not only create knowledge and consider the problems of climate change in their educational courses but also set cultural values and ensure the inclusiveness of processes of tackling climate change. Despite the increased attention to the problem of climate change and the results of educational establishments in the minimization of the negative influence of its consequences, the efforts made are not yet of the desired coverage and intensity. Moreover, there are significant differences in the educational and scientific spheres regarding the determination of opportunities and overcoming of certain skepticism concerning the effectiveness of the available climate-neutral technologies. Under such conditions, new trends in education, which are based on digital technologies and the concept of climate-smart green innovations, must ensure the maximum agreement of the approaches to climate change and their development at a new level.

2 Materials and Methods

This paper’s methodology is based on the combination of different views and approaches in scientific spheres that study digital technologies, organizational behavior, psychology, educational process, the policy of climate sustainability, etc. The system approach is used to draw links between the provisions of the mentioned spheres. It allows determining the elements and spheres of the interaction of technology, behavior, interaction, and development. Thus, the notion of climate-smart green innovations is linked with educational processes and supplemented by cultural and behavioral specifics. It is implemented through the technological component within the concept of educational technologies (EdTech).

According to the multidisciplinary character of this research, the chosen topics are considered through different research areas, which are aimed at the study of the impact of climate change on science, economy, policy and human behavior [4], attitude toward climate change in universities [6], the role and features of education aimed at climate change in society [1, 8, 9], specifics of green innovations [10, 14], the activity of universities directed at climate change [11], the interconnection between education and technologies [3], specifics of EdTech in the system of education [2], identification of the main trends and tendencies in EdTech [5, 13], and influence of educational technologies on the environmental behavior of students [11].

Thus, the main goal of this research is to describe the potential of EdTech in terms of its influence on overcoming cultural differences and ensuring inclusiveness in universities involved in the innovative activities related to smart green innovations.

3 Results

The professional activity of most people in the world is based on their education. The obtained qualification and specialty define the basic skills and competencies people use in their work. Thus, it is possible to state that the basis of the current state was set in the educational environment of previous years. The traditional approach to education characterizes it as a process of teaching certain skills and competencies, which are necessary for the professional activity of performing specialized tasks or solving problems. According to this approach, students acquire only technical skills without their connection to the system of values. This limits the potential of education and its mission. Without the foundation of universal human values, any process might be a threat to an individual, social group, or mankind on the whole. A low level of popularization and promotion of the culture of climate responsibility over recent centuries led to significant problems, one of which is climate change.

In the twenty-first century, the role of universities is substantially transformed. They cease to be purely educational or scientific environments and become social and cultural hubs that distinguish and acquire only the most topical tendencies and produce and popularize cultural values that conform to available or prospective challenges. Therefore, universities – through traditional and modern tools – together with global and national institutes form and develop a culture of environmental responsibility, climate innovations, and cultural inclusiveness.

Such an approach to education is implemented through different approaches and concepts, the main role among which belongs to Climate Change Education (CCE), the concepts of social responsibility, and sustainable development. Universities around the world realize their responsibility for teaching students and society to stimulate the alleviation of the consequences of climate change and adapt to them [7]. This role lies in universities' accepting and facilitating carbon-neutral goals and practices through three main mechanisms:

1. Functioning as carbon-neutral establishments;
2. Teaching students and disseminating the values of climate responsibility;
3. Ensuring the development of smart green innovations that are aimed at tackling climate change.

In the first case, their influence is the smallest, because it is limited by the establishment's walls. In the second and third cases, it goes far beyond the university, expanding the understanding of carbon footprint and stimulating its reduction.

Impact Rankings Times Higher Education ranks universities from the position of their achievement of the Sustainable Development Goals. Within SDG 13 (Climate action), it is determined whether educational establishments conduct research on the topic of climate change, how they implement alternative energy or introduce educational programs that are aimed at the achievement of zero emissions. The top five universities with climate-responsible activity are as follows:

- University of Tasmania. It publishes numerous studies on climate action, tracks low-carbon energy, and uses educational courses and subjects aimed at carbon

neutrality. The university's Climate Futures Research Group cooperates with Australian companies to develop forecasts for determining the influence of climate change on different economic sectors.
- University of Victoria. It actively involves students in supporting the principles of sustainable development and offers them opportunities to develop their knowledge in environmental practices.
- Wageningen University & Research. It offers several educational courses in the sphere of climate action and sustainable development, including a master's program, with scientific research in the sphere of sustainable development.
- Arizona State University. Apart from educational programs, it offers students the possibility of long-term influence on climate processes through interaction with public organizations, post-graduate programs, and a special platform that covers a wide range of specialists in sustainable development.
- King Abdulaziz University. It has research centers that study climate change and interact with government establishments, offering sustainable practices in climate action [12].

In many cases, universities strive toward the development of green innovations and the long-term impact through student education, which may be realised in more eco-friendly decisions during the purchase of a car, choice of food or energy sources, everyday behavior and professional career [1].

The above universities are at the vanguard of tackling climate change. To solve the corresponding tasks, their activity must utilize all capabilities, including technology. The study of trends in the educational environment showed substantial changes over 2020–2021. The key trends in education are as follows:

- Remote work and education
- Increase in the use of educational technologies
- Climate change
- Globalization online

The main technological trends in education and science include the use of AI, educational analytics, open educational resources, quality online education, etc. [13]. Despite the fact that the driver of these changes was the COVID-19 restrictions, the active implementation of digital technologies in the educational process started some time ago. An example of this can be the Scholarship of Technology Enhanced Learning (SOTEL) [1, 2]. SOTEL is based on the concept of teaching and is a form of a dialogue between the results of educational research and real teaching in the context of teaching and technologies [3].

Currently, the key technological capabilities of education and science are closely connected with digital technologies and have a generalized title: Educational Technologies (EdTech). Implementation of such technologies in education has a large potential in improving students' education and allowing them to become competent and get ready for a future career [11].

The comprehensive character of the problem of climate change and a large differentiation of the tools of its resolution are objective factors of differences in thoughts. Thus, from the position of perception and understanding of the problem of climate change, there are large differences among groups of people. This is largely predetermined by objective conditions and accompanying circumstances. Thus, citizens of island countries with a dry climate feel the influence of climate change in a larger way. The same is true for residents of areas with frequent natural catastrophes. However, apart from differences formed by objective reasons, an important factor in the differentiation of attitudes toward the problem posed is large cultural differences, which reflect the subjective factor.

Culture – as a totality of norms, rules and values – is formed primarily based on the personal and social self-identification of a person. Thus, a person shares certain values and identifies their role in society. Most people understand the importance of the problem of climate change and the necessity to solve it. At the same time, understanding the role of each person in this process is much differentiated, from tacit consent to a radical fight for climate values.

Technological tools in this case can be the means of overcoming cultural differences and ensuring the inclusiveness of universities in the process of creating climate-smart green innovations. The most actual tendencies of EdTech and their potential influence on climate change, including the cultural component and inclusiveness, are given in Table 1.

A part of the key trends of EdTech is directly connected with digital technologies: e.g., computational thinking, AI, AR, VR and mixed reality, global training and digital citizenship; while other trends combine digital tools with the change in the focus of the educational process. This allows structuring the thinking of students and scholars, taking into account their cultural and social features, expanding educational opportunities, etc. The expected result of such change is a better understanding of the role and capabilities of each person in tackling climate change, better development of the educational and scientific potential and the achievement of a higher level of inclusiveness of students and research in the development of green innovations based on universities.

4 Discussion

The sphere of EdTech develops very quickly and presents significant opportunities to solve the current problems, including those connected with the activities of universities. Most smart green and climate innovations involve the mandatory use of digital technologies. The potential of these technologies has not been studied comprehensively. Thus, an important scientific problem is the search for spheres and directions for using digitalization in the interests of society, to deal with global challenges, including climate change.

A similar situation is peculiar to the sphere of provision of educational services, in which EdTech tools are often used without any system or strategic planning. This does not allow for the full use of their capabilities. A significant barrier in this context is the insufficiently high level of mastering of technological tools by university

Table 1 The influence of actual trends in EdTech on overcoming cultural differences and ensuring the inclusiveness of universities under the conditions of green innovative activity

Trends of EdTech	Characteristics	Influence on overcoming cultural differences and ensuring inclusiveness
Computational thinking (CT)	Focus on algorithms, decomposition and generalization allows raising the ability to solve problems	Structures thinking, offering algorithms to solve climate problems regardless of cultural features
Professional training	Uses the differentiated individual approach to training, forms individual and inclusive competencies	Takes into account cultural differences of students, uses their specifics or adapts their vision to general conditions
AR, VR and mixed reality	Allows creating additional artefacts to make the perception and acquisition of information easier	Forms additional opportunities for expressing specific features and agreement of the visions
AI	Complements the teacher	Allows for a better understanding of cultural specifics and differences
Global training	Raises the level of understanding the influence of climate change in different regions of the world	Allows for a better understanding of the risks of other people and the support of inclusiveness
Students' profiles	Allows raising the level of citizens' climate responsibility	Forms conditions for better understanding of cultural features of other people
Studying sciences	Determines the means for raising the level of studies	Adapts the educational process to the student
Digital citizenship	Offers additional opportunities for social interaction of different people that are aimed at the prevention of risks and problems	Serves as a platform for the inclusiveness of processes that are connected with climate change
Student-oriented education	Allows unlocking each student's potential	Focuses on the cultural features of each student, ensuring their inclusiveness

Source: Compiled by the authors based on Ref. [5]

academic staff. Sometimes, the absence of the required level of training prevents them from using the latest opportunities that can substantially supplement the theoretical and practical aspects of teaching, ensure a quick exchange of data, and information processes, provide support in the creation of models, etc.

The absence of a unified approach and vision of the described problems is also a result of cultural differences between educational process participants. That is why the use of EdTech tools, apart from solving climate and educational problems, should also be aimed at studying cultural differences in the interests of green innovative projects. This will allow achieving a better convergence of views and visions and improving the level of students' and scholars' inclusiveness. The results obtained will also allow disseminating climate knowledge and values with further increase of the level of climate awareness and improvement of the environmental behavior of large groups of people.

5 Conclusions

Universities play an important role in shaping the future. Setting correct values and models in the educational process allows creating substantial preconditions for solving future problems. One of these problems is climate change, which is a result of the long intensive use of the planet's resources. Using universities as social and cultural hubs, it is possible to reach much better results in tackling climate change. It is also important to strive toward a high level of inclusiveness of people who are involved in the educational and scientific process and to reduce cultural differences between them.

Given the current trends in the educational sphere, the development of universities more and more depends on digital technologies. Therefore, the use of EdTech tools to solve climate problems and reduce the cultural differences between the participants of the educational and scientific process conforms to the requirements of our time and allows raising awareness of their role and capabilities in tackling climate change.

References

1. Cordero EC, Centeno D, Todd AM (2020) The role of climate change education on individual lifetime carbon emissions. PLoS One 15(2):e0206266. https://doi.org/10.1371/journal.pone.0206266
2. Cowling MA, Crawford J, Vallis C, Middleton R, Sim K (2022) The EdTech difference: digitalisation, digital pedagogy, and technology enhanced learning. J Univ Teach Learn Pract 19(2):1–13. https://doi.org/10.53761/1.19.2.1
3. Haynes D (2016) Introducing SOTEL. Int J Scholarsh Technol Enhanc Learn 1(1):1–2
4. Incropera FP (2015) Climate change: a wicked problem – complexity and uncertainly at the intersection of science, economics, politics and human behaviour. Cambridge University Press, Cambridge
5. ISTE (2018) 9 samykh goriachikh tem v edtech [The 9 hottest topics in edtech]. Mezhdunarodnoe obshchestvo tekhnologii v obrazovanii (ISTE) [International Society for Technology in Education (ISTE)]. http://iste.org/ru/explore/Education-leadership/The-9-hottest-topics-in-edtech. Accessed 8 Feb 2023
6. Leal Filho W, Mifsud M, Molthan-Hill PJ, Nagy G, Veiga Ávila L, Salvia AL (2019) Climate change scepticism at universities: a global study. Sustainability 11(10):2981. MDPI AG. https://doi.org/10.3390/su11102981
7. Leal Filho W, Sima M, Sharifi A et al (2021) Handling climate change education at universities: an overview. Environ Sci Eur 33:109. https://doi.org/10.1186/s12302-021-00552-5
8. Molthan-Hill P, Worsfold N, Nagy GJ, Leal Filho W, Mifsud M (2019) Climate change education for universities: a conceptual framework from an international study. J Clean Prod 226:1092–1101. https://doi.org/10.1016/j.jclepro.2019.04.053
9. Salonen A, Lehtonen A, Cantell H (2018) Climate change education: a new approach for a world of wicked problems. In: Cook JW (ed) Sustainability, human well-being, and the future of education. Palgrave McMillan, pp 339–374. https://library.oapen.org/bitstream/handle/20.500.12657/22996/1/1007165.pdf#page=349. Accessed 7 Feb 2023
10. Schiederig T, Tietze F, Herstatt C (2012) Green innovation in technology and innovation management. R D Manag 42:180–192. https://doi.org/10.1111/j.1467-9310.2011.00672.x

11. Tamang RS, Tshering, Rinchen S (2021) The impact of educational technology tools on the learning achievement of B.Ed. science students in ecology class at Samtse College of Education. Contemp Educ Teach Res 2(1):1–11. https://doi.org/10.47852/bonviewCETR2021020101
12. THE (2022) Top universities for climate action 2022. Times Higher Education. https://www.timeshighereducation.com/student/best-universities/top-universities-climate-action. Accessed 7 Feb 2023
13. Thompson K, Corrin L, Hwang G-J, Lodge JM (2021) Trends in education technology in higher education. Australas J Educ Technol 37(3):1–4. https://doi.org/10.14742/ajet.7396
14. Tietze F, Schiederig T, Herstatt C (2011) What is green innovation? – a quantitative literature review. https://www.researchgate.net/publication/228320089_What_is_Green_Innovation_-_A_Quantitative_LiteratureReview. Accessed 6 Feb 2023
15. UN (1992) United Nations framework convention on climate change. United Nations, General Assembly, New York. https://unfccc.int/resource/docs/convkp/conveng.pdf. Accessed 6 Feb 2023

Sustainable Development of the Tourism Sector of the Republic of Armenia in the Context of an Innovative Economy

Tatul M. Mkrtchyan, Gayane R. Tovmasyan, and Svetlana A. Dallakyan

1 Introduction

Nowadays, tourism is one of the most rapidly developing sectors of the world economy. It attracts new investments, creates jobs, and contributes to the socio-economic development of regions and countries. Besides advantages, tourism also negatively affects the social, economic, cultural, and environmental sectors of countries. It is a significant contributor to CO_2 emissions and climate change. It may use non-renewable resources and damage biodiversity. Moreover, cultural values and traditions may be distorted. Additionally, it may lead to an increase in crime. There may be poor working conditions in the tourism sector, etc. Thus, the world is currently moving toward sustainable tourism development, trying to solve the aforementioned issues.

Tourism should also develop the social, economic, cultural, and environmental sectors of destination places. Many countries adopt policies for sustainable tourism development that reflect the main pillars mentioned above.

The sustainable development of destinations is now more important in the context of innovative economies. This is an emerging sector, and innovative solutions may contribute to solving issues connected with sustainable development.

In today's world, innovative activity is the engine of economic and social development that determines the economy's competitiveness and its long-term development prospects. The industrial revolution and innovative and digital economy disrupt the traditional models of industry markets in the world economy, allow countries to achieve impressive rates of economic growth and increase the

T. M. Mkrtchyan (✉) · G. R. Tovmasyan
Armenian State University of Economics, Yerevan, Armenia

S. A. Dallakyan
Institute of Economics after M. Kotanyan of NAS of the Republic of Armenia, Yerevan, Armenia

E. G. Popkova (ed.), *Smart Green Innovations in Industry 4.0*, Springer Climate,
https://doi.org/10.1007/978-3-031-45830-9_43

competitiveness of their participants. Innovation is considered the main source of competitive advantage for organizations. Innovation has demonstrated many successes. They are explained by many factors: spending on R&D, technological progress, the market orientation of the organization, opportunities, and much more [8]. The high level of development of the innovation sector, technology, and knowledge is becoming a priority and determining factor in economic growth. The creation of the institutional foundations of an innovative economy is the most important condition for a new quality of economic growth. The fourth industrial revolution, known as Industry 4.0, is the concept of achieving faster manufacturing innovation with improved supply chain efficiency. In the twenty-first century, sustainable development in the context of an innovative economy is becoming increasingly relevant. Practices characterized as environmentally sustainable innovations have been grouped into four main categories of innovation (i.e., product, process, organizational, and marketing innovations) and, more specifically, into subcategories of each type of innovation [13].

Tourism is one of the main sectors of the economy in the Republic of Armenia. The sustainable development of tourism in Armenia is vital. Thus, the research aims to study the problems of sustainable tourism development in Armenia in the context of an innovative economy.

2 Materials and Methods

In 1987, the Brundtland Commission of the UN defined sustainable tourism in the "Report of the World Commission on Environment and Development: Our Common Future," mentioning that people can make development sustainable by ensuring that it meets the present generation's needs without distressing the ability of future generations to meet their own needs [31, p. 16]. According to UNWTO, sustainable tourism is tourism that estimates its present and future economic, social, and environmental impacts, directing the needs of tourists, the industry, the environment, and local communities [32, p. 13]. UNWTO and UNEP have pointed out 12 aims for sustainable tourism: economic viability, local prosperity, employment quality, social equity, visitor satisfaction, local control, community well-being, cultural richness, physical integrity, biological diversity, resource efficiency, and environmental purity [33, pp. 25–46].

After the definitions mentioned above, researchers began to explore the issues of sustainable tourism development. Starting from 1990, many researchers have investigated sustainable tourism issues, including authors such as Butler [2]; Cronin [3]; Eccles [6]; Fossati and Panella [8]; Landorf [12]; Mondal and Samaddar [14]; Nash and Butler [16]; Navarro, Martínez, and Jiménez [17]; Postma, Cavagnaro, and Spruyt [20]; Sharpley [24]; Tovmasyan [28, 30]; and Wall [34].

The term "innovation" has been used since 1934 in Schumpeter's works as an important component of entrepreneurship [25]. Other researchers have also studied innovation, including Boer and During [1]; Drucker [5]; Fischer [7]; Kimberly [11];

Myers and Marquis [15]; Porter and van der Linde [19]; Rowley, Baregheh, and Sambrock [21].

Some authors also reflect on sustainable innovation, including Hellström [9]; Iritani et al. [10]; Niinimäki and Hassi [18].

This research uses different data from the Statistical Committee of the Republic of Armenia, the World Travel & Tourism Council, and the World Economic Forum, as well as data from state agencies and other organizations. The Travel & Tourism Development Index 2021, published by the World Economic Forum, evaluates the Travel and Tourism Sustainability subindex. Armenia ranked 61st among the 117 countries in the Travel & Tourism Development Index 2021 report. It was the 37th in the report with the Travel and Tourism Sustainability subindex [35, p. 11].

The authors employ various methods, including analysis, synthesis, statistical methods, and correlation analysis.

3 Results

3.1 Analyzing Tourism in Armenia

In recent years, Armenia has seen an increase in inbound and domestic tourism. This sector considerably suffered due to the COVID-19 pandemic. However, in 2021 and 2022, it started to recover. According to official data, 1,894,377 tourists visited Armenia in 2019 [26, p. 139]. In 2020, the number dropped to only 360,338. In 2021, the number of inbound tourists increased to 875,772, increasing even more in 2022 to 1,665,658 tourists [27, p. 163]. However, the increase in 2022 is mainly connected to the flow of Russian tourists (790,397 tourists) caused by the special military operation in the east of Ukraine.

Domestic tourism also saw growth in recent years. In 2019, the number of domestic tourists was 1,544,600. In 2020, it decreased to 1,045,756 due to the COVID-19 pandemic. However, it increased to 1,595,826 in 2021, which is 3.3% higher than in 2019. In January–September 2022, the number of domestic tourists was 1,696,052, which is even higher than the number of tourists in the previous years.

Tourism significantly contributes to the GDP and employment of the Republic of Armenia. Thus, the total contribution of tourism to GDP in 2019 amounted to 923.4 billion drams or $1.8 billion (12.6% of GDP), providing 13.8% of total employment or 137,600 jobs [30]. In 2020, the total contribution of tourism to the GDP in the Republic of Armenia was 251.6 billion drams or $493.5 million (3.7% of the GDP), decreasing by 73% compared to the previous year. Tourism provided 10.2% of total employment or 97,000 jobs, which is 30% less than in the previous year. In 2021, the total contribution of tourism to GDP was 448.8 billion drams or $880.5 million (6.1% of GDP), increasing by 78.4% compared to the previous year.

Tourism provided 12.2% of total employment or 120,200 jobs, increasing by 24% [29].

Sustainable tourism aims to ensure proportional territorial development of regions, create jobs, attract investments, develop infrastructures, etc.

3.2 *Analyzing the Innovative Economy in Armenia*

The current level of economic development in the whole world and Armenia is characterized by growing competition. The impact of innovations on economic development rates is expanding. These changes are associated with the high level of innovation development, which promotes and engages new investments in knowledge.

To measure the innovative potential of developed countries in today's conditions, numerous indices are used, covering various indicators: the Global Innovation Index (GII), the Global Competitiveness Index (GCI), which includes 12 sub-indices, of which the 12th – "Innovation capability" – characterizes the innovative economy.

The GII of Armenia has been improving since 2019. In 2018, Armenia ranked 68th, 64th in 2019, and 61st in 2020. In 2018, Armenia was among the top ten lower-middle-income countries. However, compared to the previous reporting year (2019–2020), Armenia has already ranked 15th among 34 countries with upper middle income, and one of the positions of this progress is "Human capital and research" (107 to 94), "Business sophistication" (89 to 69), and "Knowledge and technology outputs" (54 to 45). This is no coincidence, as an improvement in positions was also observed in the 1st (Innovation Input) and 2nd (Innovation Output) sub-index counts from 85 to 83 and 50 to 47, respectively. In 2021, the position of the GII and several sub-indices linked to the 2020 war and COVID-19 were deteriorating.

The Global Competitiveness Index (GCI), which measures the potential for innovative development of the countries worldwide, consisting of 12 sub-indices, is combined into one index that characterizes the competitiveness of the national economy to reflect the classification of countries better. Armenia has a rather encouraging indicator confirming the capacity for innovation (the score is 4.2). The availability of scientists and engineers is also considered encouraging (the score is 4.3). However, the PCT patents indicator, which refers to the commercialization of intellectual property, is 2.6, showing that there is no connection between science and the economy, as a result of which the created innovative, scientific, and engineering abilities are inefficiently commercialized and do not enter the economy. Thus, although Armenia shows promise in certain areas, it lags behind in terms of translating research knowledge into commercial opportunities.

Following the publication of the 2018 Global Competitiveness Report, the GCI is calculated using a new methodology in revision 4.0 (GCI 4.0) [22]. In 2019, Armenia was ranked 69th out of 141 countries in the Global Competitiveness Report [23], improving its position compared to the previous year; in 2018, it was ranked

70th among 140 countries [22]. Regarding the innovation capability sub-index, Armenia was 62nd in 2019 [23, pp. 63–65].

Given the results of the analysis, it is necessary to note several issues that impede the development of emerging high technologies and an innovative economy in the Republic of Armenia:

- Despite the presence of competitive research in certain areas, the commercialization of research work is low.
- The absence of an interconnected chain of education–science–economy (especially in terms of knowledge transfer and commercialization).
- The lack of alignment between demand and supply of high-tech and ICT labor force.
- The presence of the main problem of expanding the volume of innovative research, ensuring its quality, and improving the relevant capabilities.

3.3 *Opportunities for Developing Tourism in the Context of an Innovative Economy*

Nowadays, the formation of clusters is considered one of the most critical areas of innovation activity as it makes it possible to create successful systems for commodifying new ideas. Clusters give the opportunity to decrease the total expenses on R&D. Clusters are considered to be very effective approaches for increasing economic competitiveness and promoting the development of innovations. An innovation cluster is a group of interconnected specialized companies in relevant industries and scientific and government organizations concentrated in one geographical area and working together, despite competition [4].

Additionally, tourism clustering will contribute to the sustainable development of regions. Tourism clusters will have more potential for sectoral development. It will provide more opportunities for engaging investments, creating jobs, protecting the environment, etc.

3.4 *Evaluating the Influence of Tourism on Socio-economic Development of the Regions of the Republic of Armenia*

For this evaluation, the authors performed a correlation analysis between the following indicators: the volume of tourism services (VTC) (which includes the performance volume of hotels, restaurants, and tour companies), GDP, GDP per capita, and employment. The data were obtained from the Statistical Committee of the Republic of Armenia for 2009–2020 since the full data for 2021 and 2022 is not available yet. The method of differences was used for processing the time series. Table 1 shows the results of the correlation analysis.

Table 1 Results of the correlation analysis

	Indicator		
	Pearson correlation coefficient *R*		
Location	VTC – GDP	VTC – GDP per capita	VTC – employment
Yerevan city	**0.88**	**0.86**	**0.52**
Aragatsotn	**0.70**	**0.71**	**−0.08**
Ararat	0.46	**0.92**	0.34
Armavir	**0.71**	**0.70**	**0.46**
Gegharkunik	**0.50**	**0.53**	0.12
Lori	0.30	0.33	***−0.38***
Kotayk	**0.65**	**0.66**	***−0.04***
Shirak	**0.66**	**0.70**	0.25
Syunik	0.44	0.45	0.01
Vayots Dzor	0.37	0.42	***−0. 37***
Tavush	0.41	0.43	**−0.04**
Republic of Armenia	**0.88**	**0.88**	0.25

Source: Analyzed by the authors using SPSS software

Thus, the correlation coefficient between tourism services and GDP is high in Yerevan, followed by Armavir, Aragatsotn, Shirak, Kotayk, and Gegharkunik. The correlation coefficient between GDP per capita and tourism services is high in Ararat, Yerevan, Aragatsotn, Armavir, Shirak, and Kotayk. The correlation between employment and VTC is high in Yerevan and Armavir. In Lori, Kotayk, Vayots Dzor, Tavush, and Aragatsotn, it has a negative effect, which means there is no correlation between the change in VTC and the change in the number of employed. In these regions, tourism is a developed and developing sector. Thus, the reason for the obtained result may be the low number of registered employees. However, there is no separate statistical data on the number of people employed in the different sectors of the economy. Thus, it is unclear how many people are employed in the tourism sector.

This analysis demonstrates that tourism is essential for the social and economic development of the region and the capital, Yerevan. It is necessary to create programs for the sustainable development of tourism. It is necessary to link the development of tourism with the increase of employment and contribute to the implementation of investments in the regions and the development of SMEs, which will increase the living standards and the income level of the population.

3.5 *Evaluating Sustainable Tourism in Armenia*

For this evaluation, the following pillars were considered: tourism assets and activity, tourism-related linkages and leakages, environmental and social sustainability, infrastructures, and attractiveness [28]. Armenia possesses tourist resources and makes a significant contribution to the economy. However, challenges exist concerning environmental and social sustainability, infrastructures, and tourism attractiveness.

3.6 *Measuring the Aims of Sustainable Development of Tourism*

To evaluate the extent of achieving the aims of sustainable tourism development, the authors have developed a specialized tool in this research. It consists of a list of indicators that consider the 12 aims of sustainable tourism development. Each indicator can be evaluated on a scale of 1–100, where a high value indicates good performance. For evaluating economic viability, the following indicators should be measured: direct and total contribution of tourism to GDP, incoming and domestic tourism growth rate, tourism export volume, the level of investments in tourism, and the existence of favorable conditions for attracting new investments. To evaluate local prosperity, it is necessary to measure the following indicators: the length of stay and level of spending by incoming and domestic tourists in local destinations. The indicators of employment quality may be tourism jobs locally (by age, gender, and sectors) and wage rate, training for workers, and workers' skills and qualification level. The indicators of social equity are tourism direct and total contribution to GDP and employment by regions, funding schemes for small businesses in tourism, and the number of poor people engaged in tourism. The indicators of visitor fulfillment are the satisfaction level of tourists, safety level for tourists, crime, and other local problems. The indicators of local control are tourism plans at the local level and the participation of local people in decision-making processes concerning the issues of tourism planning and development. The indicators of community well-being are the welfare of local people, existing infrastructures, and the use of the main resources in local areas. The indicators for the cultural richness and physical integrity are the conservation of C&H values and their use in tourism locally, plans for conservation of C&H values, quality of physical landscapes and conservation plans, and environmental degradation caused by tourism activities. The indicators of biological diversity are the harm of tourism on nature, wildlife, and existing conservation strategies. Finally, the indicators of resource efficiency and environmental purity are the usage of non-renewable resources in tourism activities by tourism entities, waste disposal systems, air pollution levels, climate change control systems, green certifications, and energy efficiency mechanisms by tourism entities. The evaluations make it possible to assess the degree of implementation of the aims, see the main areas of intervention, and process or implement changes in the existing sustainable tourism development programs.

4 Conclusion

Thus, the research analyzes tourism, innovative economy, and sustainable tourism development in Armenia. The analysis proves that tourism has great development opportunities; it has high significance for the economy. Innovations and innovative solutions should contribute to tourism development. The creation of tourism clusters in Armenia can be a macro innovation. Besides, innovations should be used in microspheres as well, such as using sustainable construction mechanisms by hotel industries, using climate control mechanisms, sustainable transportation, waste and water management, and energy efficiency mechanisms in tourism.

The list of indicators was proposed for evaluating sustainable tourism development aims in Armenia. Sustainable tourism development is a must, and innovative solutions should contribute to that.

References

1. Boer H, During WE (2001) Innovation, what innovation? A comparison between product, process and organisational innovation. Int J Technol Manag 22(1–3):83–107. https://doi.org/10.1504/IJTM.2001.002956
2. Butler RW (2007) Sustainable tourism: a state-of-the-art review. Tour Geogr 1(1):7–25. https://doi.org/10.1080/14616689908721291
3. Cronin L (1990) A strategy for tourism and sustainable developments. World Leis Recreat 32(3):12–18. https://doi.org/10.1080/10261133.1990.10559117
4. Dallakyan S (2022) Innovative economy of the Republic of Armenia: prospects, opportunities, trends. Institute of Economics after M. Kotanyan of the NAS RA, Yerevan
5. Drucker PF (1985) Innovation and entrepreneurship: practice and principles. Heinemann, London
6. Eccles G (1995) Marketing, sustainable development and international tourism. Int J Contemp Hosp Manag 7(7):20–26. https://doi.org/10.1108/09596119510101895
7. Fischer MM (2001) Innovation, knowledge creation and systems of innovation. Ann Reg Sci 35:199–216. https://doi.org/10.1007/s001680000034
8. Fossati A, Panella G (2000) Tourism and sustainable economic development. Springer, New York. https://doi.org/10.1007/978-1-4615-4321-3
9. Hellström T (2007) Dimensions of environmentally sustainable innovation: the structure of eco-innovation concepts. Sustain Dev 15(3):148–159. https://doi.org/10.1002/sd.309
10. Iritani DR, Lopes Silva DA, Saavedra YMB, Grael PFF, Ometto AR (2015) Sustainable strategies analysis through life cycle assessment: a case study in a furniture industry. J Clean Prod 96:308–318. https://doi.org/10.1016/j.jclepro.2014.05.029
11. Kimberly JR (1981) Managerial innovation. In: Nystrom PC, Starbuck WA (eds) Handbook of organizational design. Oxford University Press, Oxford, pp 84–104
12. Landorf C (2009) Managing for sustainable tourism: a review of six cultural World Heritage Sites. J Sustain Tour 17(1):53–70. https://doi.org/10.1080/09669580802159719
13. Marcon A, de Medeiros JF, Ribeiro JLD (2017) Innovation and environmentally sustainable economy: identifying the best practices developed by multinationals in Brazil. J Clean Prod 160(2):83–97. https://doi.org/10.1016/j.jclepro.2017.02.101
14. Mondal S, Samaddar K (2021) Responsible tourism towards sustainable development: literature review and research agenda. Asia Pac Bus Rev 27(2):229–266. https://doi.org/10.1080/13602381.2021.1857963

15. Myers S, Marquis DG (1969) Successful industrial innovation: a study of factors underlying innovation in selected firms. National Science Foundation, NSF, Washington, DC
16. Nash D, Butler R (1990) Towards sustainable tourism. Tour Manag 11(3):263–264. https://doi.org/10.1016/0261-5177(90)90051-A
17. Navarro J-LA, Martínez MEA, Jiménez J-AM (2020) An approach to measuring sustainable tourism at the local level in Europe. Curr Issue Tour 23(4):423–437. https://doi.org/10.1080/13683500.2019.1579174
18. Niinimäki K, Hassi L (2011) Emerging design strategies in sustainable production and consumption of textiles and clothing. J Clean Prod 19(16):1876–1883. https://doi.org/10.1016/j.jclepro.2011.04.020
19. Porter ME, van der Linde C (1995) Toward a new conception of the environment-competitiveness relationship. J Econ Perspect 9(4):97–118. https://doi.org/10.1257/jep.9.4.97
20. Postma A, Cavagnaro E, Spruyt E (2017) Sustainable tourism 2040. J Tour Futur 3(1):13–22. https://doi.org/10.1108/JTF-10-2015-0046
21. Rowley J, Baregheh A, Sambrook S (2011) Towards an innovation-type mapping tool. Manag Decis 49(1):73–86. https://doi.org/10.1108/00251741111094446
22. Schwab K (2018) The global competitiveness report 2018. Retrieved from http://www3.weforum.org/docs/GCR2018/05FullReport/TheGlobalCompetitivenessReport2018.pdf. Accessed 20 Feb 2023
23. Schwab K (2019) The global competitiveness report 2019. Retrieved from https://www3.weforum.org/docs/WEF_TheGlobalCompetitivenessReport2019.pdf. Accessed 20 Feb 2023
24. Sharpley R (2021) On the need for sustainable tourism consumption. Tour Stud 21(1):96–107. https://doi.org/10.1177/1468797620986087
25. Shumpeter JA (1934, 1980) The theory of economic development. Oxford University Press, London
26. Statistical Committee of the Republic of Armenia (2019) The socio-economic situation in the Republic of Armenia in 2019 January–December. Retrieved from https://www.armstat.am/file/article/sv_12_19a_421.pdf. Accessed 20 Feb 2023
27. Statistical Committee of the Republic of Armenia (2022) The socio-economic situation in the Republic of Armenia in 2022 January. Retrieved from https://armstat.am/file/article/sv_12_22a_421.pdf. Accessed 20 Feb 2023
28. Tovmasyan G (2019) Assessing some indicators of tourism sustainability: case study from Armenia. Soc Econ Chall 3(1):127–136. https://doi.org/10.21272/sec.3(1).127-136.2019
29. Tovmasyan G (2021) Capital investments, tourist tax and tourism development: the case study of Armenia. Econ Sociol 14(1):199–213. https://doi.org/10.14254/2071-789X.2021/14-1/13
30. Tovmasyan G (2022) Evaluating the role of tourism in the economic development of the Republic of Armenia and other member states of the Eurasian Economic Union. J Lib Int Aff 8(3):83–98. Retrieved from https://e-jlia.com/index.php/jlia/article/view/747. Accessed 20 Feb 2023
31. UN (1987) Our common future: report of the World Commission on Environment and Development. Oxford University Press. https://sustainabledevelopment.un.org/content/documents/5987our-common-future.pdf. Accessed 20 Feb 2023
32. UNWTO (2013) Sustainable tourism for development guidebook. Madrid, Spain. Retrieved from https://www.e-unwto.org/doi/pdf/10.18111/9789284415496. Accessed 20 Feb 2023
33. UNWTO and UNEP (2005) Making tourism more sustainable – a guide for policy makers. Retrieved from https://wedocs.unep.org/bitstream/handle/20.500.11822/8741/-Making%20Tourism%20More%20Sustainable_%20A%20Guide%20for%20Policy%20Makers-2005445.pdf?sequence=3&isAllowed=y. Accessed 20 Feb 2023
34. Wall G (2018) Beyond sustainable development. Tour Recreat Res 43(3):390–399. https://doi.org/10.1080/02508281.2018.1475880
35. World Economic Forum (2022) Travel & tourism development index 2021: rebuilding for a sustainable and resilient future [Insight Report]. WEF, Geneva. Retrieved from https://www3.weforum.org/docs/WEF_Travel_Tourism_Development_2021.pdf. Accessed 20 Feb 2023

Strategic Vision for the Decarbonization of the World Economy: Climate Action 4.0 Through 2050 (Conclusion)

Elena G. Popkova

Climate economy 4.0 is a forward-looking socioeconomic response to the worsening problem of global climate change using advanced Industry 4.0 technologies. For this response to be effective, stakeholders' efforts must be coordinated and their participation in the climate agenda balanced. To achieve this, this book has shaped a systemic vision of the entrepreneurial, technological, regulatory, and societal manifestations of climate economy 4.0.

With this, the book advanced climate horizon 4.0 all the way to the end of the Decade of Action. However, there is no doubt that even the most ambitious and global initiatives will not completely overcome the risks of climate change before 2030. In this regard, pushing the climate horizon 4.0 even further is necessary. Most national strategies for achieving a carbon-neutral economy by the world countries are designed for the period up to 2050.

The middle of the twenty-first century is the presumed starting point of carbon-free economic growth. Therefore, it is up to this period to seek to advance climate horizon 4.0 to bring clarity to the strategic vision for the decarbonization of the world economy. On the one hand, this book has expanded and clarified existing knowledge in the field of climate economy 4.0. On the other hand, the book has raised new research questions.

One of the issues that have arisen is the involvement of everyone in the operation and development of climate economy 4.0. As this book has shown, the climate economy 4.0 is attracting increasingly more supporters. Nevertheless, it continues to remain within the strict boundaries of the digital economy and Industry 4.0 market segments and cannot yet be fully extended to contemporary economic systems.

E. G. Popkova
RUDN University, Moscow, Russia
e-mail: elenapopkova@yahoo.com

E. G. Popkova (ed.), *Smart Green Innovations in Industry 4.0*, Springer Climate,
https://doi.org/10.1007/978-3-031-45830-9

Another new question is what new perspectives and new threats lie in the evolution of artificial intelligence and its socialization, which is happening at a rapid pace and may accelerate once the Decade of Action is over. In particular, the question of how artificial intelligence will resolve the contradiction between the growing energy needs of Industry 4.0, and the ambitious climate agenda 2050 is of particular scientific and practical interest.

The questions raised remain open and define further perspectives for exploring climate change risk management through smart green innovations in Industry 4.0.

Index

E. G. Popkova (ed.), *Smart Green Innovations in Industry 4.0*, Springer Climate,
https://doi.org/10.1007/978-3-031-45830-9

The manufacturer's authorised representative in the EU is Springer Nature Customer Service Centre GmbH, Europaplatz 3, 69115 Heidelberg, Germany. If you have any concerns regarding our products, please contact ProductSafety@springernature.com

Printed and bound by CPI Group (UK) Ltd, Croydon, CR0 4YY
01/10/2026
02225902-0012